PROFESSIONAL SURVEYOR'S MANUAL

U.R.S. Consultants, Akron, Ohio formerly: Beiswenger, Hoch, Arnold & Associates Preliminary Interchange design by Robert C. Vernon

PROFESSIONAL SURVEYOR'S MANUAL

Robert C. Vernon, Sr.

McGraw-Hill
New York San Francisco Washington, D.C. Auckland Bogotá
Caracas Lisbon London Madrid Mexico City Milan
Montreal New Delhi San Juan Singapore
Sydney Tokyo Toronto

McGraw-Hill

A Division of The ***McGraw-Hill*** *Companies*

1 2 3 4 5 6 7 8 9 0 FGR/FGR 9 0 2 1 0 9 8 7

ISBN 0-07-067419-1

The sponsoring editor for this book was Zoe Foundotos, the editing supervisor was Patricia Amoroso, and the production supervisor was Tina Cameron. It was set in Times Roman by Jana Fisher through the services of Barry E. Brown (Broker—Editing, Design and Production).

Printed and bound by Quebecor.

CONTENTS

Acknowledgments vii
Preface ix

Part 1 Background

Chapter 1. Highway Planning 3

Chapter 2. Ohio and U.S. Land Surveying System 11

Chapter 3. Indiscriminate Surveys 25

Part 2 Highway Plans

Chapter 4. Title Sheet 41

Chapter 5. Centerline Plat 61

Chapter 6. Typical and Superelevated Sections 67

Chapter 7. Topographic Maps 95

Chapter 8. Cross Sections 113

Part 3 Mathematics

Chapter 9. Vertical Curves 125

Chapter 10. Slope Staking 147

Chapter 11. Simple Curves 161

Chapter 12. Simple Spirals 209

Chapter 13. Transitions Spiral between Two Curves 261

Appendix Sample Survey Forms 285
Bibliography 295
Index 297

ACKNOWLEDGMENTS

Technical editing and suggestions have been provided by Dawne Vernon, who has more than 15 years experience with Right-of-Way Engineering mathematics and the supervision of engineering technicians. With a background education in both science and mathematics, Dr. Dawne Vernon is a graduate of Ohio State University with a B.S. degree in genetics, and a Ph.D. degree in molecular genetics. Without Dawne's complete understanding of the art and science of surveying and all related mathematics, this text would not be possible.

A special thanks to my wife, Roni, who has gently nudged me to the studio each day for the past three years with a warm "good luck."

Computer guidance was graciously provided by Richard Staats, a great friend and neighbor, who guided me through word processing without complaint.

Thanks to my son, Robert, Professional Surveyor who, when asked to verify the results of certain field procedures and office mathematics, was always obliging.

To all the surveyors, engineers, and architects who provided my tutoring throughout the years, I give my gratitude.

Other Residential Construction Books of Interest

BIANCHINA • *Forms & Documents for the Builder*

BYNUM, WOODWARD, RUBINO • *Handbook of Alternative Materials in Residential Consruction*

DOMEL • *Basic Engineering Calculations for Contractors*

FRECHETTE • *Accessible Housing*

FRECHETTE • *Bathroom Remodeling*

GERHART • *Everyday Math for the Building Trades*

HARRIS • *Noise Control Manual for Residential Buildings*

HUTCHINGS • *CPM Construction Scheduler's Manual*

HUTCHINGS • *National Building Codes Handbook*

MILLER, BAKER • *Carpentry and Construction—2nd Edition*

PHILBIN • *The Illustrated Dictionary of Building Terms*

PHILBIN • *Painting, Staining and Refinishing*

POWERS • *Kitchens: Professional's Illustrated Design & Remodeling Guide*

SCHARFF AND THE STAFF OF ROOFER MAGAZINE • *Roofing Handbook*

SCHARFF AND THE STAFF OF WALLS & CEILINGS MAGAZINE • *Drywall Construction Handbook*

SKIMIN • *The Technician's Guide to HVAC Systems*

WOODSON • *Professional Modeler's Manual*

WOODSON • *Be a Successful Building Contractor—2nd Edition*

PREFACE

Thirty-nine years ago, a college teacher of Fine Arts took leave from the hallowed halls of Park College, Parkville, Missouri, to begin a new career in engineering and surveying. College courses in engineering and surveying, new work experiences with consulting and state highway engineers, and twenty-six years of surveying in private practice have all formed the basis for this textbook. Portions of the material have been accumulated over the years, taught in various work and professional assignments, and refined during the past three years.

Several chapters have been recently prepared for purposes of covering a full range of difficult material that together make-up the surveyor's training. The emphasis within the textbook continues to search for elementary ways to describe surveying problems.

This text is intended as a source of renewing or adding to engineering-related skills using several of the most significant drawings associated with highway engineering and right-of-way plans, while covering a full range of mathematical and field problems experienced by today's surveyors.

Contractors, Attorneys, Appraisers, Negotiators, Students and, yes, even Surveyors often have difficulty computing and understanding the drawings historically used by consulting engineers and state highway design personnel during the development of new or improved roadways.

This textbook retraces the last thirty-nine years, remembering those who offered their assistance, those who were patient during their instructions, and those whose training provided the expertise that has led to this Professional Surveyor's Manual.

Whatever your level of expertise, this text will improve your ability to deal with the various aspects of your present profession (or future profession), that relate to the design, construction layout, and acquisition of rights-of-way, as well as all surveys or retracements of property lines near or along highways.

The text will bring you to a new level of mathematical ability, whether your expertise is in the field of surveying, construction, the legal profession, appraiser, negotiator, student, or other work-related experiences where a better mathematical background is desirable. The teaching methods, while solving all problems with the simple use of hand-held calculators, also encourages and outlines advantages of using computer programmable software programs with your P.C.

Robert C. Vernon's surveying, engineering and design experience spans thirty-nine years, including twenty-nine years in teaching. He began at the consulting engineering firm Beiswenger, Hoch, Arnold & Associates (now U.R.S. Consultants) Akron, Ohio. He was then employed as right-of-way engineering plan reviewer and plan reading instructor for the Ohio highway department's central office College Training program in Columbus, Ohio, and as assistant right-of-way coordinator for Division 10, Marietta, Ohio. Mr. Vernon commenced private surveying practice in 1969 and presently continues this practice as President of Right-of-Way Consultants & Associates/Robert C. Vernon, Inc., and its successor corporation, Robert C. Vernon Enterprises, Incorporated.

Teaching has always been a part of Mr. Vernon's career. Following receipt of a master's degree, Mr. Vernon was instructor at the Kansas City Conservatory (now part of the University of Missouri); at Park College, Parkville, Missouri; and taught in the Willard City Schools, Willard, Ohio. He received his Registered Professional Surveying license in 1963.

In addition to teaching the plan reading and surveying courses while employed at the Ohio Department of Transportation, he also taught a route surveying course (curves and spirals) annually for twenty consecutive years. This course was part of the yearly Surveyor's Refresher Program, originally sponsored by the Ohio Society of Professional Engineers, then later sponsored by the Professional Land Surveyors of Ohio.

Mr. Vernon has published several articles in the American Right-of-Way Association magazine and penned numerous course books for the Ohio Department of Transportation. He has also been affiliated with the National and Ohio Society of Professional Engineers, Professional Land Surveyors of Ohio, the West Virginia Association of Land Surveyors, the American Congress on Surveying & Mapping, and is licensed to practice in Ohio, Pennsylvania, and West Virginia.

P · A · R · T · 1

BACKGROUND

CHAPTER 1
HIGHWAY PLANNING

THE HISTORICAL IMPACT OF HIGHWAYS

Many years ago, I recall reading an interesting comment relating to the timing of the development of the U.S. roadway system and how this related to the Civil War. Clarkson (1952) concluded that, "had our roadway system and the automobile arrived in the 1860s, some forty years before the actual advent of motorized travel, the Civil War would never have occurred." At the time of the Civil War, our country was young, not yet a century old. Roads were still largely limited to local travel. Between urban areas, roads were just dusty trails or cleared paths through forests, and communication was difficult. Is it so unusual to imagine that the Civil War was fought, in part, because people in different regions of the country had not yet been drawn together via our more modern modes of travel and our various means of instant communication (telegraph, telephone, newspaper, and television)? In other words, before the world could be connected in seconds.

Improvement of the U.S. roadway system proceeded in several stages. Near the end of the 1700s, public demand for better roadways between urban areas in the eastern states resulted in the improvement of some roads that were built as, or converted to, toll roads. These improvements were paid for with private money. The first time federal money was used for road improvement was a 35-year project in the early 1800s that produced the National Pike, from Maryland to Illinois. However, expansion of railroad systems in the early 1800s decreased public interest in road building. This stage continued through the Civil War years to the end of the 1800s.

At the beginning of the 1900s, public demand again increased for improved roadways. Farmers demanded better ways to deliver goods to railroads. With the development of the automobile also came the demand for better and more direct roadways between cities. The modern impact of federal money on highway improvements began in 1916, with the first Federal Road Act to improve rural highways.

The benefits of early road construction and improvement were largely limited to people living near the improvement, but the first half of the 20th century can be credited with the development of our present roadway system, which has resulted in more widespread benefits (Wright and Paquette, 1979).

THE COMMENCING OF A CONCEPT

The construction of railroads in the 1800s, the development of motorized vehicles, the trucking industry, and the airplane in the early 1900s provided sources of transportation that forced the United States into a modern roadway concept. We have become a nation devoted to "people and product moving," especially with the design and construction, over the last 50 years, of new state highways and the Interstate Highway System. Our country now contains many large and small businesses, each making commodities that need to be delivered to purchasers. These products are "processed" hourly on our present highway system, which can be labeled the "conveyor-belt" of a tremendous assembly line. The population as a whole now has the benefits of a continentwide system of interconnecting roadways.

HIGHWAY LEGISLATION

Highway design, construction, and maintenance in the United States developed into a joint effort between the federal and state governments. The federal government does not plan, design, build, or maintain any federal highways, but it provides federal aid to help each state do this work. The Federal Highway Act of 1921 required each state highway department to select a set of primary, interstate, and intrastate highways that included not more than 7 percent of the rural mileage in the state. All federal-aid funds were to be spent improving these highways. The Federal Highway Act of 1944 required the states to design roads to be the most direct routes to connect the principal U.S. cities and the principal industrial centers in the United States for the purpose of national defense. This Act also planned the construction and improvement of secondary roads (to connect previously constructed major routes), of farm-to-market roads, and of the urban portions of federal-aid highways.

Several categories of highway now receive federal aid as part of the Federal-Aid Highway System. The highways that have continental importance are the primary system. These include the National System of Interstate and Defense Highways, begun in 1956. Also in this category are the highways in the U.S. system which are marked on road signs by the characteristic shield with the state name plus U.S. route number. These two categories together comprise the primary highways, and these are the highways that are limited to roughly 7 percent of the total rural mileage in each state. The Federal Aid Secondary System and Federal-Aid Urban System are major feeder and connector roads connecting the primary system highways that have been identified as warranting federal involvement (Clarkson, 1952).

More details about the development of the Federal-Aid Highway System are available in Clarkson (1952) and Wright and Paquette (1979).

Most of the engineering design and construction of federal-aid highways took place in the 1960s and 1970s. Why were these decades the peak, when Federal legislation was enacted as early as 1921? The answer is that there were other economic and political considerations than road building in the United States until the 1950s, and it took until the 1960s to see any appreciable results of the response in the 1950s to a considerable "lost-ground demand" (Woods et al., 1960).

In the 1930s, money was not available to build roads because of the Depression. In the 1940s, workers and equipment were devoted to war production, not road building, because of World War II. At the end of World War II, demand was finally able to be addressed, and road-building programs commenced. The 1950s are generally considered the beginning of our present Interstate Highway System, with much credit given to then-President Dwight D. Eisenhower. The interstate highway construction program leveled off in the 1980s and 1990s, but there will always be new connectors to design or new building construction located along highways that requires the surveying retracement of existing highway centerlines or right-of-way lines. This text should help make highway plans and the related mathematics more understandable.

ROADWAY BENEFITS

Multiple benefits emerged from improvements in these various highway systems. A comparison of the United States of today to the United States of 1940 reveals the changes that have occurred in all segments of society:

1. Commercial development
2. Industrial development
3. Agricultural development
4. Natural resources development
5. Appreciation in land values
6. Availability of fire and police protection
7. Access to hospital and medical facilities
8. Availability of public utilities and sanitary systems
9. Access to recreational facilities
10. Strengthening of the national defense

SELECTING THE ROUTE

The selection, design, and construction of a new highway involves many facets that must be reviewed before a state or federal agency will approve a road construction project. An example of the engineering preparation preceding the approval of a new highway is shown in Figure 1.1, showing a route study linking Ohio with West Virginia.

The recommended route by the designers is shown in a solid line from a city in southeast Ohio to the Ohio River. Alternate routes in Ohio are shown in a dashed line, and alternate routes in West Virginia are shown in a dotted pattern.

After the beginning and end points, in general concept, for a proposed roadway have been fixed, several roadway corridors that connect these beginning and end points are analyzed. Frequently, one corridor may be initially preferred, for one of many reasons, and it is designated the recommended design. Then, the recommended route and several alternate routes are analyzed for their respective cumulative costs based on many factors. The factors considered include, but are not limited to, the following:

1. The land area through which each route will pass must be considered for potential land usage, considering the present costs of the property in today's market versus a projected cost of the same property if its highest and best use is attained.
2. The land must also be considered for its potential industrial value. If a route is located through areas that are zoned for manufacturing, future industrial improvements will not only increase the land value but also increase the costs to purchase the industrial improvement or to relocate the improvement to other areas, if the roadway is constructed along that route.

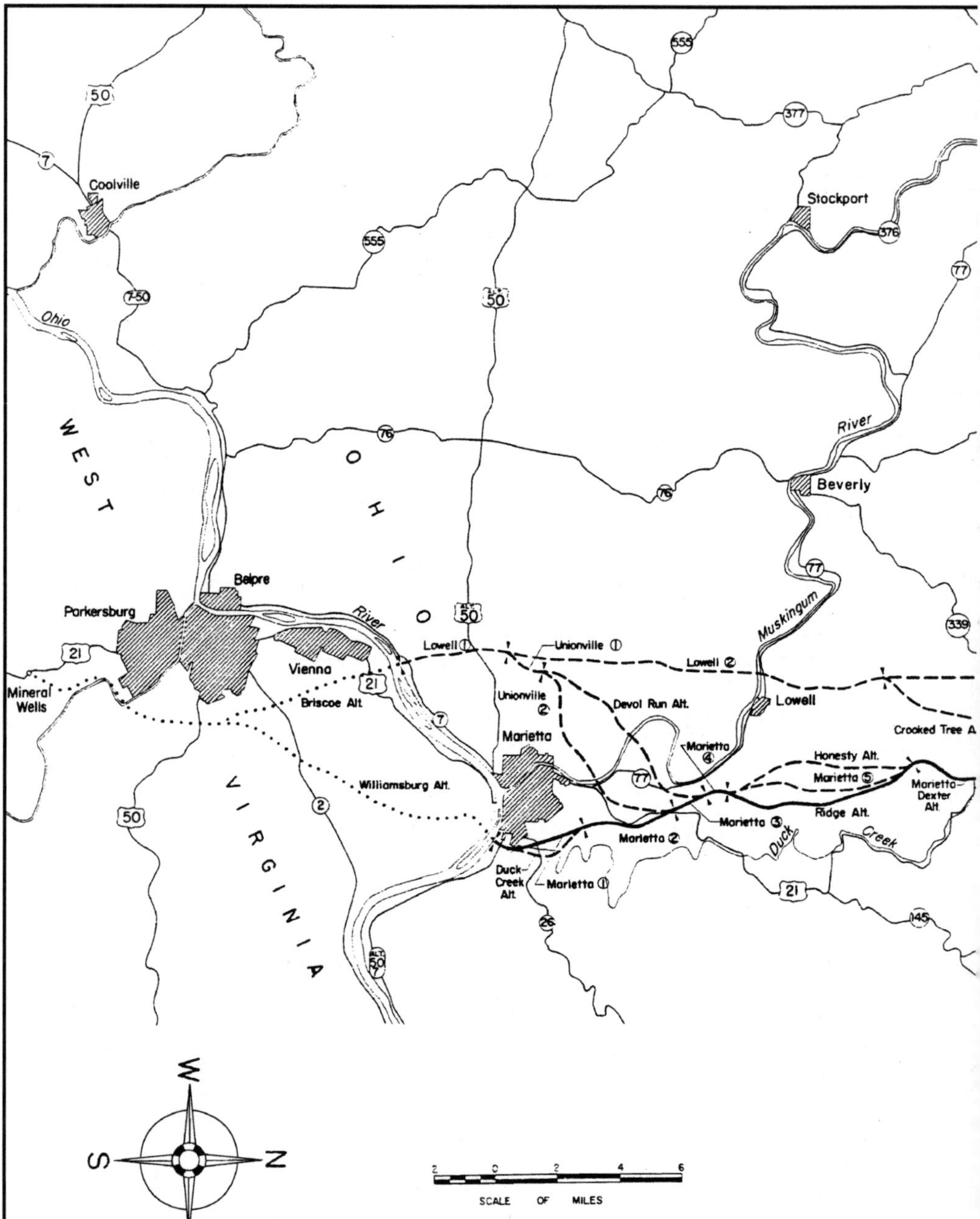

FIGURE 1.1 Preliminary route studies *(Ohio Department of Transportation. Interstate Highway 77. Route Studies. Cen[…] Office. Columbus, Ohio.)*

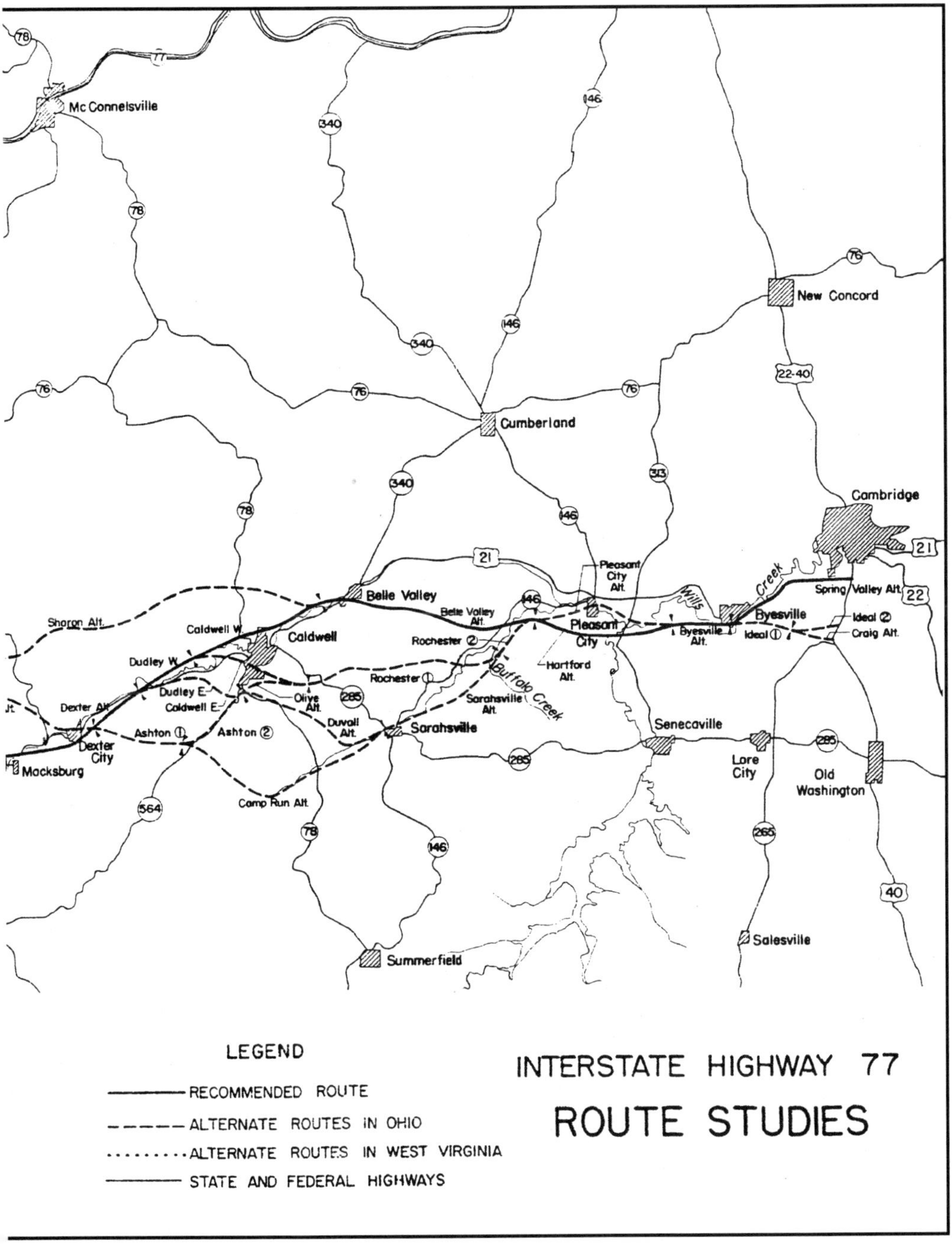
Mc Connelsville
New Concord
Cumberland
Cambridge
Pleasant City Alt.
Belle Valley
Belle Valley Alt.
Spring Valley Alt.
Byesville
Ideal ②
Ideal ①
Craig Alt.
Byesville Alt.
Wills Creek
Sharon Alt.
Caldwell W.
Caldwell
Rochester ②
Pleasant City
Hartford Alt.
Dudley W.
Rochester ①
Dudley E.
Caldwell E.
Olive Alt.
Sarahsville Alt.
Buffalo Creek
Dexter Alt.
Duvall Alt.
Sarahsville
Senecaville
Ashton ①
Ashton ②
Dexter City
Macksburg
Lore City
Old Washington
Camp Run Alt.
Salesville
Summerfield
LEGEND
RECOMMENDED ROUTE
ALTERNATE ROUTES IN OHIO
ALTERNATE ROUTES IN WEST VIRGINIA
STATE AND FEDERAL HIGHWAYS
INTERSTATE HIGHWAY 77
ROUTE STUDIES

3. Existing commercial, city, state and national parks; other recreational facilities; industrial production centers; shopping centers; etc., are all important considerations for their proximity to the proposed route. The closer these facilities are to the proposed route (without actual acquisition), the more serviceable the proposed route will be to the community.
4. A thorough site study of the ground is made to determine general conditions and to locate any obstructions that might interfere with the selection. A ground review is not only conducted in the office using aerial photography and U.S. Geological Quadrangle maps but also involves a review made by walking the route. Many existing objects that would lie next to or be crossed by a proposed roadway can greatly affect the engineering design and cost. These could include existing highways, waterways, and railroads. Many soil and topography conditions such as wetlands or potential slide areas also affect design and cost. The topography of each corridor affects whether horizontal and vertical alignments can be planned that result in similar amounts of excavation relative to fill, a situation that frequently minimizes earth-moving costs.
5. Comparative costs for various alternate routes must be made for utility changes, land acquisition for rights-of-way, soil conditions, major and minor bridges, interchanges, length of the improvement, and engineering fees.
6. A traffic analysis must be conducted between one locality and another to determine whether the roadway improvement is warranted. Local planning commissions, as well as state and federal funding agencies, require an analysis of existing traffic. A daily traffic count is used to calculate the appropriate current traffic volume and a 20-year future prediction of traffic volume, since that time span represents the average life of a modern roadway improvement.

Route studies were formerly a part of Needs Studies, mandated at 10-year intervals by federal legislation. The present Federal Highway Administration (formerly, The Bureau of Public Roads) has oversight on all federal highway funds, disbursing the moneys to the state, but deligating the process of highway selection to each individual state. Consulting engineers and/or state highway engineers prepare studies showing the need for new highways, either connecting major municipalities or to provide greater vehicular mobility for national defense. Rough engineering reports contain certain recommendations for new routes, interchanges, bridges, etc., and include a cost analysis for each proposal.

FINALIZING THE ROUTE

The second stage of highway development involves finalizing a proposed route as part of the preliminary engineering alignment design. During preliminary design, route proposals prepared by consultants, are presented to the state departments of transportation where, for each proposal, cost comparisons are made, traffic analyses are prepared, land use studies are prepared, soils and geology tests are made, environmental impact studies are prepared, and a recommendation of one particular route is suggested. Federal Highway matching funds are requested at this time, if applicable, and approvals to continue the design are requested. Finalizing the route involves a thorough review of the following:

1. Using recent aerial photographs and U.S. Geological Survey quadrangle maps to determine commercial, industrial, subdivision, and individual home density along the proposed route.
2. Locating detailed information of the present streets (sizes, pavement types, and traffic volume analysis), as well as railroad rights-of-way, stream or river navigational information, cemeteries, and other topographic restrictions.

3. Locating detailed information of the existing above- and below-ground utilities, including future upgrading of the facilities by the utility companies.
4. Determining control points that will affect the grade, such as railroads that cannot be raised or lowered, rivers and streams that cannot be grade altered, cross streets that cannot be adjusted vertically, and other physical objects that control the elevation of the new improvement.
5. Reviewing, in conjunction with federal and state agencies, all existing or proposed interchanges or other major arteries that must be connected to the new improvement.
6. And finally, many hours must be devoted to determining the existence, the extent, and the location of any proposed facility improvements being planned by private industry and commerce for future growth.

The third stage of highway development begins when one of the proposed routes is chosen, and appropriation of funds are secured. The alignment design for the route can then be finished, and the roadway design, rights-of-way surveys, and land acquisition can proceed. Usually, for projects less than \$1 million, the state generally provides the engineering, surveying, land acquisition, and negotiation, and then submits the construction stage to bid from private construction contractors. Projects in excess of \$1 million are generally contracted to private engineering design firms, with the construction stage again submitted to bid.

Larger highway improvements are separated into smaller units and awarded to several contractors so that the total cost per project does not exceed \$5 million, excluding bridges. Completion dates are determined by the state department of transportation, subject to the complexity of the project.

Improved distance-measuring and photographic mapping technology and improved coordinate-geometry software now allow several approaches to design and plan preparation. The traditional time for field-staking of the chosen centerline alignment can now be postponed until the beginning of the construction stage. This can be desirable and cost-beneficial when a proposed route passes through congested urban areas. Newer engineering software can now produce computer-generated horizontal alignment, vertical profile, and cross-section plan sheets during the roadway design stage, in addition to showing alternative alignments during the earlier stage of route location study.

The final design stage for highway development starts with the completion of the horizontal and vertical alignment of the centerline. After the roadway width and typical section are chosen, the roadway design proceeds. Earthwork and drainage problems are solved. Bridges are designed. Plan, profile, and cross-section sheets are drafted or produced by computer, and quantities and costs of material and earthwork are computed. Many of these plan sheets and mathematics are explained in greater detail in later chapters in this text.

After the right-of-way width is chosen, property surveying proceeds. Deeds and survey plats are obtained from the local court house for each property owner along the route. The deeds are checked for mathematical accuracy of closure and, in combination with property line evidence located by the surveyor in the field, are used to reconstruct the boundaries for each property owner. Using a right-of-way base map that contains topographic details, existing property lines, and the location of all engineering design features, and with the help of the cross sections, the limits of construction are determined and plotted. The proposed property lines for land to be acquired are planned, corner coordinates and acreage are calculated, and legal descriptions are written. Certified appraisers estimate the fair market value of the land purchases. Negotiators then discuss with owners the impact of the proposed land acquisition on their remaining property and successfully arrive at an agreement to purchase. In cases where no agreement can be reached with the property owner, suits are automatically filed in court for legal determination of the land value.

For the surveyor, the last part of highway development includes much surveying during and after construction. If the horizontal alignment of the centerline, including all points at the beginning and ending of curves and transition spirals, has not already been staked in the field and referenced, that is done prior to construction. Slope stakes are set to show the construction contractor where to make the initial earth cuts and fills for the rough grading of the surfaces underneath the roadway. All bridge and drainage structures are staked horizontally and vertically.

Another round of precise staking helps the contractor place the subgrade roadway layers and final pavement layers in the proper horizontal and vertical locations. Cross sections are needed periodically to help estimate the amount of construction progress for payment purposes. Finally, if applicable, an as-built survey is completed after construction, documenting all approved deviations from the original design.

CHAPTER 2
OHIO AND U.S. LAND SURVEY SYSTEM

OHIO LAND SYSTEMS

I have spent a large part of my career surveying in Ohio, which has the most complex land surveying history of any state. For some projects, that complexity caused many technical problems, but, for the most part, the result was that I learned a lot of interesting history. Parts of Ohio were surveyed using a rectangular system of surveying (which will be described in this chapter); other portions of the state were surveyed using a metes and bounds system (which will be described in Chapter 3); and both systems were used with several variations in different parts of the state. The portions using a rectangular system, newly designed by the Continental Congress after the Revolutionary War, were, in fact, very successful test cases for how to approach surveying all the land west of the original 13 states. Sherman (1925) described Ohio's role in the origin of the U.S. Rectangular System in the following unique way:

> Eternal Rome, whence was quarried the ideas for some of our grandest architecture! Imperial Rome, whence come first principles of the world's most sweeping legal system! How striking that she (eternal Rome) should also have supplied a world-system of surveys! And how strange that this system should have lain dormant through the sleeping centuries of the Dark Ages to be revived and applied anew on a magnificent scale in America, beginning in Ohio, pioneer state of our public domain!

ORIGIN OF THE RECTANGULAR SYSTEM

The origin of the rectangular system can be traced to the Roman Empire. In addition to constructing the most extensive military road system of ancient times throughout Europe, Great Britain, North Africa, and parts of Asia, there is evidence that some Roman cities, military encampments, roads, and land surveys were laid out using rectangular coordinates. One example of such evidence is the markers or monuments found in the interior of Tunis in Africa that show a rectangular layout (Clewley, 1910). There is also much evidence that the ancient Greeks and Egyptians used precision geometry and leveling to position and align buildings and pyramids. Land surveying was certainly practiced also, but it is unclear whether either culture applied a rectangular system to land location.

A rectangular system was used sparingly in Europe after the Roman Empire. The practice was revived in some settlements in the New England colonies of North America in the 1600s and 1700s. Many towns in New England were planned as 6- to 10-mile-square tracts with regularly shaped interior lots so there would be no gaps of land between neighboring lot owners. The original plans said lots were to be staked in the field before occupancy, plats were to be drawn and recorded, and the boundary lines were to be walked and restaked periodically. Some towns achieved these objectives admirably. Other attempts included irregularly shaped inner lots and interior corners that were never staked or were lost due to no later retracement and no restaking. The norm in other colonies was to do "indiscriminate" (metes and bounds) surveys of land after it was chosen for its desirability and occupied, using natural boundaries where possible as property lines and leaving undesirable lands between adjacent properties unclaimed and undescribed.

As North America was being colonized in the early through late 1600s, the wording of land descriptions in the original charters from the various kings of England, or from settlement agencies authorized by the kings, profoundly affected the way subsequent land descriptions were worded when the United States became an independent country after the 1783 treaty with England, Spain, and France. The English charters for land from New England through Carolina defined the north and south boundaries of each tract as specific lines of latitude relative to the equator. Two other aspects of wording in the charters also affected the U.S. land description system. Some, but not all, of the original charters covered land from the Atlantic Ocean all the way west to the Pacific Ocean. Unfortunately, many charters cited latitude lines that overlapped with tract descriptions of charters to the north and/or south of them. Even though few English settled west of the Alleghenies until after 1750, by the time of the Revolutionary War in the later 1700s, all land east of the Mississippi River and north of 31° north latitude (Florida) was claimed in overlapping charters by more than one of the original east coast colonies. All this unoccupied English "Crown" land was claimed by the newly formed United States at the beginning of the Revolutionary War.

Fortunately, the bond formed between colonies by the war provided for a solution to these overlapping land claims. Six of the original 13 states had claims to land west of the colonized east coast; the other seven states had no western claims. After much debate in the Continental Congress, states with western claims began giving their western lands to the Congress to be sold for the common good of all states.

New York was the first to cede its western lands, in 1780; the last state was Georgia, in 1802. The New York cession mollified most of the seven states with no western land claims, some of which had refused to sign the Articles of Confederation until the western lands issue was resolved. Maryland refused to sign until Virginia relinquished its western land claims in 1781; and with Maryland's signature, the Articles of Confederation became legal with two-thirds of the required states' signatures. Two states retained claim to small tracts of western land. Connecticut kept two tracts (the Connecticut Western Reserve and the Firelands) in what would become the state of Ohio. Virginia kept what became the state of Kentucky plus one tract (the Virginia Military Reserve) in Ohio.

Debate began again in the new U.S. Congress concerning how these newly acquired public lands would be divided, surveyed, and conveyed. The New England states wanted to design and survey tracts using a rectangular system; the southern states wanted to continue using metes and bounds descriptions. Both factions were allowed design control over land in different portions of Ohio, so Ohio contains some land surveyed in various versions of a rectangular system and some land surveyed using the indiscriminate method. The test case in parts of northern Ohio of what is now the U.S. Rectangular System was such a success that all other states after the original 13 were surveyed using the rectangular system. The exceptions were West Virginia, Kentucky, Tennessee, and Texas, which were never part of the federal government lands (White, 1983).

THE NATIONAL RECTANGULAR SYSTEM

The U.S. Rectangular System consists of tracts called "quadrangles" (or blocks) that are approximately 24 miles square. Each quadrangle is divided into 16 townships that are approximately 6 miles square; and each township is divided into lots (now called sections) that are approximately 1 mile square. It is interesting to note that early planning for this system, as proposed by Thomas Jefferson in the Continental Congress, suggested 100 one-mile-square lots in 10-mile-square townships. Congress revised this to 49 1-mile-square lots in 7-mile-square townships. It was revised again to the 6-mile-square townships described above before implementation on the ground.

As in the earliest English charters, the boundaries of tracts in this system were set by lines of latitude and longitude. Principle meridians (true lines of longitude) were established as reference lines from which to establish the east and west boundary lines of quadrangles. Baselines (true latitude lines) were established as reference lines from which to establish the north and south boundary lines of the quadrangles. A small copy of C. E. Sherman's map showing the principal meridians and baselines established in the United States is shown in Figure 2.1. A larger-size color map was included in Sherman's third volume (Sherman, 1925). If you don't have access to Sherman's original books or the reprints, you should try to obtain a copy for your reference library, perhaps through a rare book dealer. They are invaluable, even now, for retracement or subdivision surveying of land located within the rectangular system.

Using the principal meridians and baselines, guide meridians and standard parallels were to be staked at 24-mile intervals. Then, range lines and township lines were to be staked at 6-mile intervals. Finally, section lines were to be staked at 1-mile intervals.

Figures 2.2 and 2.3 illustrate the naming of the successive levels of reference lines and the successive subdivision steps for the variously sized tracts in the rectangular system, from quadrangles through sections. Figure 2.4 then shows the next level of subdivision, for an individual section. Townships or sections located east and west of each other are called "tiers;" townships or sections located north and south of each other are called "ranges."

Surveys that are referenced to any of the 34 principal meridian and baseline intersections are described in such a way that one and only one parcel of land can be located as described using this system. The major complication of the rectangular system is that, because meridian (longitudinal) lines converge as they approach a pole, all tracts defined by this system cannot be true squares. The southerly line of each "24-mile-square" quadrangle will truly be 24 miles, while the northerly line of each quadrangle will be less than 24 miles. Therefore, the true dimensions of each "6-mile-square" township will vary depending upon which tier they fall in, with townships in tiers closer to the southerly line of each quadrangle having larger east-west dimensions than townships near the northerly line of each quadrangle. Ignoring measurement discrepancies, the easterly lines should be the same dimensions as the westerly lines for all quadrangles, townships, and sections.

However, at the subdivision level of 1-mile-square sections, the dimensional discrepancies caused by longitudinal convergence are not shared uniformly among all sections.

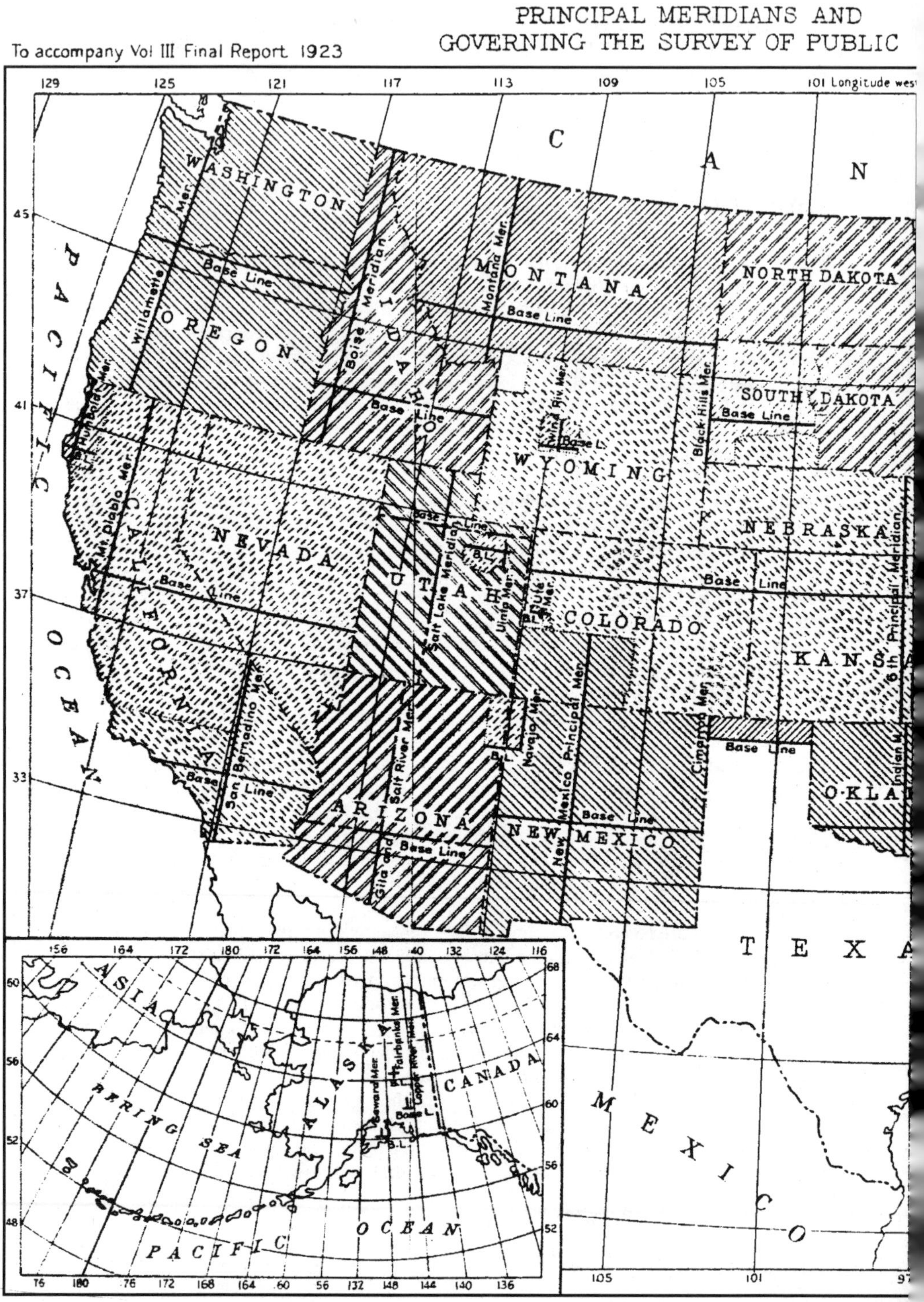

FIGURE 2.1 Public lands of the United States *(Sherman, C.E.* Ohio Lands Subdivision. *1925. Columbus, Ohio. The Ohio S University Press.)*

ACCOMPANYING BASE LINES
LANDS OF THE UNITED STATES

Ohio Topographic Survey
C. E. Sherman, C.E.

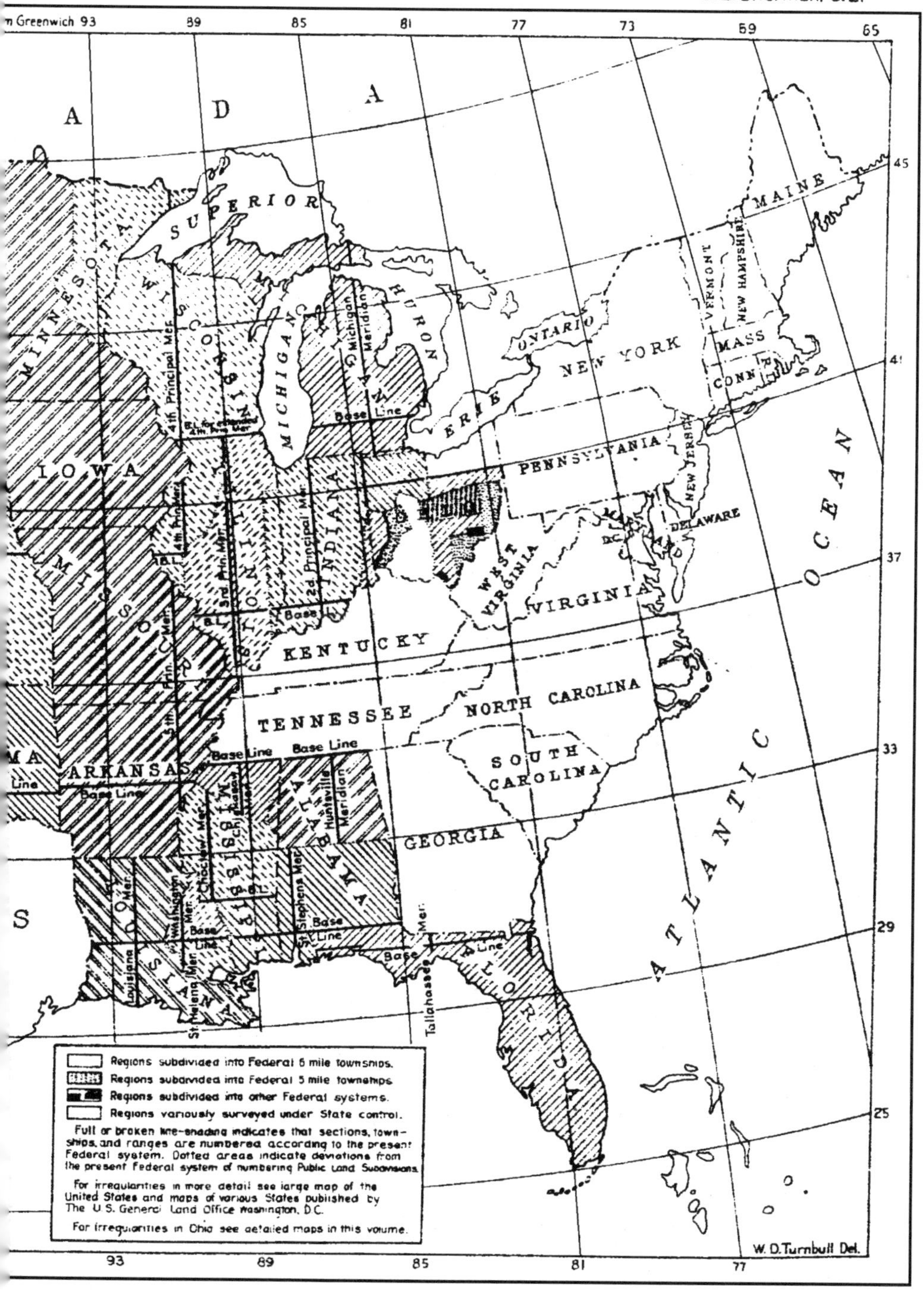

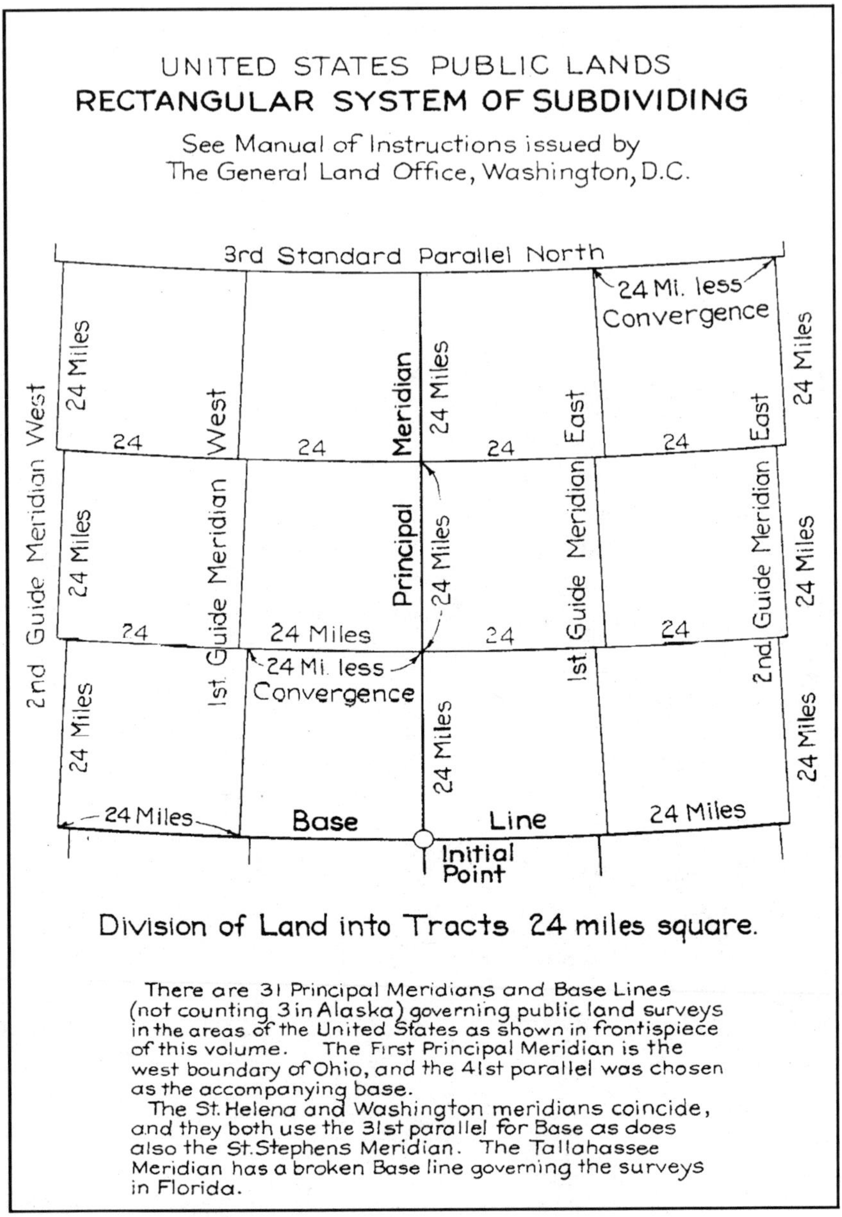

FIGURE 2.2 United States rectangular system *(Sherman, C.E.* Ohio Public Lands: *1925. Vol. III. Columbus, Ohio. The Ohio State University Press.)*

U. S. PUBLIC LAND SURVEYS

SUBDIVISION OF 24-MILE BLOCK

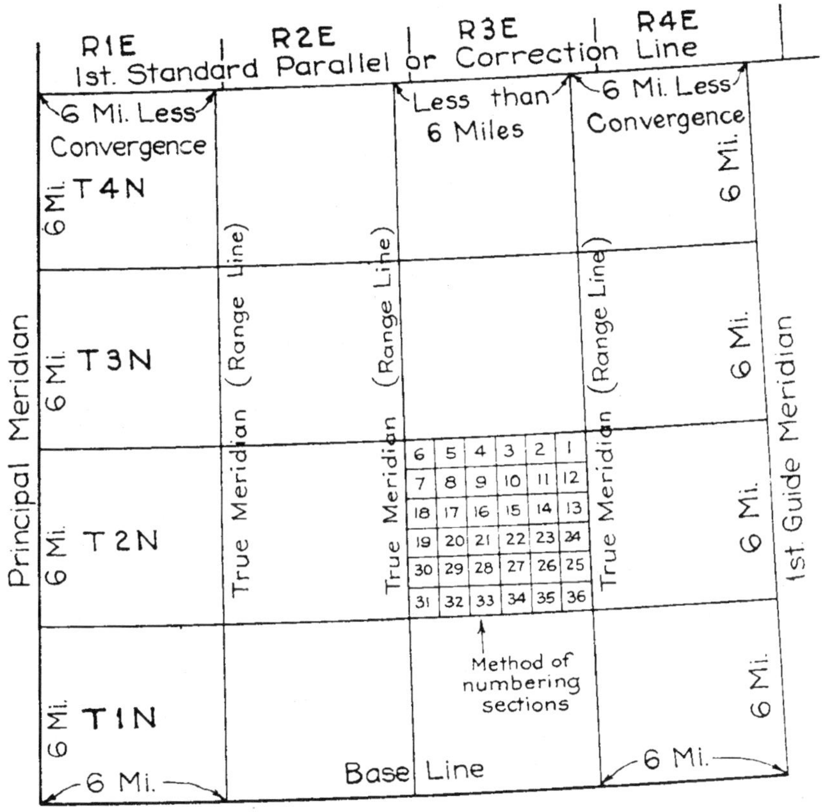

Showing the Division of Land into Townships

Range lines are run as true meridians for a distance of 24 miles. On them township, section and quarter-section corners are set as the range lines are run. Thru the six-mile posts set on range lines, curving lines are run westerly to complete the township boundaries. On these curving lines, following parallels as nearly as may be and yet striking the 6-mile posts on range-lines, mile and half-mile marks are set on the lines when in final position.

FIGURE 2.3 Public land surveys, subdivision of 24-mile blocks *(Sherman, C.E.* Ohio Lands Subdivision: *1925. Vol. III. Columbus, Ohio. The Ohio State University Press.)*

Beginning at the southeast corner of each township, the meridional section lines are run parallel to whichever township line forms the township's southerly line. All convergence discrepancies, measurement errors, and acreage adjustments are forced into the west half of the range of sections lying closest to the west line of the township and into the north half of the tier of sections lying closest to the north line of the township. Figure 2.3 shows one 24-mile-square block or quadrangle divided into 6-mile-square townships, where distance adjustments are made along the northerly parallel or so-called correction line. This figure also shows the method of numbering the 36 sections within each township, where section 1 always commences in the northeast corner of the township, and numbering increases from east to west (Sherman, 1925; White, 1983).

Typical Description In The Rectangular System

The following is a typical rectangular system description:

> "Situated in the State of . . . , Defiance County, Milford Township and being all of the southeast Quarter of the southwest Quarter of Section 6, Township 2 North, Range 1 East, of the 1st Principal Base and Meridian, and containing 40 acres, more or less."

Figure 2.4 shows the graphic location of the above description, where 1-mile-square section 6 is located in the second block north, called "township 2 north," and in the first block east, called "range 1 east." This figure further details the subdivision of a quarter and quarter-quarter section. Each section can be divided into quarter sections, quarter-quarter

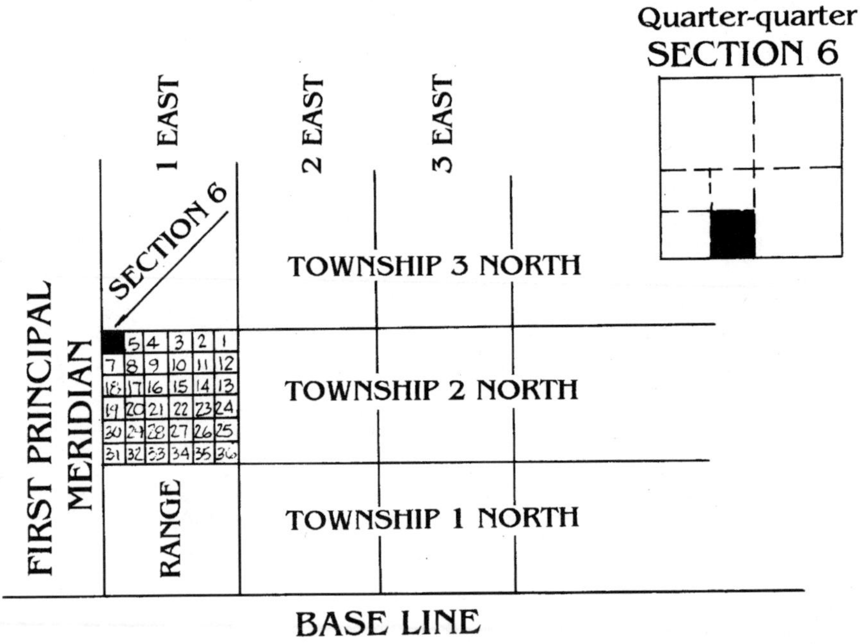

FIGURE 2.4 Rectangular system of townships, ranges, and sections

sections, or even half quarter-sections (law enacted in 1832) (Sherman, 1925). The full section contains 640 acres, while the quarter section contains 160 acres, and the quarter-quarter section contains 40 acres. A helpful hint that can be used when reading a typical description and attempting to locate the parcel of land is to read and draw the description backward. In other words, first visualize and sketch the southwest quarter of a section, then sketch the southeast quarter of that quarter, and so on.

Definitions In The Rectangular System

The following data outline the various parts of the U.S. Rectangular System and their definitions:

1. **Quadrangle = block.** A tract of land defined by latitude and longitude lines nominally 24 miles square.
2. **Townships.** Each quadrangle is divided into 16 townships and each township is nominally 6 miles square.
3. **Section.** Each township is divided into 36 sections and each section is nominally 1 mile square.
4. **Meridian.** Converging longitudinal lines.
5. **Baselines.** Parallel latitudinal lines.
6. **Town = tier.** All townships east or west of a given meridian line.
7. **Range.** All townships north or south of a given baseline.
8. **Fractional townships.** Townships that are less than 6 miles square.
9. **Greenwich, England.** The reference line for longitudinal lines.
10. **Equator.** The reference line for latitudinal lines.
11. Acreage
    a. Quadrangle. Nominally 24 miles square, or 368,640 acres.
    b. Township. Nominally 6 miles square, or 23,040 acres.
    c. Section. Nominally 1 mile square, or 640 acres, 80 chains, or 5280 feet in each direction.
    d. Quarter section. Nominally ½ mile square, or 160 acres, 40 chains or 2640 feet in each direction.
    e. Quarter-quarter section. Nominally ¼ mile square or 40 acres, 20 chains or 1320 feet in each direction.
12. Angular measurement in degrees, minutes and seconds.
    a. 360° in a circle.
    b. Each degree has 60'.
    c. Each minute has 60".
    d. Each degree represents 69.17 statute miles at the equator.
    e. 178° 58' 05" represents 178 degrees, 58 minutes, 5 seconds.
    f. Acre. 43,560 ft^2, 10 square chains.
13. Linear measurement.
    a. Surveyor's chain = 66 ft in length.
    b. One link in a surveyor's chain = 7.92 in, with 100 links per chain.
    c. Surveyor's tape. 100 feet, 200 feet, and 300 feet in length are the most common.
    e. Perch = 16.5 ft
    f. Pole = 16.5 ft
    g. Statute mile = 5280 ft

THE OHIO RECTANGULAR SYSTEM

The first principal meridian for all land in the United States is the west line of Ohio, located at longitude 84° 48' 50" west of Greenwich, England. The closest baseline for Ohio is 41° 00' 00" north latitude from the equator. These are Ohio's only principle meridian and baseline. Most western and northern parts of Ohio were surveyed later than the southern or eastern parts and were subdivided using the National Rectangular System of 6-mile squares that was then applied in most states west of Ohio. The remaining parts of Ohio were subdivided as follows:

1. Using various early versions of a rectangular system, where the land was divided into square townships but not necessarily 6 miles on each side
2. Using various square or rectangular tract or lot subdivisions
3. Using descriptions from indiscriminate surveys

OHIO'S 18 LAND GRANTS

Ohio is unique among the states by virtue of the number and variety of its land subdivisions (Ohio Auditor of State, 1993, and Sherman, 1925). No other state has so many different kinds of original surveys, with Ohio being divided into 18 different "land grants," the first of which is the Virginia Military Survey District of 1784, some 19 years before Ohio was admitted to the Union in 1803. Figure 2.5 shows Sherman's (1925) map, which was the first map ever prepared that showed the location of all 18 land grants. The following lists chronologically the 18 land grants, showing the year(s) of introduction, and provides a few details of each land subdivision system:

1. Virginia Military Survey District (Used metes and bounds descriptions.) 1784
2. Connecticut Western Reserve (Had only 5-mile-square townships, irregularly and nonuniformly subdivided into rectangular tracts, then into many nonrectangular lots.) 1786
3. The Old Seven Ranges (The only tract surveyed by government surveyors for the Continental Congress, but the requirements were waived for the use of true meridional lines, and magnetic bearings were used; compensation for longitudinal convergence was not handled as in the present-day system; numbering system for sections is unconventional.) 1786
4. The First and Second Ohio Company Purchase (Used an earlier version of the rectangular system and had unusual sizes and shapes of fractional sections because lots within sections needed to be planned so that all shareholders obtained lots according to the number of shares owned.) 1787
5. The Fire Lands (Had only 5-mile-square townships, which were then subdivided into only four quarters.) 1792
6. The Donation Tract (Was not part of the Ohio Company Purchases but was surveyed by them. Townships were not divided into sections but into irregularly shaped "allotments" chosen by settlers, then subdivided into many shapes and sizes of rectangular lots.) 1792
7. Symmes Purchase (One of only two portions of the United States where ranges and towns were named and numbered using an orientation that is 90° opposite from the convention used in the U.S. Rectangular System, corners were very poorly surveyed

by nongovernment surveyors, and the original owners were required to arrange for surveying on their own.) 1794

8. The French Grants (Consisted of several shapes and sizes of mostly rectangular lots.) 1795 and 1798
9. United States Military District (Had only 5-mile-square townships, which were then subdivided into only four quarters, then into several sizes of rectangular lots.) 1796
10. The Refugee Tract (Used the same townships and sections as the Congress Lands south of it, but then the sections were divided in half with a north-south line to obtain 320-acre lots.) 1798
11. Congress Lands (Compensation for longitudinal convergence was handled unconventionally.) 1798–1801
12. Congress Lands (Compensation for longitudinal convergence was handled unconventionally.) 1798–1802
13. Congress Lands (Compensation for longitudinal convergence was handled unconventionally.) 1799–1804
14. Congress Lands (Used the same opposite orientation for towns and ranges as in the Symmes Purchase, but surveying was better than in Symmes because it used government surveyors.) 1802–1805
15. Two and Twelve Mile Square Reserves (Consisted of lots surveyed by the government. Then surrounded two military forts, and were excluded from the U.S. Rectangular System that surrounds them.) 1805
16. Congress Lands (The most uniform surveys in all of Ohio, where government surveyors used a version of the U.S. Rectangular System that is the closest to the current version used in all later states and was the last and best test in Ohio of the U.S. Rectangular System.) 1819
17. Congress Lands (Same as 16.) 1820–1821
18. The Michigan Survey (Land disputed between Ohio and Michigan; surveys were done by extending meridians and baselines from Michigan into the area.) 1836

PROPERTY LINE RETRACEMENT DIFFICULTIES

The perimeters of most of these land grants were originally located by federal government surveyors, but then did not necessarily locate the interior lines. Much confusion and litigation occurred when the interior lines of the land grants were later surveyed under nongovernmental guidelines. The confusion was generally caused by inaccurate surveys, where the interior lines of townships or sections as originally staked in the field contained errors of different magnitudes. In the retracement of these lines, adjustments for the errors were attempted. Ohio law has long recognized that these were ill-advised adjustments. The courts now request that the lines retraced in all later surveys be replaced as originally surveyed, even if originally staked in error. Therefore, lines of occupation carry strong evidence of property division, and disputes are becoming less of a burden on the courts.

Even with these errors associated with the rectangular system, land described relative to the reference lines in the U.S. Rectangular System most often results in fewer boundary disputes or retracement difficulties than either a description of a rectangular-shaped lot that is described relative to some civil subdivision under local laws or a description of an irregularly shaped parcel defined by natural boundaries.

FIGURE 2.5 Ohio's 18 Land Grants *(Ohio State Auditor's Office. Reprint 1995.* The Building of Ohio. *Columbus, Ohio. Jim Petro, Auditor of State. Sherman, C.E.* Ohio Lands Subdivisions: *1925. Vol. III. Columbus, Ohio. The Ohio State University Press.)*

The U.S. Rectangular System of surveys, as opposed to indiscriminate surveys (see Chapter 3), is mathematically controlled by using astronomical observations of star or sun positions as a means of establishing accurate bearings, preparing legal descriptions, and locating the properties. Great positional accuracy can be obtained by commencing a survey from these government-established control intersections, while metes and bounds surveys are usually oriented by their relation to some natural land boundary. Different surveyors may choose different commencing lines and corners, thus leading to slightly different positions for the same property on the ground.

This has resulted in much litigation where metes and bounds surveys are prevalent. It has been estimated that more money has been spent by individual property owners for surveys in Ohio than in all the other states combined.

MUNICIPAL, COUNTY, AND STATE BOUNDARIES

All who live in growing cities are familiar with the rapid changes in the location of corporation and local township lines. And yet, these lines have been used in some legal descriptions to describe property boundaries. Government-surveyed reference lines, on the other hand, can be perpetually used in the transfer of property because the control lines and intersections never change. Can you imagine trying to locate a property line that followed a county line, if the legal description was written when Ohio was admitted to the Union with 20 counties in 1803? Since 1848, Ohio has contained 88 counties and has experienced many changes in those lines through the years. So, many old deeds can be misleading if county lines are used to describe property. Not only are county lines subject to change, but so are state lines. As an example, until the year 1863, West Virginia was a part of Virginia; and Ohio, until the year 1803, was also a part of Virginia.

For some properties, it may even be difficult to find what county or state has saved the older deeds and survey records. In addition to your local city and/or county auditor, recorder, and engineer, many state and federal offices contain records you may need. Local to my area are the federal Historical Archives in Frankfort, Kentucky; the State Auditor's offices in several states, and the library at Marietta (Ohio) College. The information in this chapter is in a brief outline, and many conditions arise in the field that require special consideration. If you need more details pertaining to retracing federal government surveys, you should request the most recent "Manual of Instructions for the Survey of the Public Lands" issued by the General Land Office in Washington, D.C. Much more fascinating historical information about the U.S. Rectangular System is available in White (1983) and information about Ohio's land grants can be found in Sherman (1925).

CHAPTER 3
INDISCRIMINATE SURVEYS

INTRODUCTION

The land conveyance system in the United States that is acceptable to the general public, the courts, title companies, insurers of property, attorneys, and surveyors involves a "deed" that is prepared based on a "legal description," and the recorded document is used to transfer property. The deed, prepared by an attorney, is recognized as the total legal document used to transfer ownership of land from one party to another, while the description is that portion of the deed prepared by the surveyor to describe the physical boundaries of the property.

Indiscriminate surveys refer to land that has not been surveyed using the U.S. Rectangular System referenced to meridian and baseline intersections, but where deed descriptions are prepared based on random boundary locations, which may or may not include natural boundaries. The lighter area of the United States map in Figure 2.1 represents the states where indiscriminate surveys dominate, while the darker area represents states using the U.S. Rectangular System of surveys.

Indiscriminate surveys are described in at least two ways:

1. By reference to the boundary of an adjacent owner by name
2. By means of the survey method identified as "metes and bounds"

By definition, the word "mete" means to allot by measure, and the word "bound" means to form the boundary of. To make these measurements, early surveyors in the United States used a compass and chain. Today's surveyors use steel tapes, precise transits, electronic distance measuring (EDM) devices, and/or "total station" devices that include the transit and EDM device housed in the same frame. Tomorrow's surveyors will probably use global position satellite systems for some projects.

The considerations described below apply when surveying U.S. land for conveyancing regardless of the devices used to measure the boundary lines and locate the boundary corners, at whatever accuracy.

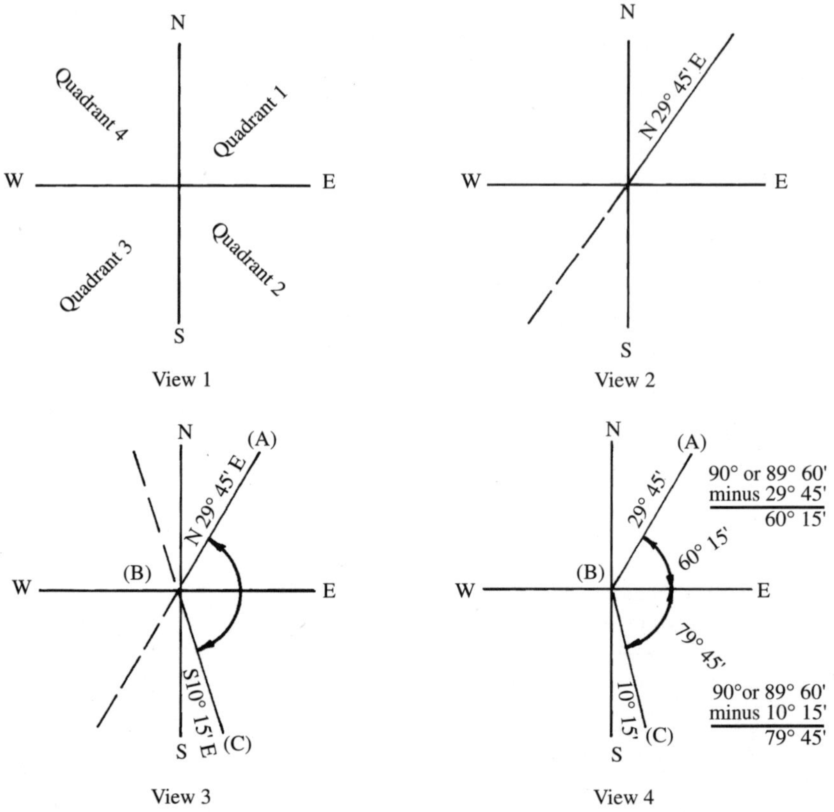

FIGURE 3.1 Manual method of computing angles

ELEMENTS OF A DEED DESCRIPTION

There are five considerations used by surveyors, attorneys, and the court systems of the United States when recorded instruments (deeds) are being reviewed for property ownership and description determination. These are intent, identity of title, location, geometric shape, and size.

Intent

A good legal description must unambiguously locate the land. The seller and buyer must be able to read and understand the description, so the seller knows this is what he or she intends to convey and so the buyer knows what he or she is buying. The wording used should be clear enough that anyone looking at this description, as part of the public record, can also understand the location of the property.

The surveyor accomplishes this by using a narrative in the description that combines language plus a mathematical identity of the boundary lines.

If a dispute arises concerning a land title, the dispute goes to a judge or jury. The court will review the deed and the property description, as the legal profession says, "from the four-corners thereof." This means that the court will search for words within the whole of the deed that will lead to a conclusion concerning the intent. If the court is still unable to find the intent from the description, it will consider evidence (called extrinsic testimony) offered by various individuals on behalf of each party. Extrinsic evidence can include information found in other adjacent deeds or plats and testimony by disinterested parties who have personal knowledge that might contribute to the solution. Most frequently, when testimony is presented, witnesses relate property lines to natural objects such as trees, streams, and rivers, and use descriptions such as down the hollow, top of the ridge, and near the hole in the ice. When the natural object can be found and it is determined the object is original, the court generally accepts this as primary evidence of the boundary, to the exclusion of mathematical metes and bounds (courses and distances). If the intent can be clearly stated, the court will uphold (sustain) the description based on natural objects, regardless of the bearings and distances contained in the survey.

Note that this is similar to the court conclusions mentioned in Chapter 2, concerning original surveying errors in the rectangular system surveys. The corners as staked, even with errors, are given a greater weight of evidence than a mathematical description of a line.

Title Identity

The title must identify a certain parcel of ground where no conflict exists with other senior properties (those properties recorded earlier). A deed which expresses the location, the geometric shape, and the size of the parcel is not sufficient if the description includes part of the neighbor's property. Therefore, a thorough boundary review of all senior property rights must be finalized before the junior parcel can be described. (Brown, 1957).

When conflicts do occur between junior and senior property rights, such as overlaps (deficiency) and spaces (excess) between properties, such conflicts must be resolved before the legal description can be finalized. Perhaps each deed taken separately describes a perfect geometric shape, expresses the area of its respective parcel, and refers by reference to a government or subdivision control, but when the parcels are surveyed as a unit, spaces or overlaps exist. When overlaps occur, the senior parcel (earliest recorded deed) has priority, but when spaces occur (assuming the space is not previously described), the former grantor (owner of the parent tract) technically has title to the land.

When subdivisions are recorded, on the other hand, they are considered "simultaneous conveyances" and there are no junior or senior rights. If errors exist in the platting of a subdivision, those errors are prorated according to the lineal feet of frontage or backage in each lot as related to the error.

When title identity is in question, the court, attorney, and surveyor should come together to solve this conflict. Occasionally, two property owners can exercise their rights and resolve property disputes by mutual agreement without court action, but whatever decisions are made by the combatants, these agreements must be surveyed and recorded in the public records. To do otherwise will result in the disputed property lines being subsequently surveyed in their original positions, with the conflicts continuing.

Many years ago a private property owner threatened the state by suggesting that traffic would be halted on a completed section of freeway, and that admission would be charged over the land in question, because the highway department had not appraised, negotiated, and acquired his property. The private property owner knew of the problem very early in the project but purposely neglected to inform the state highway department or the contractor of his property ownership during design or construction. Title identity in this situation became acute, and the lack of full and complete acquisition could have resulted in a complete freeway shutdown. The problem was resolved due to the good nature of the wayward owner, but only after much embarrassment to the engineering designers who prepared the plans and to the highway department in general.

Location

Every description must include the state, county, township, town and range, section, principal meridian and baseline (if applicable), and other municipality (if pertinent) in which the land is located. The description requires a point of beginning that is referenced to other adjacent properties, and in many states, the beginning of the description and survey must be field-located in relation to a section, half section, quarter section, quarter-quarter section, or some other prominent locatable point. A good point of beginning is a fixed physical monument, such as an iron pin or pipe, new concrete corner, old sandstone corner set during government or county surveys, centerline of interstate or county roads, or other points that can be physically touched, as opposed to "paper" corners that may have no accurate or unambiguous location.

Natural objects used to describe a location, such as beginning at "the hole in the ice" (referred to earlier) can be humorous, and thought to be an impossible point of beginning. The use of extrinsic evidence to help in locating a parcel (as described in title identity) was an important solution to an actual "location dilemma" involving "the hole in the ice."

Discussions with adjacent neighbors, senior citizens, and disinterested parties often solve the most bizarre deed notations with ease. In this situation, extrinsic testimony revealed that the hole in the ice referred to a point caused by a spring under the floor of the lake that continued to flow in summer and winter, causing the hole in the ice to occur. The hole was located and it was a perfectly acceptable "point of beginning."

Geometric Shape

Every description must define a closed geometric figure that has a particular (regular or irregular) shape that can be graphically plotted by using geometry, trigonometry, and other drawing skills. The mathematical description of each line of the closed figure must result in an "error of closure" for all sides of the total figure that is within an acceptable accuracy. Details on closures and acceptable accuracy are covered later in this chapter.

Size

Every description must contain the number of square feet and acres encompassed within its boundaries to an acceptable accuracy. The words "more or less" are generally included following the declaration of acreage in a description. These words were originally included in early surveys where some boundary lines (e.g., creeks) may not have had a mathematical description. Determination of acreage, in these cases, could only be approximated. To the surveyor today, it is more accurate to interpret the words "more or less" as representing the number of square feet lost or gained based on the number of digits shown to the right of the decimal point for the acreage.

Example:

When converting 3.6543 acres to square feet (3.6543 times 43,560, which is the number of square feet per acre), the product is 159,181 square feet (ft^2). Suppose, instead, the digits to the right of the decimal point for this same acreage were rounded to two places. In that case, when converting 3.65 acres to square feet (3.65 times 43,560), the product is 158,994 ft^2, showing a difference of 187 ft^2. In this situation, no field or mathematical error has occurred. The difference is just the result of rounding off the number that was calculated for the acreage.

TYPES OF LAND DESCRIPTIONS

There are many ways of describing land when transferring title from one property owner to another. Of the five methods described below, only the metes and bounds description will be discussed in this chapter. The U.S. Rectangular System was previously discussed in Chapter 2.

1. Title can be transferred using descriptions referencing the U.S. Rectangular System.
2. Title can be transferred by registering the property with the courts as Torrens land.
3. Title can be transferred in total (without survey), if there are no existing restrictions set by the city or county.
4. City and/or subdivision lots can be transferred by referring to all or part of a lot as recorded in a subdivision of record. Many restrictions to this method exist, and the attorney and surveyor must be completely aware of the land survey requirements in their local area.
5. Title can be transferred using descriptions from indiscriminate surveys, which need a course (a direction) and a distance, and perhaps a physical description, for each boundary line.

An example of a description using the indiscriminate method follows:

Situated in the Southeast Quarter, Section 7, Township 12,
Range 8, County of Stark, State of Ohio

Sublot No. 1

Being a part of a tract of land presently owned by George Davis Jones, et al., and recorded in Volume 2967, page 962 of the Stark County records and bounded and described as follows: Commencing at a point on the west line of said Quarter Section, which point is south a distance of 865.50 feet from the center of said Section 7; thence N° 86 10' 00" E a distance of 461.11 feet to a point (the above portion of the lead-in is recorded in Volume 2731, page 642 of the Stark County records); thence continuing along the same course a distance of 88.20 feet to a point; thence N 2° 55' 00" E a distance of 42.78 feet to the True Place of Beginning; thence continuing along the same course N 2° 55" 00" E a distance of 508°. 72 feet to a point; thence S 88° 18' 00" E a distance of 149.67 feet to a point; thence S 67° 01' 00" E a distance of 137.20 feet to a point, said point being on the westerly Right-of-Way of Amber Drive, said Amber Drive is recorded in Plat Book 13, page 41 of the Stark County records; thence S 57° 43' 00" W along the westerly line of Amber Drive a distance of 20.00 feet, said point being the PC of a circular curve on Amber Drive; thence along the line of the circular curve and the Right-of-Way of Amber Drive a distance of 165.60 feet, whose radius is 87.45 feet, whose delta angle is 108° 30' 00", whose chord length is 141.94 feet and whose chord bearing is S 3° 28' 00" W to a point, said point is also the PT of the circular curve on Amber Drive; thence S 42° 52' 29" W a distance of 406.18 feet to the True Place of Beginning, and containing 87,078 square feet (1.999 acres), more or less, as surveyed by John Jones in May, 1964.

The description above introduces many technical terms and abbreviations you may not yet recognize, but they should become more familiar as you continue to read this book.

GRAPHIC PRESENTATION OF LAND DESCRIPTIONS

Plotting deed descriptions is an important tool for the professional. Knowing and visualizing sizes, shapes, and areas of properties can be invaluable for planning field searches for property evidence, comparing the descriptions of neighboring parcels, preparing graphic presentations for many purposes, and even understanding your own deed better as a private land owner. Few counties require graphic presentations of individual deeds as a prerequisite for recording. However, the property owner may record, at his or her discretion, any document (property plat, letter, lease agreement, etc.) that is deemed necessary. Graphic visualization is such a powerful way to convey information that many property owners want to include a drawing in their legal conveyance. On the other hand, subdivision design drawings, called "plats," of divisions of large properties into a few or many lots are required to be recorded in the county within which the subdivision is located.

ANGULAR ORIENTATION

The exterior boundary lines of property designed into subdivisions, as well as almost all roadway designs, need to be oriented to some angular reference in order to establish bearings on the property lines. The universal angular reference used most frequently is "true north" (astronomical). Individual deed descriptions can also be referenced in this manner, but individual properties will most often use bearings from earlier surveys or bearings found in neighboring deeds. These bearings will probably be compass bearings, not astronomical bearings.

The most common surveying methods to establish a bearing related to true north involve precise astronomical observations, usually of the sun or of a star like Polaris. Both the solar and star observations can be made with standard surveying equipment, and a bearing can be transferred to any property line, traverse line, or other surveyed line where this degree of accuracy is required. Angles obtained from astronomical observations are mathematically corrected to a true bearing based on the exact year, day, hour, minute, and second of the observation. The federal government and/or commercial companies publish a text, generally covering a 2-year span, called *Celestial Observation Handbook and Ephemeris*, which can be used. The surveyor needs special night lighting equipment when observing Polaris and special safety filters when observing the sun. The angle-to-bearing calculation is simple enough for either programmable computers or hand-held calculators.

PLOTTING DEED DESCRIPTIONS

When computers are used in the plotting process, two-dimensional coordinates are calculated and stored in the computer for the beginning and end points of each property line, and the boundary corners of the property are plotted from these rectangular coordinates. Computer-aided drafting (CAD) programs can provide much assistance in many aspects of surveying and for subdivision and roadway design. With some graphic design programs presently in use, the highway design can be viewed on the computer monitor rather than on paper. Cross-section and profile views can be plotted from the computer's coordinate geometry memory of stored topography to assist in choosing a design, before any conventional physical plan sheets are ever drawn by a drafter or computer. The possibilities are unlimited, but the programs are expensive.

Hand-plotting by coordinates is also a viable alternative for viewing deed descriptions. An assumed rectangular coordinate grid system is established on graph paper and each individual point is plotted after coordinates are obtained. Another type of hand-plotting (by angle and distance) uses either meridian reference lines, interior angles, deflection angles, or "field-observed" angles (when plotting lines of a traverse rather than boundary lines from a deed description). For these types of plotting and for many other graphic presentations, a protractor and engineer's scale are still used when expensive computer software and plotters are not available. Trouble shooting the software when problems arise is much easier if you know the logic behind the procedure and if you can do the procedure by hand. The examples in this text are presented so you can see how the procedures and/or mathematical calculations are done by hand. You can then transfer that knowledge to a more confident use of software and drafting aids if they are available to you.

Bearings

Bearings are usually used to define the direction of a line in legal descriptions. The convention for a bearing is that the angle used is the angle that the line makes relative to either a due north or due south meridian reference line, plus the direction of the angle relative to north or south. The primary direction of the line is written before the angle and is always either N or S. The secondary direction of the line is written after the angle and is always either E or W.

For example: The line defined by S 10° 15' E shown on (Figure 3.1) has an angular difference from due south of 10 degrees and 15 minutes. Common ways of writing this are N 59-47-05 W or S 59° 47' 05" E for a bearing of south 59 degrees 47 minutes 5 seconds east. All bearings have reversible directions, depending on the direction one is traveling. The angle of the bearing will not change, but the directions can reverse, e.g., NE becomes SW. All bearings fall into one of four quadrants, which are numbered clockwise from due north as: the northeast quadrant, quadrant 1; the southeast quadrant, quadrant 2; the southwest quadrant, quadrant 3; and the northwest quadrant, quadrant 4. No bearing angle can be greater than 89° 59' 59". When this occurs, the bearing moves into the next adjacent quadrant. When the bearing is 90° from due north or due south, the bearing becomes due east or due west. A set of bearings for a property line or a traverse can be referenced to astronomical north, to magnetic north, or to an assumed grid system.

Azimuths

Azimuths are defined by angles based on a circle of 360°, beginning with 0-00-00 for due north and continuing in a clockwise direction until the azimuth equals 359-59-60, thus returning to due north.

Plotting Angles by Meridian Method

Plotting a deed (or a traverse) using angles and distances can be done in several different ways. Meridional reference lines can be drawn as phantom lines at the point at the end of each boundary line or traverse line. The two phantom lines should intersect at 90° to each other, forming a baseline and meridian. As shown at all four points of the tract or traverse shown on (Figure 3.2), the convention for most plotting is to draw the reference lines so that the top of the page is due north.

With a protractor, an angle equal to the required bearing can be measured from the N-S meridian line at each point, with an accuracy commensurate with the equipment being used. This method establishes the direction of each line, whereas the length of each line is measured using an engineer's scale. At each angle break in the survey, an additional baseline and meridian are drawn, and the next line is plotted using the angle from the N-S reference line and the distance. This continues around the deed perimeter, with the final

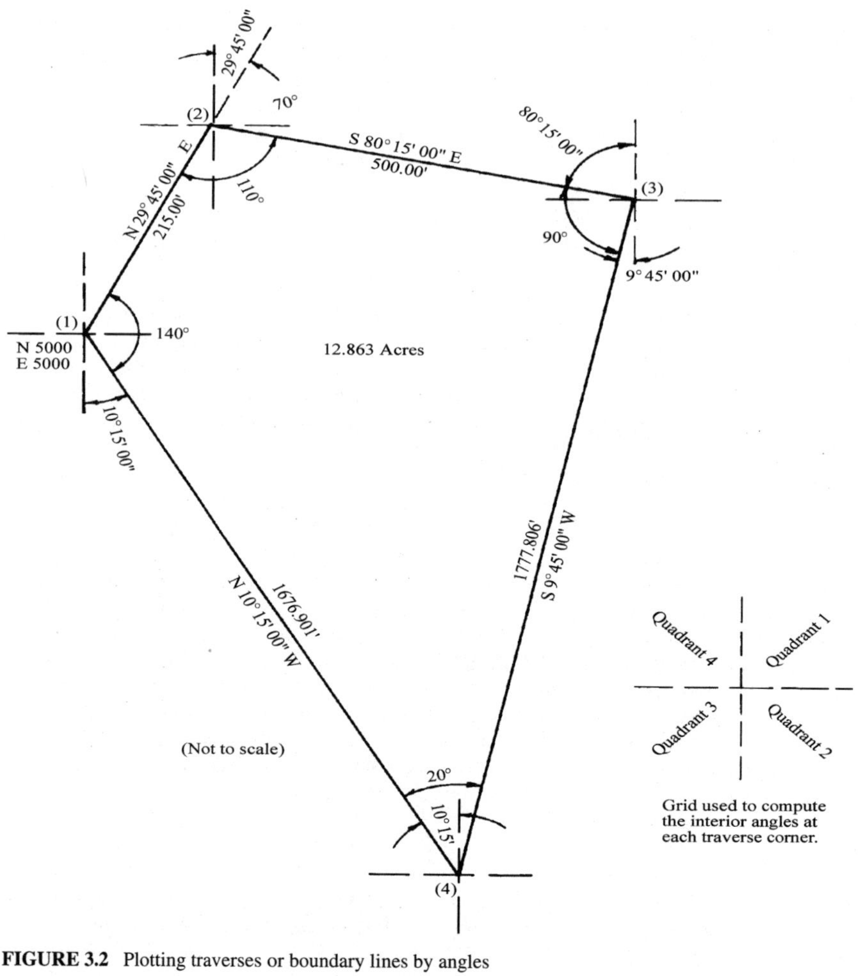

FIGURE 3.2 Plotting traverses or boundary lines by angles

closing line returning to the place of beginning, the point at which you started plotting. If your plotting skill is good and the equipment used allows adequate accuracy, this type of layout will produce a good approximation of the size and shape of the deed description. If you made no errors in plotting, but the final line does not end at the beginning point, there may be an error in one or more bearings or distances in the deed description. The mathematical procedures to check a deed description or a closed traverse for accuracy will be described later in this chapter.

Plotting with Interior Angles

Deed descriptions or closed traverses can also be plotted using the angles at each angle break in a closed figure that are interior to the figure. For a closed traverse, plotting with interior angles has the advantage of forcing the determination of interior angles at an early stage, which needs to be done anyway in order to verify the mathematical accuracy of the traverse before plotting commences. For a traverse, the interior angles may be the angles actually observed in the field, so no additional computation would be necessary. Deflection angles (the "supplement," or angle that brings the sum of two angles to 180°) can also be used for plotting. The deflection angle of 70° relative to the interior angle of 110° is shown at point 2 in Figure 3.2.

Figure 3.1 shows examples of two angles to plot from the N or S reference meridians if you use the meridian method for plotting. One example shown is an angle of 10°15' to plot from due south to obtain a bearing of S 10°15' E. Figure 3.1 also shows the calculations needed to determine the interior angle at point 1 of the tract or traverse shown in Figure 3.2.

From Bearings to Angles

Plotting by angle requires that the angle be computed between adjacent traverse lines where the bearings are known. A simple example of this is found on Figure 3.1 and more complex examples in Figure 3.2. The bearings for the example in Figure 3.1 are as follows:

1. N 29° 45' 00" E, second view
2. S 10° 15' 00" E, third view

To complete an angle computation, prepare a sketch showing two lines intersecting each other at right angles, forming four 90° quadrants, such as shown in Figure 3.1 (first view). Sketch the approximate location of the first bearing, N 29° 45' 00" E, in the northeast quadrant of this intersection (second view). Extend the bearing in a back direction, using a dashed line, to distinguish the direction of your survey from the same bearing but in an opposite direction. The dashed lines are quite often used to help you visualize the problem. Continue by graphically locating the second bearing, S 10° 15' 00" E, as shown in Figure 3.1 (third view), extending the line in the back direction as before, and compute the angle shown by arrows at point *B*.

Each quadrant contains a total of 90°. Therefore, the remaining angle within any quadrant (the "complement," or angle that brings the sum of two angles to 90°) is simply a subtraction of the angle of the bearing (in degrees, minutes, and seconds) from 90°. Figure 3.1 (fourth view), shows the remaining angles after the bearings are subtracted from 90°. Angle *ABE* equals 60° 15' 00" in quadrant 1, and angle *EBC* equals 79° 45' 00" in quadrant 2.

At this point in your computation, review the sketch to determine the interior angle required and decide if the bearings can be used in their exact positions in quadrants 1 and 2 to make the computation, or if you must use one or more of the dashed bearing lines in their back directions in quadrants 3 and 4 to more easily visualize the angle being computed. The example in Figure 3.1 shows an interior angle that can be computed using the bearing lines in quadrants 1 and 2. Therefore, angle *ABC* is computed by adding the two remaining (or complement) angles in both quadrants, and the interior angle of these two bearings thus becomes 140°.

Complex Bearing-to-Angle Computation

Figure 3.2 could represent either a closed survey traverse or the boundary lines of a recorded deed description. The interior angles are computed for plotting as follows: To compute each interior angle before plotting, draw the phantom reference grid at each corner on a sketch, as in Figure 3.2, where you can visualize the interior angles and determine whether the bearings of the lines, when converted to angles, must be added or subtracted.

Traverse corner no. 1. This interior angle was explained and computed above (see Figure 3.1) and is 140°.

Traverse corner no. 2. The corner at point 2 in Figure 3.2 consists of the bearing S 80° 15' 00" E, going away from the point, and the bearing N 29° 45' 00" E, coming toward the point. An observation of these two bearings relative to the south line of the grid reference lines shows that the 80° 15' angle in quadrant 2 and the 29° 45' angle in quadrant 3 can be added to obtain the interior angle. Therefore, the interior angle at point 2 is 80° 15' 00" plus 29° 45' 00" for a total of 110°.

Traverse corner no. 3. The interior angle at point 3 in Figure 3.2 is somewhat more difficult to visualize. The bearing of line 2-3 (S 80° 15' 00" E) represents the large angle in quadrant 4, and the small angle in the quadrant is required to compute the total interior angle. Therefore, the angle required in the fourth quadrant is the complement, or 90° 00' 00" minus 80° 15' 00", resulting in a complement angle of 9° 45'. The bearing of line 3-4 (S 9° 45' 00" W) represents the small angle in the third quadrant, whereas the large angle is required to compute the total interior angle.

The complement of this angle is 90° 00' 00" minus 9° 45' 00", resulting in a complement angle of 80°15'. Therefore, the interior angle at point 3 is 80° 15' 00" plus 9° 45' 00" for a total of 90°.

Traverse corner no. 4. The interior angle at point 4 in Figure 3.2 is less complicated than the previous point, if you relate the bearings in the traverse to the north line of the grid reference lines. The bearing angle in quadrant 1 and the bearing angle in quadrant 4 result in an interior angle of 20° (9° 45' 00" plus 10° 15' 00" equals 20°).

Angle Summation Check

When this stage in the computation has been reached, you probably feel ready to plot. However, certain angular checks must be made first, thus eliminating unnecessary drafting time if the angles have been incorrectly computed from the bearings in the deed description. For a closed traverse, this angular check will show whether the angles were observed in the field to a sufficient level of accuracy. This check involves the addition of all interior angles in the closed figure to determine whether they equal the total sum expected for the interior angles in a figure with the same number of sides. A simple formula for the sum of the interior angles in a figure with any number of sides is (n minus 2) 180°. In other words, the sum of the angles in any closed figure is equal to the total number of sides, diminished by 2, with the result multiplied by 180.

Example:

Assume that a seven-sided figure has been surveyed and the summation of the interior angles is calculated to be 899° 58' 42". Using the computation outlined above, diminishing the number of sides in the figure by 2 leaves 5 sides, and multiplying this result by 180° gives a product of 900° for the interior angles of a seven-sided figure. The interior angles in the seven-sided figure in our example do not match 900°. Although the difference between the expected sum of angles and the actual sum of angles is relatively small (0° 01' 18"), the mathematics must be reviewed. For a deed description, perhaps either one or more interior angles were computed incorrectly or the interior angles were added incorrectly. For a traverse, one or more of the angles may have been observed using incorrect technique or the angle written incorrectly in the field notes. Angular errors of an unacceptable magnitude must be reviewed and corrected before continuing.

Correcting Angular Errors

Angles can be adjusted in several ways, depending on circumstances. Some circumstances warrant apportioning the error equally to all angles. In other cases, you may review technique, terrain, and the length of the backsight and foresight lines for particular angular observations before deciding to assign parts of the error unequally to the most suspect angles.

Today's survey instruments are sufficiently accurate for angular observations that angular corrections rarely exceed 3 to 4" per angle. Instruments that measure directly to 1" produce the best results. Corrections of greater magnitude probably indicate that one or more angles have been observed incorrectly, and the work must be retraced. If the instrument being used for angular observation measures directly to 6" or less, a simple doubling of the horizontal angle and closing the horizon for vertical angles is sufficient.

Double an Angle Horizontally

Doubling a horizontal angle means turning the angle commencing with 00 on the plate and ending with the observation. Clamp the upper motion after the angle is read and loosen the lower motion, turning back to the first point and observe the angle again. This second reading should approximate twice the first observation, but divide the second reading by 2 to arrive at the "mean" (average), which is the usable angle.

Close the Horizon Vertically

Closing the horizon on a vertical angle (after adjusting all level vials) means to observe the point, read the vertical scale and note the angle, sight the point again after inverting the scope, and read the vertical scale and note the angle. The two angles, one direct, and the other reverse, should have a summation of 360°.

Final Angle and Distance Adjustments

If final adjustments still need to be made to the horizontal closure after you revisit the field, and if the angular adjustments fall within the range of 3 to 4" per angle, you can feel satisfied. However, if the angles require greater adjusting, the work should be reviewed for a possible error. Computer software is available to compute, balance, and adjust angles (plus distances, if necessary) within the figure to produce on paper what looks like an acceptable survey. When angular adjustments become necessary to balance the survey, you need to remember that the original survey corners still remain in the field, and an adjustment on paper may not match the true positions of these corners. Subsequent surveys will reveal these problems, and the error(s) will be moved around depending on your starting point.

Assuming that you have corrected the angular errors, one additional verification of the accuracy of the survey must be made. This second mathematical check involves combining each measured distance around the perimeter of the survey with the angles observed at each perimeter break point. A perfect set of field-measured distances and field-observed angles can never be attained and, therefore, the surveyor strives for an acceptable degree of accuracy.

Formerly, errors were considered adequate if meeting the following guidelines:

First order:	1 ft of error in every 15 to 25,000 ft traversed
Second order:	1 ft of error in every 10 to 15,000 ft traversed
Third order:	1 ft of error in every 5 to 10,000 ft traversed

FIGURE 3.3 Ever-changing surveying methods *(Scott, Bob. Artist/Architect/Engineer. 1996. Marietta, Ohio.)*

First-order surveys were originally associated with commercial properties; second-order surveys were originally associated with residential property; and third-order surveys were originally associated with farm land.

Surveying accuracy has been changing over the past 30 years due to the great advances made in laser technology for measuring distances. For decades, measurement of horizontal and vertical angles could be done using transits that provided much greater accuracy than the accuracy of distance-measuring equipment (the tape or chain). The infrared light beams (laser) of the late 1950s began to change and equalize the accuracy ratio between angles and distances, and surveying took on new standards. Using EDM devices, today's surveyors are expected to maintain an accuracy ratio of 1 ft of error in 100,000 lineal ft (or better) of traverse.

Such a ratio is mathematically computed with software that combines the horizontal angles and laser distances, computes the overall error in the survey, and then provides a method(s) to adjust these errors. There are also several ways to compute the error of the survey and make adjustments using simple hand-held calculators, but the processing time is much greater.

EDM devices allow the use of highly accurate horizontal control in projects where taping or chaining would be impossible. In the example shown in (Figure 3.3), a crane lifted us to the top of a 60-ft-high bridge pier so we could mark the horizontal centerline across the abutments and piers of the highway. Construction projects that require this type of precise control from the surveyor show that surveying has come a long way in technique from the time when the surveyor vowed that "I will level the chain . . . and well and truly plumb the tally pins. . . ."[1] However, the professional responsibility toward work performed for the public good, whether it is property corners or bridge pier corners, has always been present for those of us who make this work more than just an occupation.

Caution: No computer adjustment should ever be made to a survey when it is obvious the angles are incorrect. Also, measured distances can never be adjusted independently without first accepting the validity of the angles. When an incorrect angular survey is accepted and paper adjusted by practicing professionals, the public trust is threatened, and the surveyor's reputation and career is jeopardized. Never compromise your integrity.

---

1 Part of a government surveying crew oath, circa 1800 – 1850 (White, 1983).

P · A · R · T · 2

HIGHWAY PLANS

CHAPTER 4
TITLE SHEET

INTRODUCTION

This chapter commences a series of chapters in which major design sheets within a set of construction and right-of-way engineering drawings are reviewed. Study of these chapters should make the surveyor, attorney, appraiser, negotiator, realtor, and members of the general public more capable of highway plan interpretation. The next five chapters describe five major design sheets and how each can be interpreted, understood, and used.

Much has been written about the engineering plan development of roads and the many design features contained therein, but little has been written concerning the analysis of the individual sheets within the plan. The gathering of engineering data brings the conceptual design, mathematics, and drawings together to form a complete engineering analysis known as the highway plan. Highway plans can contain only a few sheets for small projects or many sheets for larger projects. Regardless of the length, each project contains a "title sheet," and that is the emphasis for this chapter. Figures 4.1 and 4.2 represent two examples of title sheets found in state highway construction plans, and they contain many important facts necessary to thoroughly understand the project, such as:

1. Location of the project (state, county, and township)
2. Identity of the project (county, state, or interstate)
3. Route number (such as County Road 2, or Interstate 470)
4. Length of the project in lineal feet and miles
5. Type of construction (such as grading, drainage, paving, structures, signing, and lighting)
6. Stationing of the project
7. Occasionally, the individual sheet layout along the centerline

Figure 4.1 is a construction title sheet for an older interstate highway project approved in fiscal year 1972, and Figure 4.2 represents a title sheet for a state highway improvement whose design was still in progress in 1995, when the sheet was provided to me to copy.

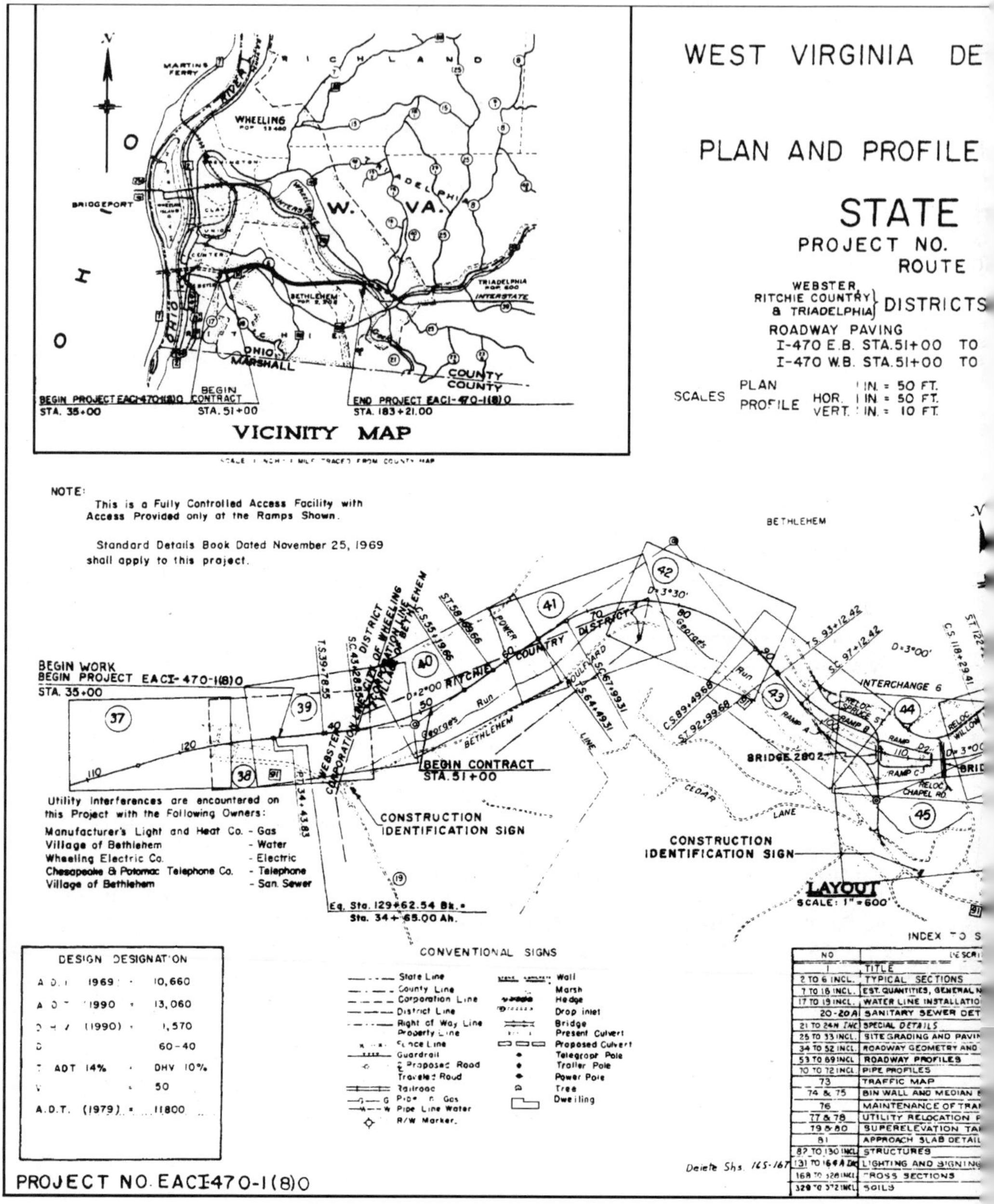

FIGURE 4.1 Typical title sheet from the construction drawings *(The West Virginia Department of Transportation, 1972.)*

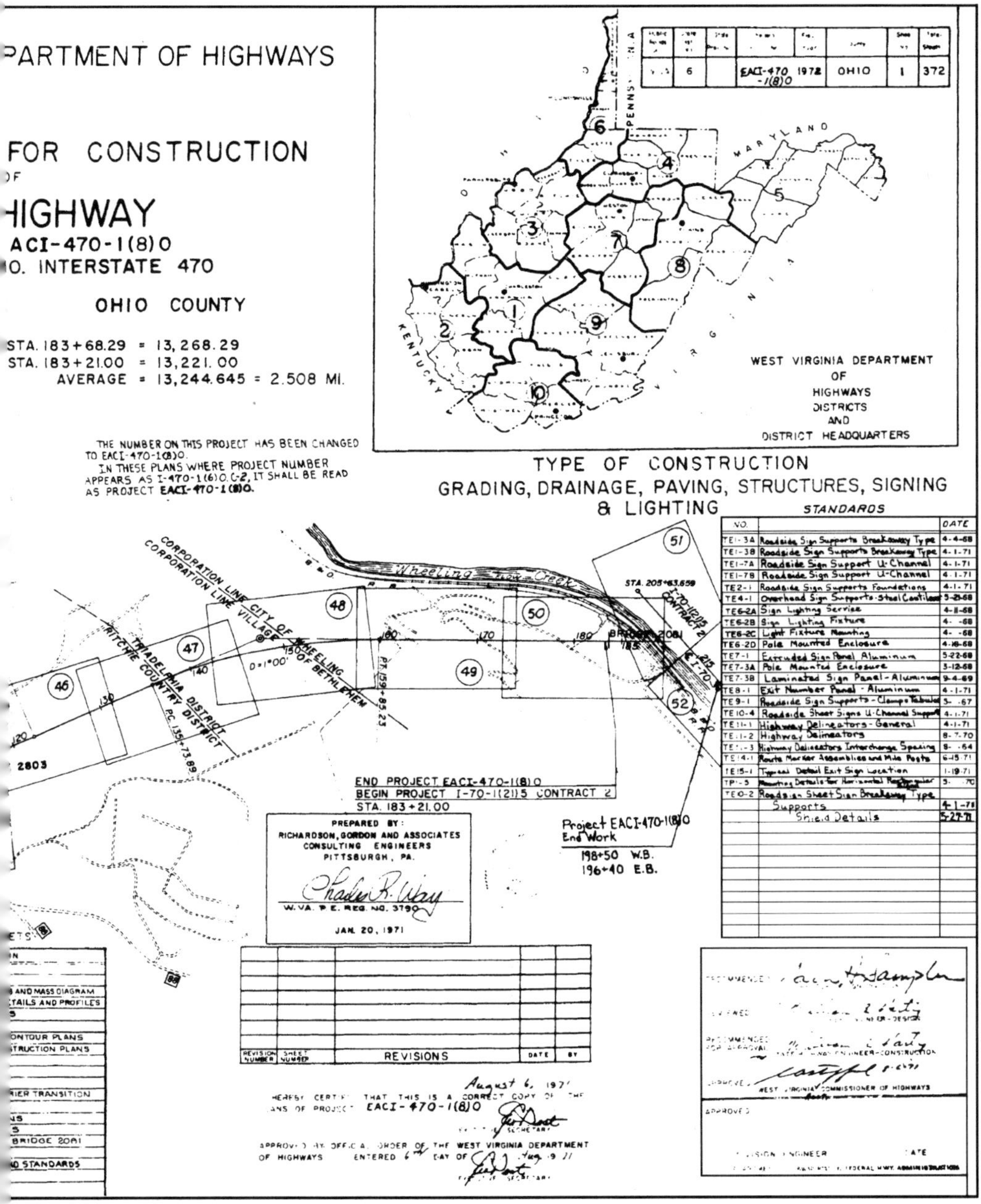

NO.		DATE
TE1-3A	Roadside Sign Supports Breakaway Type	4-4-68
TE1-3B	Roadside Sign Supports Breakaway Type	4-1-71
TE1-7A	Roadside Sign Support U-Channel	4-1-71
TE1-7B	Roadside Sign Support U-Channel	4-1-71
TE2-1	Roadside Sign Supports Foundations	4-1-71
TE4-1	Overhead Sign Supports-Steel Cantilever	5-23-68
TE6-2A	Sign Lighting Service	4-11-68
TE6-2B	Sign Lighting Fixture	4- -68
TE6-2C	Light Fixture Mounting	4- -68
TE6-2D	Pole Mounted Enclosure	4-18-68
TE7-1	Extruded Sign Panel Aluminum	5-22-68
TE7-3A	Pole Mounted Enclosure	3-12-68
TE7-3B	Laminated Sign Panel-Aluminum	9-4-69
TE8-1	Exit Number Panel - Aluminum	4-1-71
TE9-1	Roadside Sign Supports - Clamps Tabulated	5- .67
TE10-4	Roadside Sheet Signs U-Channel Support	4-1-71
TE11-1	Highway Delineators - General	4-1-71
TE11-2	Highway Delineators	8-7-70
TE11-3	Highway Delineators Interchange Spacing	8- .64
TE14-1	Route Marker Assemblies and Mile Posts	6-15-71
TE15-1	Typical Detail Exit Sign Location	1-19-71
TP1-5	Mounting Details for Horizontal Rectangular	5- .70
TE0-2	Roadsign Sheet Sign Breakaway Type	
	Supports	4-1-71
	Shield Details	5-27-71

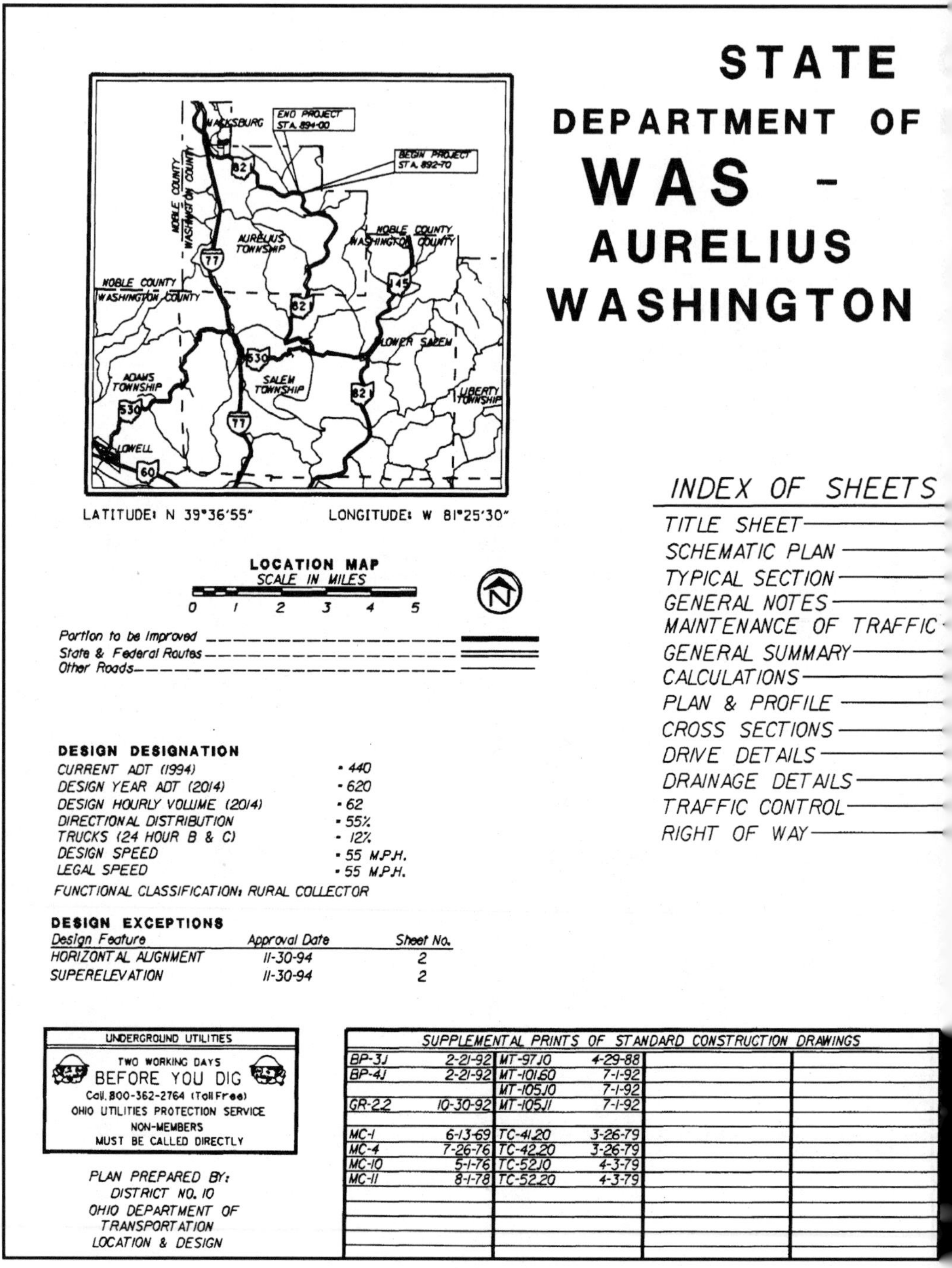

STATE
DEPARTMENT OF
WAS -
AURELIUS
WASHINGTON

LATITUDE: N 39°36′55″ LONGITUDE: W 81°25′30″

LOCATION MAP
SCALE IN MILES
0 1 2 3 4 5

Portion to be Improved
State & Federal Routes
Other Roads

INDEX OF SHEETS

TITLE SHEET
SCHEMATIC PLAN
TYPICAL SECTION
GENERAL NOTES
MAINTENANCE OF TRAFFIC
GENERAL SUMMARY
CALCULATIONS
PLAN & PROFILE
CROSS SECTIONS
DRIVE DETAILS
DRAINAGE DETAILS
TRAFFIC CONTROL
RIGHT OF WAY

DESIGN DESIGNATION

CURRENT ADT (1994)	• 440
DESIGN YEAR ADT (2014)	• 620
DESIGN HOURLY VOLUME (2014)	• 62
DIRECTIONAL DISTRIBUTION	• 55%
TRUCKS (24 HOUR B & C)	• 12%
DESIGN SPEED	• 55 M.P.H.
LEGAL SPEED	• 55 M.P.H.

FUNCTIONAL CLASSIFICATION: RURAL COLLECTOR

DESIGN EXCEPTIONS

Design Feature	Approval Date	Sheet No.
HORIZONTAL ALIGNMENT	11-30-94	2
SUPERELEVATION	11-30-94	2

UNDERGROUND UTILITIES
TWO WORKING DAYS
BEFORE YOU DIG
Call 800-362-2764 (Toll Free)
OHIO UTILITIES PROTECTION SERVICE
NON-MEMBERS
MUST BE CALLED DIRECTLY

PLAN PREPARED BY:
DISTRICT NO. 10
OHIO DEPARTMENT OF
TRANSPORTATION
LOCATION & DESIGN

SUPPLEMENTAL PRINTS OF STANDARD CONSTRUCTION DRAWINGS

BP-3J	2-21-92	MT-97J0	4-29-88		
BP-4J	2-21-92	MT-101.60	7-1-92		
		MT-105J0	7-1-92		
GR-2.2	10-30-92	MT-105J1	7-1-92		
MC-1	6-13-69	TC-41.20	3-26-79		
MC-4	7-26-76	TC-42.20	3-26-79		
MC-10	5-1-76	TC-52J0	4-3-79		
MC-11	8-1-78	TC-52.20	4-3-79		

FIGURE 4.2 Typical title sheet from the construction drawings *(Ohio Department of Transportation. 1995. Division 10. Marie Ohio.)*

)F OHIO
TRANSPORTATION
821 - 16.91
TOWNSHIP
COUNTY

PROJECT DESCRIPTION

IMPROVEMENT OF 0.02 MILES OF STATE ROUTE 821 BY REPLACEMENT OF A STRUCTURE WITH A BOX CULVERT OVER AN UNNAMED TRIBUTARY OF THE WEST FORK OF DUCK CREEK. 130 FEET OF PAVEMENT WILL BE REPLACED AND A DRIVE CONSTRUCTED.

FEDERAL PROJECT NO.
SURFACE TRANSPORTATION PROGRAM

PID NO. **13219**

CONSTRUCTION PROJECT NO.

RAILROAD INVOLVEMENT **NONE**

WAS-821-16.91

1/20

______ 1
______ 2
______ 3
______ 4
______ 5
______ 6
______ 7
______ 8
______ 9-11
______ 12
______ 13-16
______ 17
______ 18-20

1995 SPECIFICATIONS

THE STANDARD SPECIFICATIONS OF THE STATE OF OHIO, DEPARTMENT OF TRANSPORATION, INCLUDING CHANGES AND SUPPLEMENTAL SPECIFICATIONS LISTED IN THE PROPOSAL SHALL GOVERN THIS IMPROVEMENT.

I HEREBY APPROVE THESE PLANS AND DECLARE THAT THE MAKING OF THIS IMPROVEMENT WILL REQUIRE THE CLOSING TO TRAFFIC OF THE HIGHWAY AND THAT DETOURS WILL BE PROVIDED AS INDICATED ON SHEET 5.

SUPPLEMENTAL SPECIFICATIONS		
820	3-18-92	
841	5-16-84	
843	7-29-88	
931	7-19-94	
942	3-18-92	
944	5-2-94	

Approved
Date______ ______________ District Deputy Director

Approved
Date______ ______________ Engineer, Bureau Of Bridges And Structural Design

Approved
Date______ ______________ Deputy Director, Design

Approved
Date______ ______________ Director, Department Of Transportation

During their careers, surveyors, attorneys, realtors, appraisers, and negotiators need to interpret both old and new title sheets. The older title sheet (Figure 4.1) identifies the project as in the state of West Virginia, Ohio County, and traversing through Webster, Ritchie and Triadelphia Districts and as Interstate Route 470. The newer title sheet (Figure 4.2) identifies the project as the state of Ohio, Washington County, Aurelius Township, and as State Route 821. The remainder of this chapter will concentrate on how the highway plan title sheet outlines various bits of information helpful to the public or the professional.

THE HEADING

A highway plan title sheet has a heading similar to the following:

State of Ohio
Department of Transportation
WAS.- 821- 16.91
Aurelius Township
Washington County

County

The first three letters are an abbreviation of the county (WAS. as in Washington), the middle numbers are the roadway designation (as in State Route 821), and the last set of numbers (16.91) are the length in miles from the beginning of the county to the project.

The first three letters in the county designation do not always represent the first three letters in the name of the county. Two of several exceptions from Ohio's 88 counties are:

MOT for Montgomery County
MOE for Monroe County

Route

The roadway designation number refers to county, state, or interstate routes. Your personal knowledge of the local area will help to determine whether the project is state or interstate. Occasionally, the letter I will precede the route number to indicate the project is interstate. The absence of the letter I, however, does not always mean the project is state-sponsored. You will need to review the plans more thoroughly to make a final determination.

Section

The last numbers (such as 16.91 in the example above) represent the distance, in miles, from the southerly or westerly entrance of the road into the county.

SIGNIFICANT FEATURES

County, route, and sections are discussed in greater detail on the pages that follow, but all title sheets contain several other significant features that should be addressed. As shown in Figure 4.1 and/or Figure 4.2, these are:

1. The *vicinity map* (sometimes called the *location map*). This map orients the project in proximity to a county, township, municipality, state, or interstate route and other geographical topography in the area. Look for the location map on each of the title sheets (Figures 4.1 and 4.2) and notice where the project is located, where the "begin" and "end" designations are located, where the route intersects with other roadways (if applicable), and where the project is situated in relation to land topography.

2. The *index of sheets*. This index provides the name of each series of sheets in the project. Review the index of sheets on Figures 4.1 and 4.2, noticing the number of sheets (372) in the older (and larger) project versus the number of sheets (20) in the newer project that is still in design stage.

3. The *centerline schematic*. In some instances, as shown in Figure 4.1, each "plan and profile" sheet is also shown graphically, blocked out in rectangular form along a schematic of the centerline survey. For larger projects, this type of detail on the schematic can help you quickly locate your area of concern to a particular plan sheet among many. In some plans, this schematic is not placed on the title sheet but on a separate sheet of its own, usually immediately after the construction title sheet. Schematics can also be found after the right-of-way title sheet described later in the chapter.

4. The *design designation*. This design designation outlines the traffic analysis statistics, providing the average daily traffic of the route at the time the project is designed (1969 for Figure 4.1 and 1994 for Figure 4.2), contrasted with a 20-year (1990 for Figure 4.1 and 2014 for Figure 4.2) projected average daily traffic count. This traffic analysis forms the basis for the engineer's design.

5. The *standard drawings*. These drawings consist of preapproved designs for objects that can be used on many different projects and therefore do not need to be redesigned for each project. The list of standard drawings includes the preparation date of the drawing, the title of the drawing, and the identifying number of the drawing. On Figure 4.1, the standard designs included objects such as lighting (TE6-2A), roadway sign supports (TE1-3A), overhead sign supports (TE4-1), highway delineators (TE11-1), route markers and mile posts (TE14-1), and typical exit signs (TE15-1).

6. The title sheet also includes the name of the *engineering designers, approval signatures, revision table, utility warnings, scales, conventional signs*, and *type of construction* (whether the project includes grading, drainage, paving, structures, signing and/or lighting, etc.).

BEGINNING, ENDING, AND LIMITATION OF ACCESS

In addition to the various items discussed above, the "beginning," ending," and "limitation of access" for each project is delineated, either by printed text or with graphics (such as arrows) pointing to the locations on the vicinity map or on the centerline schematic (if a schematic is included). When the access is limited, commercial development on side road intersections with the mainline is prohibited by the agencies participating in the funding. Usually 600 ft is chosen as the limiting distance between commercial properties and the mainline on new construction, but the distance is often variable on older construction.

COUNTY, ROUTE, AND SECTION

The *county* and *route* location of a project are readily understandable. However, a more detailed explanation is required to understand the significance of the *section*. In this context, section does not refer to a particular 640-acre tract of land surveyed in the U.S. Rectangular system, as described in Chap. 2. For a highway project, the section number of the project refers to the distance, in miles, from the project to the nearest county line. The physical location of the route has commonly been identified as a section number over the years by state and federal agencies. It should be noted that these rules are not mandated by law, but you can reasonably expect the section numbering of a project to be outlined as follows:

1. The highway will numerically proceed from south to north in the county.
2. The highway will numerically proceed from west to east in the county.
3. The *section* number for each improvement will commence at zero (written 0.00) at the point where the improvement enters the county and will increase on a mileage basis, in either a north or east direction until the improvement leaves the county. The process will be repeated as often as a new county line is intersected by the improvement. In "station" notation (described later in this chapter), a beginning section number of zero becomes station 0 + 00 (pronounced zero plus zero zero). (Note, however, that stationing is in feet, not miles).

The section number of a project is frequently the designation commonly used by right-of-way personnel when projects are being discussed. A particular county, route, and section number can mean different things to different employees.

To thc billing section, the number can be remembered as a project for which the clerk had recently completed a partial claim with the federal highway administration. To the records clerk, it can mean a project just filed or sold (meaning completed). To the appraiser, it can recall cemetery, industrial, commercial, and other property that were difficult to appraise. To the engineer, it can recall a project with multimillion dollar bridges, multiple cubic yards of rock excavation, subsurface coal mines, and other engineering difficulties that provided a memorable career experience.

An example of section numbering follows. A 2-mile road improvement intersects the southern boundary of Summit County. (Assume the route number is Interstate 71). The official designation of the project, at the county line, will be Sum 71- 0.00. The next project commencing where the Sum 71- 0.00 improvement ends is therefore called Sum 71-2.00. The point of beginning for this improvement is 2 miles within the Summit County line. The *section* numbers become the equivalent of mileage digits and increase progressively as the improvement is extended into the county.

When a highway originates within the county, the most southerly or westerly point of the improvement is assigned a *section* designation corresponding to the end of some previous improvement or it can be assigned a digit, equal in miles, from where this route intersects the county line.

CONSTRUCTION TITLE SHEET VERSUS RIGHT-OF-WAY TITLE SHEET

Title sheets come in different forms and often include different information. Originally, there was only one title sheet associated with and included with a complete set of highway plans. This title sheet was separated from the construction drawings to form a similar title sheet for the right-of-way engineering part of the plan. Confusion began immediately because the right-of-way portion had different needs than the construction portion.

Construction engineers were satisfied because the title sheet referred to construction items. However, the right-of-way engineers were dissatisfied because much of the standardized information on the construction drawing did not apply to the right-of-way portion of the plan. This need to match drawing information with usage prompted the highway department to initiate and include two different title sheets, one for the construction portion of the plan and the other for the right-of-way portion. The differences are significant and make the separation into two parts more convenient for everyone.

Differences

The major differences between construction and right-of-way title sheets are:

1. The conventional signs are different.
2. The index of sheets are different.
3. The standard construction drawings are omitted from the right-of-way title sheet.
4. The limitation of access shows on the right-of-way title sheet.
5. The approval signatures are often different.
6. The scale of the drawings are often different.

Conventional signs, index of sheets, and scales will be discussed in more detail below.

Conventional Signs

Conventional signs are a tabulation of the graphic symbols used throughout a project. Most symbols are the same from project to project so that there is uniformity in presentation and in understanding and you can soon remember what most symbols represent. A tabulation on the title sheet means that the definitions of lines and graphics are not required on each sheet in the plan. Figure 4.3 is representative of the title sheet in the right-of-way portion of a plan. By comparing the conventional signs in this illustration with the conventional signs in Figure 4.2 (same project), you can see that the construction title sheet (Figure 4.2) has no conventional signs, whereas the right-of-way title sheet (Figure 4.3) has several signs that will be commonly used throughout the plan. Figure 4.4 is another right-of-way title sheet from a different project that also contains conventional signs. Obviously, knowing the meaning of these conventional signs assists the professional in the interpretation of the right-of-way drawings. (Note: The exclusion of the conventional signs in Figure 4.2 is probably because the project was still being in design phase, and not all features of the plan sheets were completed.)

Index of Sheets

Without the indexing of sheets, a considerable amount of time would be lost in searching for certain aspects of design in the project. Figures 4.1 and 4.2 (both construction title sheets) strive to eliminate this problem by including an index that describes different aspects of engineering design and where that design can be located.

In the Index of Sheets in Figure 4.2, sheet 1 is obviously the *title sheet*. Sheet 2, called the "*schematic plan*," is a compilation of all the mathematics that tie the centerline of sur-

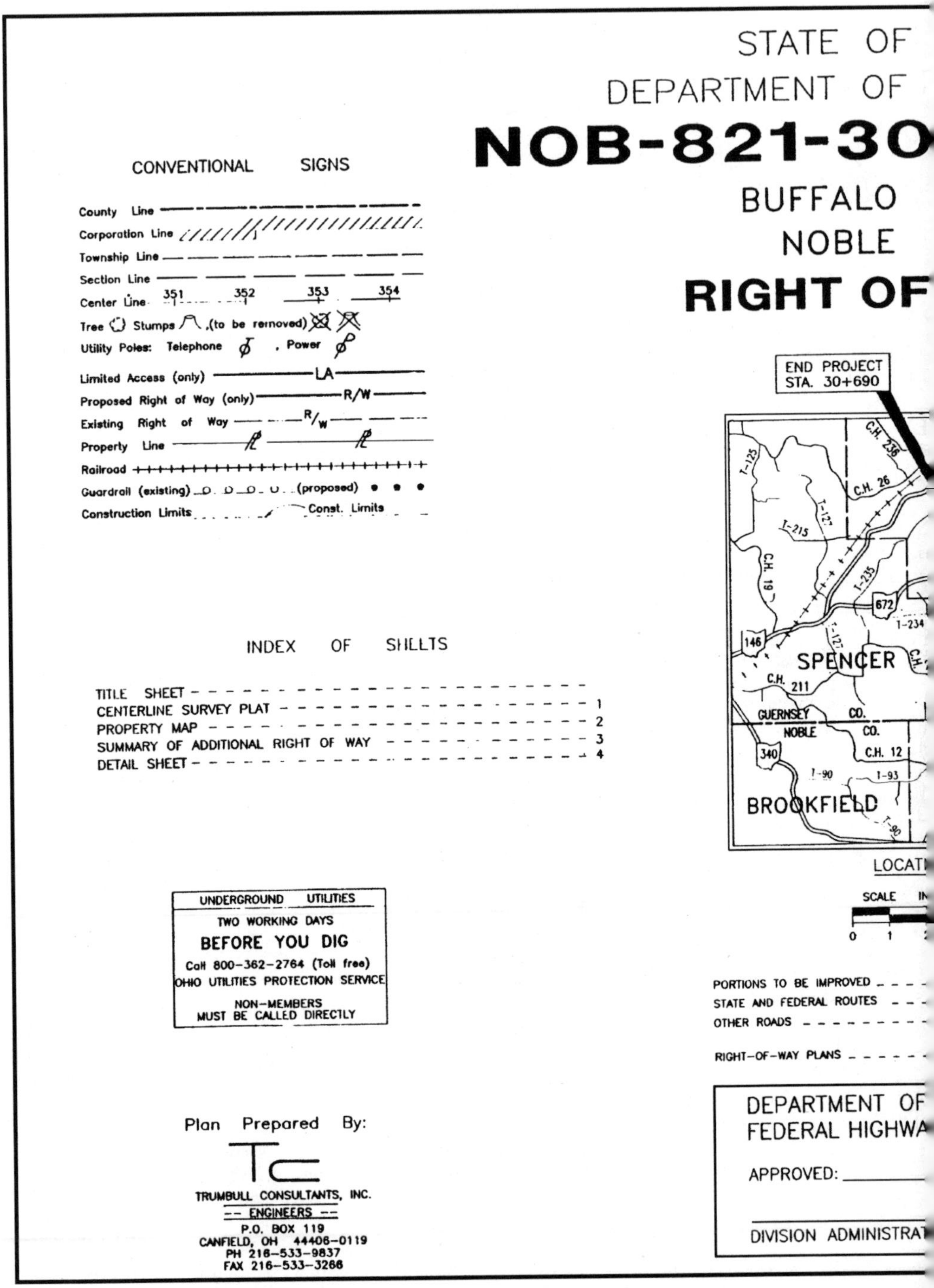

FIGURE 4.3 Typical Right-of-Way title sheet *(Ohio Department of Transportation. 1995. Division 10.)*

OHIO
TRANSPORTATION
.630(19.03)
TOWNSHIP
COUNTY
WAY

metric units

SURFACE TRANSPORTATION PROGRAM

PROJECT DESCRIPTION

IMPROVEMENT OF 0.06 km OF STATE ROUTE 821 BY REPLACEMENT OF ONE STRUCTURE OVER AN UNNAMED STREAM, INCLUDING APPROACH RECONSTRUCTION.

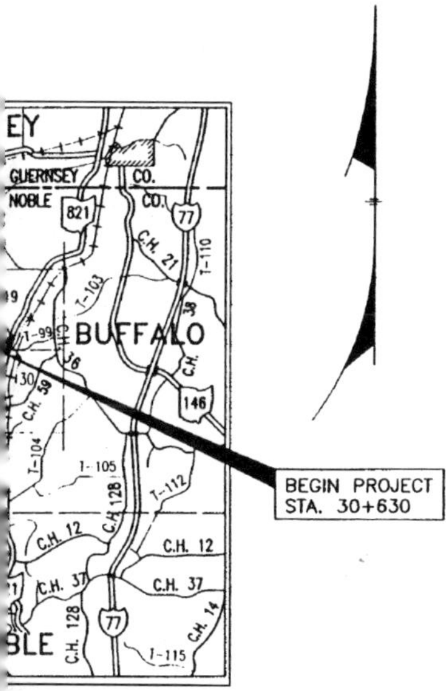

AP

METERS

4 5

0 5 m 10 m
SCALE IN METERS

ACQUIRING AGENCY: STATE OF OHIO

1995 SPECIFICATIONS

The standard specifications of the State of Ohio, Department of Transportation, including changes and supplemental specifications listed in the proposal shall govern this improvement.

I hereby approve these plans and declare that the making of this improvement will not require the closing to traffic of the highway and that provisions for the maintenance and safety of traffic will be as set forth on the plans and estimates.

ANSPORTATION
MINISTRATION

DATE

Approved ______________________

Date __________ Assistant Deputy Director, Office of Real Estate Administration

Approved ______________________

Date __________ Director, Department of Transportation

FEDERAL PROJECT NO.	PID NO.	STATE JOB NO.	RAILROAD INVOLVEMENT	
STP	12087	109550	NONE	NOB-821-30.630(19.03)

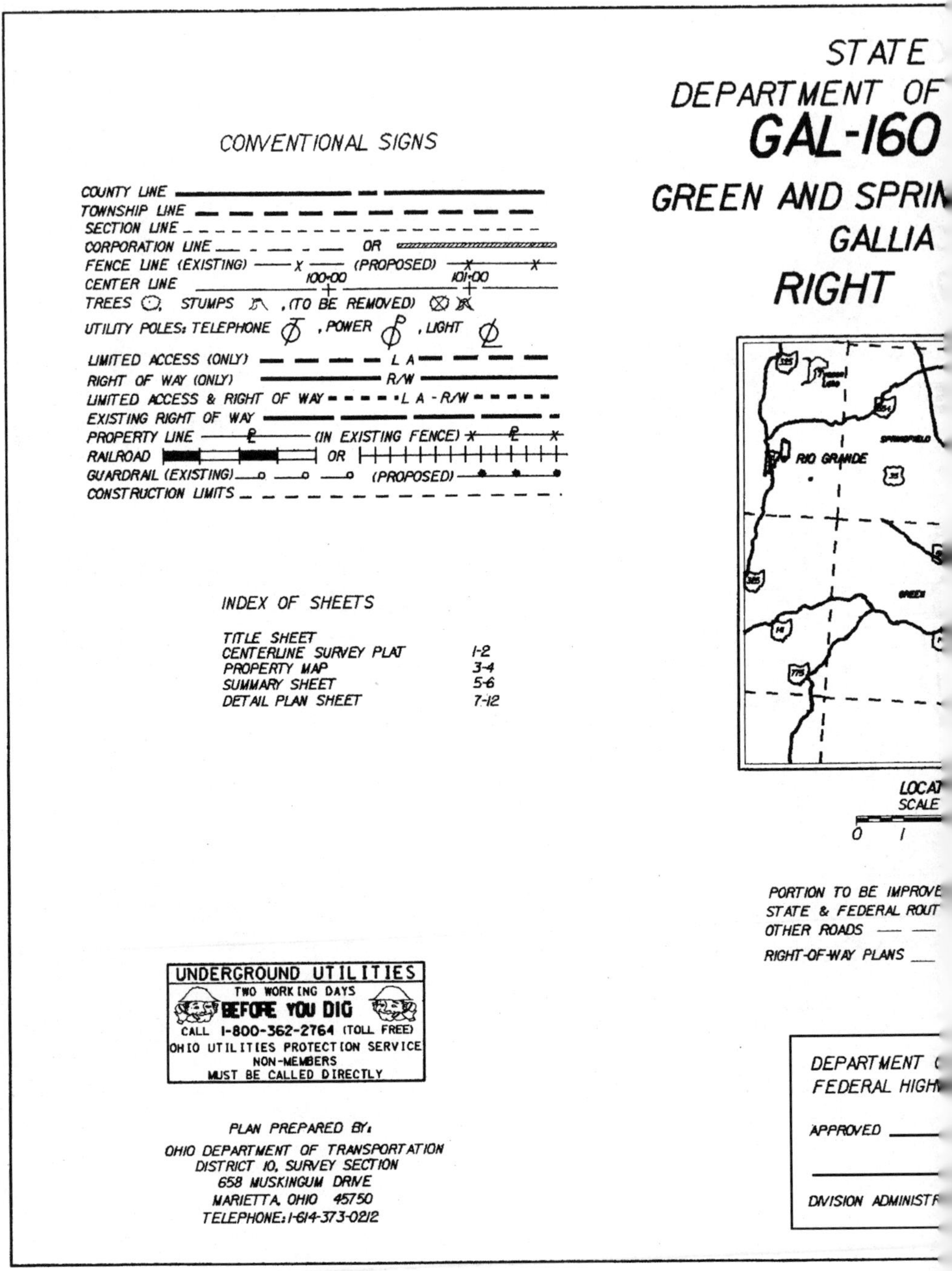

FIGURE 4.4 Typical Right-of-Way title sheet *(Ohio Department of Transportation. 1995. Division 10.)*

OF OHIO
TRANSPORTATION
-(2.77)
FIELD TOWNSHIPS
COUNTY
OF WAY

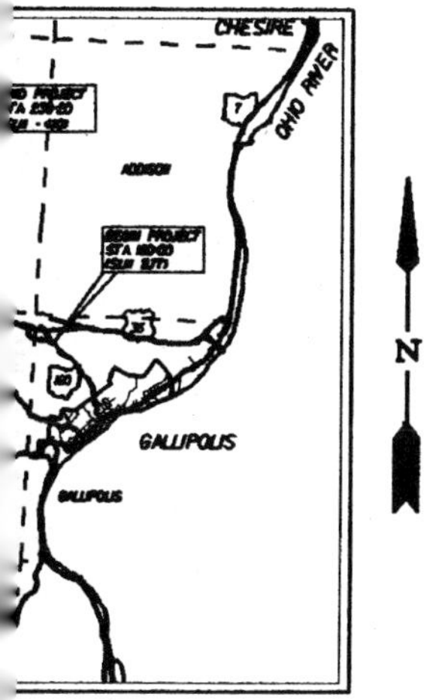

4

SCALE IN FEET

SPORTATION
NISTRATION

DATE

SURFACE TRANSPORTATION PROGRAM

PROJECT DESCRIPTION

RECONSTRUCT AND WIDEN EXISTING PAVEMENT
MINOR RELOCATION

ACQUIRING AGENCY: STATE OF OHIO

LIMITED ACCESS

A SECTION OF THIS IMPROVEMENT WILL BE LIMITED ACCESS RIGHT-OF-WAY AND ABUTS A PREVIOUSLY ACQUIRED SECTION OF LIMITED ACCESS RIGHT-OF-WAY ON U.S.R. 35, WHICH HAS BE DECLARED A LIMITED ACCESS HIGHWAY OR FREEWAY BY ACTION OF THE DIRECTOR IN ACCORDANCE WITH THE PROVISIONS OF SECTION 5511.02 OF THE REVISED CODE OF OHIO.

1995 SPECIFICATIONS

THE STANDARD SPECIFICATIONS OF THE STATE OF OHIO, DEPARTMENT OF TRANSPORTATION, INCLUDING CHANGES AND SUPPLEMENTAL SPECIFICATIONS LISTED IN THE PROPOSAL SHALL GOVERN THIS IMPROVEMENT.

I HEREBY APPROVE THESE PLANS AND DECLARE THAT THE MAKING OF THIS IMPROVEMENT WILL NOT REQUIRE THE CLOSING TO TRAFFIC OF THE HIGHWAY AND THAT DETOURS WILL BE PROVIDED AS INDICATED ON SHEET 2.

APPROVED ______________________

DATE ________ ASSISTANT DEPUTY DIRECTOR, OFFICE OF REAL ESTATE ADMINISTRATION

APPROVED ______________________

DATE ________ DIRECTOR, DEPARTMENT OF TRANSPORTATION

FEDERAL PROJECT NO. STP

PID NO. 8483

STATE PROJECT NO. 108360

RAILROAD INVOLVEMENT NONE

TITLE SHEET

GAL-160-(2.77)

vey, entrance and exit ramps, side roads and bridges together, and how it shows how this total mathematical figure relates to the existing ground. Without the schematic plan, construction surveying cannot commence. Sheet 3, called the "*typical section*," is covered in greater detail in Chapter 6. Sheets 4 and 5, called "*general notes*" and "*maintenance of traffic*," are mostly self-explanatory. The general notes are included in the plan in a narrative format. The notes explain, in one location, all construction features that must be addressed, such as governing specifications, elevations, azimuths, guardrail, select embankment, protection of bridge slopes, concrete and macadam pavement, wingwalls, delay in traffic, seeding mixtures, fill benching, disposal of unused material, utility relocations, and other special items that outline in a clear fashion the contractor's responsibility.

Sheet 6, called the "*general summary*," is an outline of all materials (called quantities) in the project, where the item is located, and the size or volume of each item. Sheet 7, called the "*calculations*," is the backup mathematics pertaining to the quantities.

Sheet 8, called the "*plan and profile*," is covered in greater detail in Chapter 9. It is the heart of the construction for any project. Sheets 9 to 11, called the "*cross-sections*," are covered in greater detail in Chapter 8. Sheet 12, called the "*drive details*," is generally a special presentation on one or more sheets that use an appropriate large scale to present the material for easier review. The data includes a graphic plan of the drive (or any other feature), showing the new feature, how it ties to the existing ground, the proposed and existing elevations of the improvement, and the mathematics (curve data, offsets, etc.) required to survey and construct the object.

Sheets 13 to 16, called the "*drainage details*," include designs for all side ditches, drive pipes (culverts), mainline culverts (concrete or otherwise), corrugated metal pipes or multiplate structures used for drainage, animal or pedestrian usage, water courses (completely relocated or realigned), and all other drainage features that require engineering design and capacity calculations. Sheet 17, called "*traffic control*," is another narrative outline. It explains the traffic maintenance the contractor must meet during certain stages of construction. Those traffic relocations from lane to lane during the "battle of the barrels," traffic detours off the Interstate onto secondary roads, and other temporary traffic configurations during highway construction might seem like extemporaneous plans but, actually, these traffic controls are designed by the engineers during plan preparation.

Finally, sheets 18 to 20, called "*right-of-way*," are the drawings required to show limits of property acquisition in the vicinity of the highway improvement. In general, plans prepared before 1962 acquired the rights-of-way in the form of easements (to be used during the existence of the improvement but to return to the original or adjacent owner once the improvement ceased to exist). Since 1962, all permanent rights-of-way (R/W parcels) and limited access (L/A parcels) have been acquired in fee title. Drainage relocations (X,Y,Z parcels) located outside the rights-of-way or limited access lines are still acquired in the form of easements. Work agreements (W/A parcels), on the other hand, are secured from property owners to grade small ditch slopes or other earthen remedial dressing or shaping of the ground when the appraised value of the work is in small dollars (generally $500 or less).

Note: The examples described do not necessarily cover all aspects of design that you may find in other plans. Review, review, and review your plans thoroughly to uncover unusual design features, unusual restrictions, unusual construction circumstances, etc.

SCALES

Engineer's Scale

A scale provides a fixed relationship between the true dimensions of objects which exist on the ground (or objects proposed to be constructed) and the graphic representation of the corresponding object on a map. In all construction and right-of-way plans, except bridge drawings, an engineer's scale is used. The engineer's scale converts true dimensional length, in feet, to a 1-inch (in) unit of measurement on a drawing.

Example:

Most engineer's scales consist of six separate scales, marked 10, 20, 30, 40, 50, and 60. Using the part marked 10, for each inch on the scale, one can use any one of an unlimited number of translated lengths. For one drawing, 1-in on the drawing could equal 1-ft of true dimension in the field. For another drawing, 1-in could represent 10, or 100, or 1000 ft. Using the part marked 20, one can interpret 1-in on the drawing to represent 20 ft on the ground or 200 ft or even 2 ft. Using these various scale translations, the actual field measurements can be scaled down to be presented on standardized engineering drawings.

Standardized plan sheet sizes sometimes vary, but a 22 by 34-in sheet is common for the "full-size" set of most highway engineering plans. The "miniature" plan will be explained later in the chapter.

Architect's Scale

The architect's scale is another method of converting actual dimensions to a representation on a drawing. The size of the work determines the need for certain scales. Consequently, the architect, who designs building-size projects rather than highway-size projects, works with scales that are more useful for representing smaller true dimensions at larger drawing sizes. An architect's scale translates true dimensions, in feet, into inches (or fractions thereof) on the drawing. Many architect's scales consist of 11 separate scales, marked 16, 3, 1½, 1, ¾, ½, ⅜, ¼, ³⁄₁₆, ⅛, and ³⁄₃₂. The common examples below reveal the different and opposite conventions used on the two types of scales. The unit of measurement on the architect's scale changes for each different size of drawing, whereas the unit of measurement on the engineer's scale remains at 1-in, no matter what the translated dimension.

Examples of common scales are:

Architect			Engineer		
1½ in	=	1 ft	1 in	=	1 ft
1 in	=	1 ft	1 in	=	10 ft
¾ in	=	1 ft	1 in	=	20 ft
½ in	=	1 ft	1 in	=	30 ft
⅜ in	=	1 ft	1 in	=	40 ft
¼ in	=	1 ft	1 in	=	50 ft
³⁄₁₆ in	=	1 ft	1 in	=	60 ft
⅛ in	=	1 ft	1 in	=	100 ft
			1 in	=	1000 ft

The scales shown above are by no means the maximum number that can be devised. Any of the above scales can be interpreted to produce various results. On the part of the engineer's scale marked 50, a drawing can be plotted at a 1-in = 50-ft scale or a 1-in = 5-ft scale, etc. Other scales in the 10 series (1-in = 10 ft, 1-in = 20 feet, 1-in = 30 ft, 1-in = 40 ft, and 1-in = 60 ft) can also be interpreted with unusual values, such as:

1-in = 2 ft, 1-in = 3 ft
1-in = 4 ft, and 1-in = 6 ft

The same portions of the engineer's scale can also be interpreted as follows:

1-in = 200 ft, 1-in = 300 ft
1-in = 400 ft, 1-in = 600 ft, and so forth

MINIATURE PLAN

Scales and their applications have long been confusing to many professionals. Adding further confusion is the presence of a "miniature" set of plans, which is a useful tool for surveyors and other professionals in occupations related to property acquisition.

The miniature plan is no longer at the true scale of the original plan drawings, but the scale (as printed in text or as depicted using a bar graph), is written on the drawings before size-reduction of the plan sheet occurs. Careful review of a miniature plan is recommended since an incorrect scale analysis can often lead to multiple errors, not only for the surveyor but for the other professions who have direct contact with property owners and who may misrepresent the plans by using an incorrect scale.

The miniature plans are a set of plans reduced photographically from a full-size sheet to a sheet one-half its original length and width. This reduction in both directions decreases an otherwise full sheet to one-quarter its original area. In some localities these plans are referred to as "half size," and under other circumstances these plans are referred to as "quarter size," but both terms refer to exactly the same plans.

ANALYZING THE SCALE

Obviously, some careful analysis must be exercised before choosing the scale from which you will make your measurements. Many methods of analyzing the map scale can be devised; however, two methods are recommended and discussed below:

1. Measure the distance between stations on the centerline.
2. Measure the bargraph on the drawing.

HIGHWAY STATIONING

Highway "stations" are represented on highway plans at 100-ft intervals, and each 100-ft interval on the full-size plan is drawn at the number of inches stated on the drawing. For example, at a 1-in = 100 ft scale, the difference between two stations 100 ft apart will measure 1-in on the full-size plans. An engineer's scale (or a conventional ruler) can be used to measure this separation.

Highway "stationing" is a way of labeling measured increments along the centerline of a roadway to simplify the description of the location of a point on the centerline. It would be difficult to describe a certain point on the centerline of a new (or old) roadway by specifying an amount of ft (or thousands of ft) from some known "base point."

Instead, the simplest method of stationing assumes that the station of the base point can be identified as zero (but written 0 + 00). Then, a point located 100 ft along the centerline from the 0 + 00 base point would be called station 1 + 00 (and pronounced one plus zero zero). A point 200 ft from the base point is station 2 + 00, and so on. Stations at very large distances from the base point do not use commas, so a station located 10,000 feet from the base point is called station 100 + 00.

Stations can be established along the centerline of a highway or along a baseline used as a substitute for the centerline.

Stations are numbered, as discussed earlier in this chapter, commencing from the south and increasing to the north, as well as commencing from the west and increasing to the east.

Bar Graphs

The second suggestion for checking the scale of a plan is to measure the bar graph on the drawing. The bar graph is a solid bar subdivided in various ways, usually into 1-in segments. Use your engineer's scale (or a conventional ruler) to measure the bar graph and compare the measurements. The example below was prepared when this page measured 8½ by 11 in:

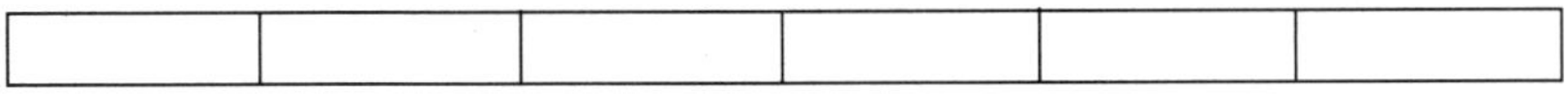

0 + 00 1 + 00 2 + 00 3 + 00 4 + 00 5 + 00 6 + 00

The simulated bar graph, above, is divided into 1-in subdivisions, and each subdivision represents 100 ft. If this bargraph were reproduced on a miniature set of plans, the 1-in subdivisions would then measure ½ in. Therefore, the total distance in 1-in is equal to 200 ft. If your analysis has determined that the information on the miniature plan is contrary to the scale identity, you must make proper scale adjustments.

Example:
Assume that your comparison of two centerline stations, as well as the bar graph, indicates that the distance between centerline stations on the drawing represents one-half of the scale identity. You can be confident that the set of plans has been reduced to one-half its original size. Therefore, a distance that was intended to represent 100 ft to 1 in, now represents 100 ft to ½ in.

To reconcile this difference, simply use a different part of the engineer's scale. In the example above, simply use the portion of the engineer's scale marked 20 instead of the portion marked 10. The correct measurements at a 1-in = 200 ft scale can then be read directly from the measuring device. The first graduation on the scale (approximately $\frac{1}{16}$ in) is equal to 10 ft, the fifth graduation on the scale is equal to 50 ft, and so forth. Whenever you have made scale adjustments for the miniature plan (11 by 17 in), remember to return to the true scale when using original sheets (22 by 34 in) within the set of drawings.

STATION OFFSETS

Stations are also used to locate existing or proposed items by identifying the object as so many feet left or right of the centerline. Objects located left or right of the centerline are named relative to the direction of stationing. As you look "up-station" along the highway (in the direction of increasing station numbers), those items that are to your left are considered left of centerline, and those items that are to your right are considered right of centerline.

Example:
Assume a designed right-of way break point is 59 feet left of station 3 + 00. The break point will be identified on the plan as $\frac{3+00}{59' \text{ Lt.}}$. Objects that are located on the right side of the centerline will also have similar notations. In this example, assume a right-of-way break point is set at 210 ft right of station 6 + 00. This point will be identified as $\frac{6+00}{210' \text{ Rt.}}$.

You will also find a similar type of notation in the field surveys discussed in Chapter 8. When cross-sectioning, elevations and distances are field-located to establish the basis for drawing the cross section and are noted in the surveyor's field book. However, the station is identified only once, on centerline; thereafter, the notation includes the distance left or right of the centerline with the elevation of the point directly beneath the distance, such as: $\frac{89'}{616.34}$. Any notation concerning left or right of centerline is generally omitted because cross-section field notes are usually arranged to record elevations and centerline offset distances on the left side of the page (denoting points left of centerline) or on the right side of the page (denoting points right of centerline).

INTERMEDIATE STATIONS

Existing or proposed points that lie somewhere between even 100 ft stations are shown in the following manner:

Example:

Assume the point is 15 ft ahead (up-station) of station 4 + 00. The point is then identified as station 4 + 15. If more accurate distances are required, such as 39. 67 ft ahead of station 11 + 00, where the tenths and hundreds of a foot are used, the point is identified as 11 + 39.67.

Fixed physical objects located left or right of centerline are also identified in a similar manner.

Example:

Assume the end of a culvert pipe is 121.75 ft left of centerline station 100 + 50. 25. The end of the pipe culvert will be identified as $\frac{100 + 50.25}{121.75' \text{ Lt.}}$.

STATION EQUATIONS

Stations can be used to determine distances between objects by subtracting one station from the other. However, some highway plans have what are known as "station equations." This term refers to a known mathematical point on a highway that has two stations assigned to it: one station "looking back" along the centerline and a second station "looking ahead" along the centerline. There are several reasons that contribute to the need for station equations. These are listed below:

- Changes in alignment which add or subtract lineal feet to the project.
- Mathematical errors becoming known subsequent to staking the alignment in the field.
- Centerline alignment crossing county or township lines. (Many county and townships also station roads differently. Others just extend the same mathematics on the centerline that their neighboring government used).
- When centerline curves are substituted for spirals, there will be a difference in length and, therefore, a station equation will be needed.

There are many other situations that can and do exist which prompt the need for station equations—perhaps you can add to the list from your experiences. In any event, to simply subtract stations to arrive at a distance between points, without thoroughly reviewing the design plan or the right-of-way plan, could lead to severe consequences.

NEGATIVE FRACTIONAL STATIONS

Negative fractional and full stations can be found in multiple locations, such as a point located behind (or back-station from) the beginning of a project.

Example:
Consider a point lying 31 ft back of the 0 + 00 station. If such a situation existed, it would be identified as 0 – 31 (zero minus 31). Rarely does this situation occur because of advanced planning by the designers who anticipate these problems.

Example:
Assume a point is located 31 ft behind a new project whose beginning station is 10 + 00. This point could be called 10 – 31 (10 minus 31) but will more appropriately be called 9 + 69. Similar situations to this last example often occur, generally in the design of drainage ditches.

CHAPTER 5
CENTERLINE PLAT

RECORDATION

Of all the drawings in a set of highway construction plans, one of the most important is the centerline plat, shown in Figure 5.1. This is due to the significance that U.S. courts place on court-recorded documents. The centerline plat represents the mathematical location of the highway improvement within the state, county, or township and to which each parcel of land required for construction has been mathematically related. This single drawing from the highway plans, according to state law, is recorded in the court house of the county in which the project is constructed. This recording is available for the public to review and is used for present and future re-establishment of state and private property rights.

Before 1962, property needed for highway construction was described and acquired by means of "station and offset" corners. This method of description temporarily replaced the customary bearings and distances to which the public had become accustomed in other types of deeds. Descriptions for land acquired using this earlier method were prepared by the right-of-way engineering staff and involved considerable reliance on the accuracy of drafted plan sheets. Generally the stations to be used as boundary corners for the new acquisitions were set (chosen) at right-of-way break points or at property intersections with the acquisition line, and the offset distances from centerline were scaled. The scaling process was based on the graphics of the proposed centerline in relation to the existing topography. This process, at best, was insufficient. Since 1962, all permanent highway acquisitions have been described by metes and bounds (bearings and distances), with deeds prepared in this manner.

PLAT INFORMATION

The centerline plat is the primary document showing the exact location of the improvement within the county and the graphical and mathematical appearance of the centerline. The centerline survey plat contains other useful and important tabulations on the drawing.

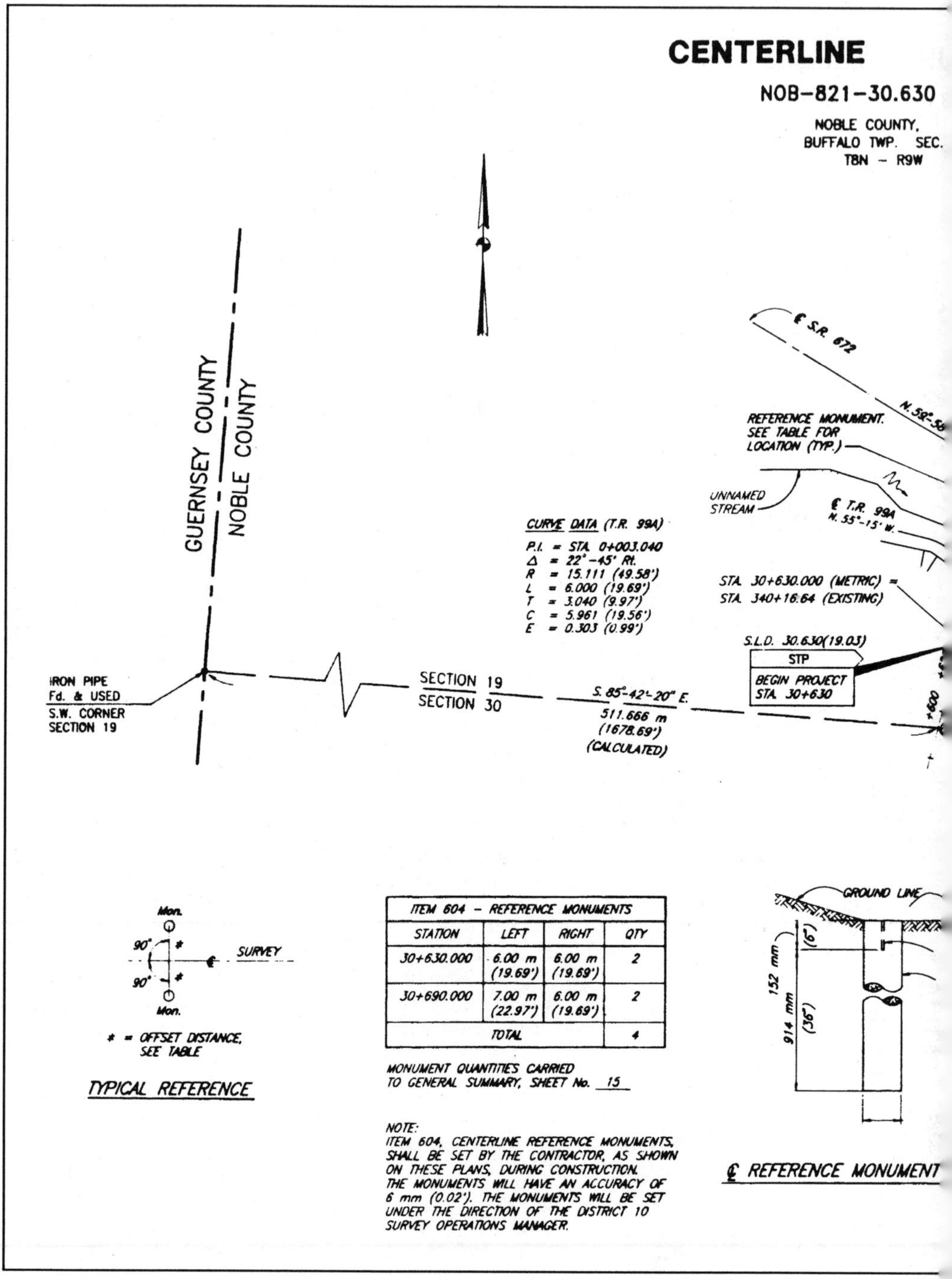

ITEM 604 – REFERENCE MONUMENTS			
STATION	LEFT	RIGHT	QTY
30+630.000	6.00 m (19.69')	6.00 m (19.69')	2
30+690.000	7.00 m (22.97')	6.00 m (19.69')	2
TOTAL			4

FIGURE 5.1 Typical centerline plat *(The Ohio Department of Transportation. Centerline Plat. 1995.)*

URVEY PLAT

.03)

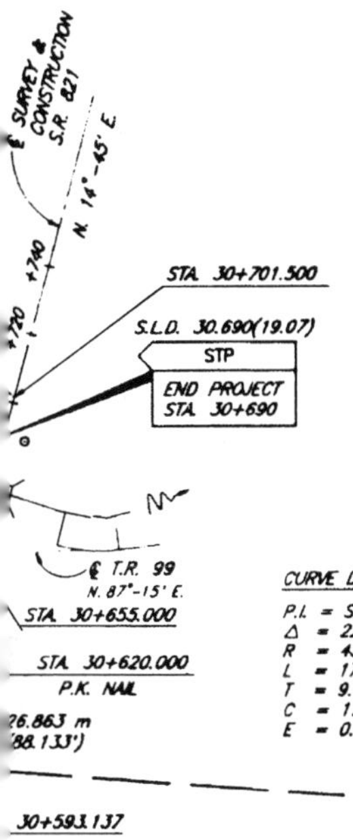

CURVE DATA (T.R. 99)

P.I. = STA. 0+056.881
Δ = 22°-47' Lt.
R = 45.250 (148.46')
L = 17.993 (59.03')
T = 9.117 (29.91')
C = 17.875 (58.65')
E = 0.909 (2.98')

metric units

℄ REFERENCE MONUMENT

CENTERLINE REFERENCE MONUMENTS ARE SHOWN ON STANDARD CONSTRUCTION DRAWING MC-1 (REV. 6-13-69) OF THE OHIO DEPARTMENT OF TRANSPORTATION. THE PLACING OF THE MONUMENTS SHALL BE UNDER THE DIRECTION OF A REGISTERED SURVEYOR AND ARE TO BE SET, AS SHOWN, BY THE HIGHWAY CONTRACTOR AT THE TIME OF CONSTRUCTION. ANY ALTERATIONS, WITH PRIOR APPROVAL OF THE DEPARTMENT OF TRANSPORTATION, SHALL BE NOTED AND O.D.O.T. SHALL BE NOTIFIED OF THE NEW LOCATIONS.

BASIS OF CENTERLINE BEARING

ALL BEARINGS SHOWN ARE BASED ON BEARINGS AS RECORDED FROM THE ORIGINAL 1941 SET OF PLANS S.H. 353 SEC. "0", SHEET 3 OF 4.

NOTE: IN CASE OF DISCREPANCY BETWEEN METRIC AND ENGLISH EQUIVALENTS, THE METRIC VALUES SHALL TAKE PRECEDENCE.

7 mm (1/2") STEEL ROD

ASS C
NCRETE

E UPPER 150 mm (6")
ALL BE FORMED

3 mm DIA.
")

RECEIVED __________, 19___

RECORDED __________, 19___

BOOK __________ PAGE ______

COUNTY RECORDER

I HEREBY CERTIFY THAT THIS PLAT IS A TRUE DELINEATION OF A SURVEY MADE FOR THE OHIO DEPARTMENT OF TRANSPORTATION IN 1994 BY VERNON SURVEYING CO.

DATE 6/26/95

REGISTERED SURVEYOR No. 6282

STATE OF OHIO PROFESSIONAL SURVEYOR

HORIZONTAL SCALE IN METERS 0 20 m 40 m

FEDERAL PROJECT NO. STP

PID NO. 12087

STATE JOB NO. 109550

CENTERLINE SURVEY PLAT

NOB-821-30.630(19.03)

1/4

31/34

Perhaps the most important of these are the centerline "*curve data.*" The mathematical curve data for each curve are placed on the drawing in the vicinity of each curve or spiral and represent all the mathematics required to field-locate a curve or spiral.

Curve symbols and their nomenclature and meaning are explained in Chapters 11, 12, and 13. Note that the mathematics necessary to reestablish the physical location of properties acquired by the state under fee title will not be found on the centerline plat; but it is on another plan sheet called the "*property plat.*"

An adequate centerline plat will also include the following:

1. The names of streets and roads
2. The names of local streams, rivers, and lakes
3. All existing railroads and the name of the carrier
4. All major topography
5. County, township, lot, and subdivision names
6. The scale of the presentation
7. The north direction on the drawing
8. The "begin" and "end" points of the project

If the project is controlled by state or federal agencies, the mathematics of the centerline will probably be tied to the appropriate *state plane coordinate system.* If so, this fact will be noted on the centerline plat. Projects controlled by the state plane coordinate system can be related mathematically to any other improvement within the state because all projects share a common coordinate grid system, established by the National Geodetic Survey.

The metric system is making additional inroads into U.S. measuring systems, and federally funded highway projects are now required to show metric distances. Figure 5.1 shows distances in both feet and meters. Notice that if a discrepancy occurs between feet and meters, the distance in meters controls.

THE CENTERLINE OF SURVEY

The *line* of single importance on the drawing is the centerline of survey, from which the improvement will be constructed and from which the properties have been acquired. This line will be prominent on the drawing, distinguishable from all other lines, and mathematically related to existing streets, county or township roads, intersecting interstate highways, section, quarter section, quarter-quarter section, and county line intersections, as well as intersecting rivers, streams, and railroads, if applicable. The centerline plat thus becomes a paper duplicate of the actual survey in the field.

Method of Location

The method of field-locating the centerline of survey is the culmination of many months or years of engineering design. During this design period, the centerline is established using aerial photographic mapping or other design methods, which are covered more thoroughly in Chapter 9. Today's highway designers have the added option of using computer software to store and retrieve U.S. Geological Quadrangle maps, state plane coordinates, and elevations for any locality.

Design can be completed in the computer and viewed on the monitor. For the following discussion, however, assume the aerial photography method was chosen to design the centerline.

After the centerline is approved and the design completed, the engineers mathematically relate this centerline to the original survey control that was established when the aerial photography was first undertaken. There are two methods of controlling aerial photography: (1) where survey ground control is set before the photographs are taken and (2) where several existing features are chosen from U.S. Geological Quadrangle maps after the photographs have been taken. When the second method is used, the existing U.S. Geological Survey (USGS) located ground features are oriented in the commercial plotters and related mathematically to each other to provide distances. The commercial plotter also establishes elevations at each of the chosen USGS located ground features, and the photogrammetrist prepares the topographic maps from this information. One existing feature often chosen as a control point for this method is the telephone pole, but it is hard to see how a surveyor would "occupy" a telephone pole (to "occupy" a point, a surveyor needs to set a transit, or hold a plumb bob, directly over the point). The first method, where *horizontal and vertical control* is set by the surveyor in the field prior to the photography, continues to be recognized as the superior aerial mapping method. The second method is often considered when the project has a lower budget, but the accuracy specifications must be more flexible. Whatever control is used, it must be verified for mathematical accuracy and probably adjusted (balanced) to eliminate mathematical errors, as described for closed traverses in Chapter 3.

Many months can elapse while the design is in progress. During this period, some of the ground control will be maliciously destroyed, while other controls will be displaced due to winter weather, rain, or earth movement, such as landslides. The surveyor commences a search for all remaining control evidence on the ground. When found, the control must be verified for its correct field location. This can be done by measuring between ground controls to verify whether the occupation distance is consistent with the documented distance. Angular checks between the ground evidence found must also be made. Angular discrepancies could indicate that a rotation has occurred.

Assuming the ground control meets all documented distances, the surveyor knows that one last opportunity remains to field-check ground controls, and the computed centerline. This opportunity comes when the engineer's centerline design is carried from control point to control point. To assist in this layout, the designer prepares a computer printout relating each original traverse corner (control from the aerial photography) to the nearest proposed centerline break point. When sufficient centerline break points are established, the surveyor commences to connect these points on centerline. In wooded conditions, this becomes physically demanding since all trees and brush along the centerline must be removed and the distances between break points on the centerline measured for their accuracy.

When curve or spiral alignments are designed in the centerline, the mainline tangents are first extended until they intersect, forming the point of intersection (PI). If this is impractical due to field conditions, the long chord is surveyed instead. Under no condition is the surveyor permitted to lay out the arc of the curve or arc of the spiral for purposes of locating the mainline tangents. A layout of sequential points on a curve will always produce incremental errors that lead to an inaccurate tangent line. Where spirals are designed on either end of mainline curves, not only are the main tangents surveyed before the curves points are set, but the long and short spiral tangents are also located before establishing the arc of the spiral. Chapters 11 and 12 provide sufficient reasons for limiting centerline layouts to approved procedures.

SCALES

The centerline plat, although showing similar information to that found in the small-scale "vicinity map" located on the title sheet of a set of highway plans, is a full-size (22 by 34 in) drawing by itself. The scale of the presentation is limited to a minimum size of 1-in equals 400 feet by specifications in the state's *Right-of-Way Manual*, with even smaller-scale numbers recommended and preferred. The smaller the scale number (such as 1-in equals 300, 200, 100, or 50 ft), the larger the drawing the more topographic features that can be shown per inch on the drawing.

Example:

If we compare the minimum scale of 1-in = 400 ft with a 1-in = 50 ft scale, we can see that 400 ft of topography must fit in 1-in on the map, versus 50 ft of topography fitting within the same 1 map inch, which is recommended. Significantly more detail can be obtained by using the 1-in = 50 ft scale. However, the final scale presentation is subject to the complexity of the project, the length of the project, the importance of showing whole physical features on a single sheet (rather than dividing the feature on two sheets), and, finally, the designer's subjective decision.

A review of Figure 5.1 shows that the scale chosen for this presentation (shown in a bar graph at top right on the plan sheet) is $\frac{8}{10}$-in equals 40 meters, (or 1-in equals 50 m) on the miniature plan (half-size sheet). (The size of the illustration as you view it in this text is even smaller than a half-size sheet.) Since the original plan sheet (22 by 34 in) is twice the miniature size, the original scale of the presentation was 1-in equals 25 m. Since meters can be converted to feet by multiplying each meter by 3.2808333 ft, a 1-in = 25-m scale is also a 1-in = 82.02 ft scale.

REFERENCING

Because the centerline survey has great significance to the future of all land transfers adjacent to highway improvements, a method has been developed to help relocate and restore the centerline by the use of "reference monuments." Figure 5.1 shows a *typical reference detail* that shows where the reference monuments are to be placed in the ground to assure that the centerline survey can be reestablished. Most commonly, concrete monuments are placed at each centerline break point (at the beginning and end of curves or spiral alignments) as well as periodically along tangent (straight line) alignments.

Where great distances occur between break points, reference monuments are generally placed not to exceed 500-ft distances or at lesser lengths where intervisibility cannot be maintained. Reference monuments are usually placed on the centerline, but in the example in Figure 5.1, the monuments are placed at locations left and right of centerline.

To complete the engineering construction requirement for reference monuments, a *standard construction detail* is located in the lower center of the figure. This standard drawing outlines the construction specifics for the monument, such as class C concrete must be used, 8-in in diameter, 36-in long, with a ½-in steel rod embedded in the upper 6-in of concrete which will be hand formed, for a total finished length of 42 in.

Figure 5.1 includes a note regarding *reference monuments* (MC-1, revised 6-13-69), which identifies who is legally qualified to place them.

Many undesired and unnecessary experiences have occurred in the past when monuments were placed by nonlicensed construction personnel and where the monuments were later shown to be located incorrectly. Much additional time and money has been consumed by state departments of transportation, and by private surveyors, to correct these errors. Most states now require the contractor to place registered professional engineers and registered professional surveyors in responsible control, in terms of their respective expertise, for each municipally financed highway project.

CHAPTER 6
TYPICAL AND SUPERELEVATED SECTIONS

INTRODUCTION

Typical and superelevated sections are plan drawings which depict detailed cross-sectional views of the pavement of the highway. These views show the thickness and types of material to be used for the various surface and subsurface layers of the pavement and shoulder. They also show the various slopes of the pavement from the centerline through the edges of pavement to the backslopes beyond the ditches. Since typical sections are shown as cross-sections drawn at 90° perpendicular to the centerline of the highway, the pavement slopes are called *cross slopes*.

A normal pavement cross-section shows the pavement in a crown condition, with the center at a higher elevation than both edges of pavement, to provide the opportunity for water to drain off the pavement. Pavement cross slopes frequently change to a superelevated condition in curved area of highways, to make turning safer for automobiles than would be provided by a constant crowned condition. In superelevated portions of the highway, the pavement in the curve is banked by raising the elevation of the outside edge of pavement higher than the centerline and lowering the elevation of the inside edge of pavement. Figure 6.5 shows the basic concept of crowned and superelevated conditions.

Transitions between crowned and superelevated conditions are also designed, to ensure a smooth appearance of the edges of pavement, with no abrupt or excessive elevation changes that might make automobile travel unsafe.

The typical section plan sheets and superelevation tabulation sheets to provide the visual and mathematical descriptions of these transitions. A typical section refers to all parts or locations along the highway that have construction uniformity. The portions of the roadway that are constructed in a similar manner are defined and identified by means of centerline stations.

Rather than design roadways with construction variables, the engineer prepares a uniform design based on guidelines obtained through years of research by counties, states, and the federal highway administration. These agencies make recommendations regarding the type and width of surface, slope of that surface, type of base material, type of shoulder material and width, location of ditches, rates of earthen side slopes in cuts and fills, guard rail placement, and many other things for highways carrying certain vehicular weights and speeds.

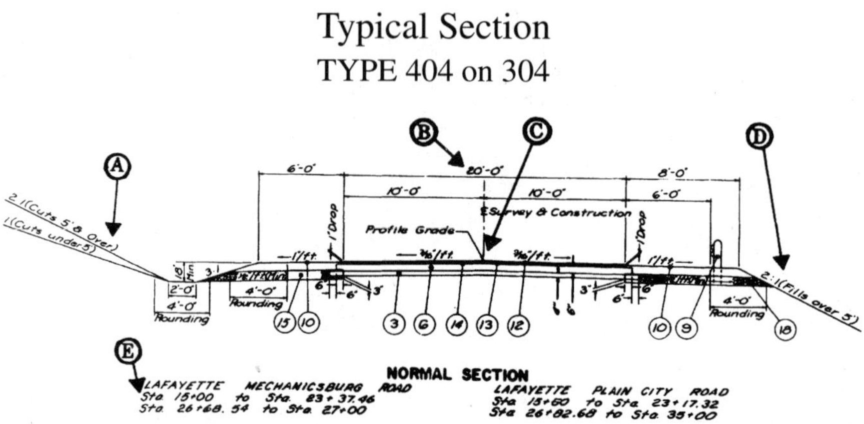

(A) Notice that the rate of slope depends upon the height of the cut.
(B) The pavement is 20 ft wide.
(C) Notice the "profile grade" is located on the centerline of survey and construction.
(D) This earthen side-slope ratio shows the highway in fill condition.
(E) This typical section applies between the stations noted and certain identifiable roads.

FIGURE 6.1 Secondary road typical-section *(The Ohio Department of Transportation.* Plan Reading Course. *Central Office, Columbus, Ohio.)*

When this statistical information is reviewed, certain design standards are chosen to correspond with the project being designed. The engineer prepares a set of highway plans that incorporates only those features appropriate for the subject project. Figure 6.1 is a typical section representing secondary roads, Figure 6.2 represents a mainline road with asphalt concrete surface on bituminous aggregate base material, and Figure 6.3 represents a more complex interstate roadway system.

TYPICAL SECTION LIMITATIONS

Except for small projects, one typical section is seldom sufficient to cover the construction needs of an entire project, and therefore, several typical sections are prepared that incorporate unusual (or different) construction needs. The divisions (or limits) between typical sections are always shown by centerline "*stations*."

Highway "*stationing*" was discussed in previous chapters and is a way of labeling measured increments along the centerline of a roadway to simplify locating a point on the centerline. It would be difficult to describe a certain point on the centerline of a new (or old) roadway by specifying an amount of feet (or thousands of feet) from some known "base point."

Instead, the simplest method of stationing assumes that the station of the base point can be identified as zero (but written 0 + 00). Then, a point located 100 ft along the centerline from the 0 + 00 base point would be called station 1 + 00 (and pronounced one plus zero zero). A point 200 ft from the base point is station 2 + 00 (pronounced two plus zero zero), and so on. Stations at very large distances from the base point do not use commas, so a station located 10,000 ft from the base point is called station 100 + 00.

Points identified with more precision use a normal decimal point (for the decimal parts of a foot), as in station 149 + 23.79. Stationing proceeds along the length of a tangent, along the circular curve portion of a curve, and along the arc of a spiral (which will be covered in later chapters).

Base-point stations do not always commence with 0 + 00. Some roadways are extensions of other numbering systems. In general, and there are exceptions, stations increase from *west* to *east* and from *south* to *north*. There are also stations that end at county lines and begin anew with either + 00 or other designations mandated by the county in which the roadway exists. Divisions between states almost universally commence stationing that applies only to that particular state. In any event, typical sections are limited by virtue of beginning and ending stations.

TYPICAL SECTION

Now that we have some knowledge of stations, and know that a certain construction uniformity (typical section) is applied using centerline stations to determine these limits, we should also expect to find sufficient typical section drawings that cover all the stationing in a project. This means that the station limits associated with one typical section will lead directly to another typical section without missing portions. Quite often, however, one typical section will describe more than one set of limiting stations, as shown on Figure 6.1, so you need to review each typical section thoroughly.

Example:

Figure 6.1 shows a typical section that controls two separate roads and also two separate portions of each road, such as from sta. 15 + 00 to sta. 23 + 37.46, and from sta. 26 + 68.54 to sta. 27 + 00 (Lafayette/Mechanicsburg Road) and from sta. 15 + 50 to sta. 23 + 17.32 as well as from sta. 26 + 82.68 to sta. 35 + 00 (Lafayette/Plain City Road).

The missing stations are always covered in a second or third typical section, where the missing limits will be outlined. Most often, when station leaps occur, other design features control and force additional typical sections to be included in the design. A few possibilities that might precipitate additional drawings are curved and spiraled roadways with superelevation, bifurcated or separated roadways (two lanes on different elevations), or a narrowing or widening of the median width.

Figure 6.3 shows a typical section, or typical used for preliminary design of Interstate 77 before final design contracts were awarded, and therefore, the drawing does not go into the details of assigning station limitations. Whether the typical represent a complex interstate roadway system or a less complicated secondary roadway system, the highway stationing will always control.

Stationing controls the lengths of typical sections, and usually the stations can be subtracted to determine these lengths. Stations can also be subtracted to calculate the length of a project. However, in both situations, there are problems when "station equations" are encountered. This has been covered in previous chapters, but a short review might be in order.

STATION EQUATION

A station equation simply refers to a known mathematical point on a highway that has two stations assigned to it: one station "looking back" along the centerline and a second station "looking ahead" along the centerline. For example:

Station 116 + 45.00, (back) equals station 110 + 00.00, (ahead)

When surveying or calculating the length of this project, all points down-station (back) from the station equation are measured using station 116 + 45.00 as the control, and all points

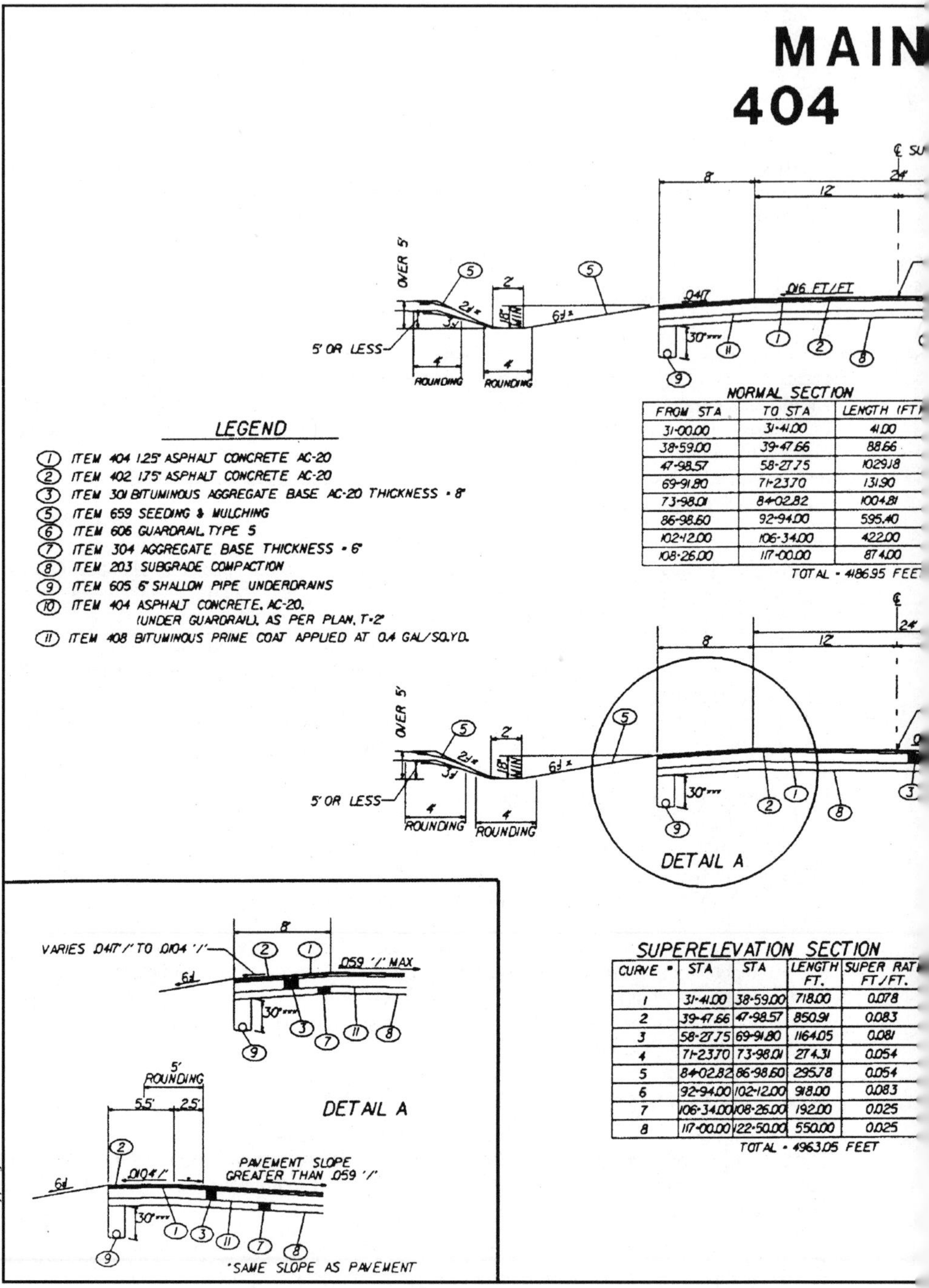

FIGURE 6.2 Primary road typical-section *(The Ohio Department of Transportation. 1995. Division 10.)*

LINE
ON 301

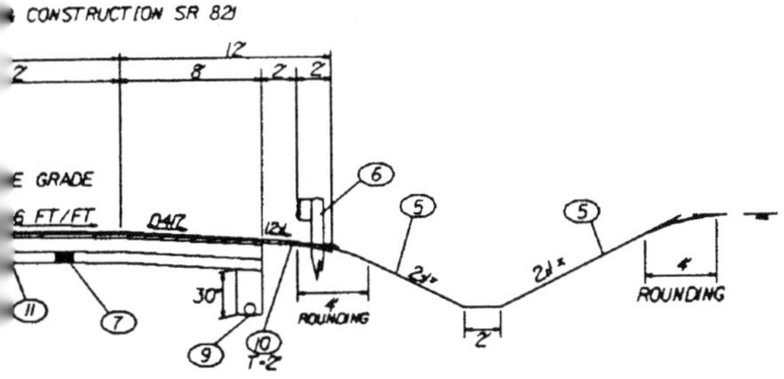

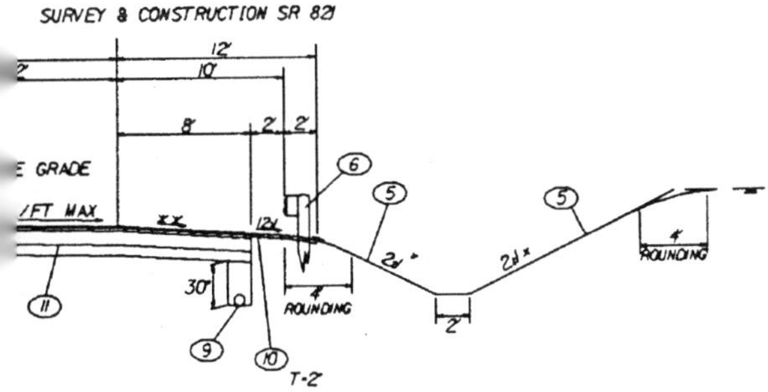

NOTES

x UNLESS OTHERWISE SHOWN ON CROSS SECTIONS

xx 0.0417 FT/FT OR RATE OF SUPER IF GREATER

xxx UNLESS OTHERWISE SHOWN ON SHEET NO. 25

TYPICAL SECTIONS

WAS-821-0.59

3

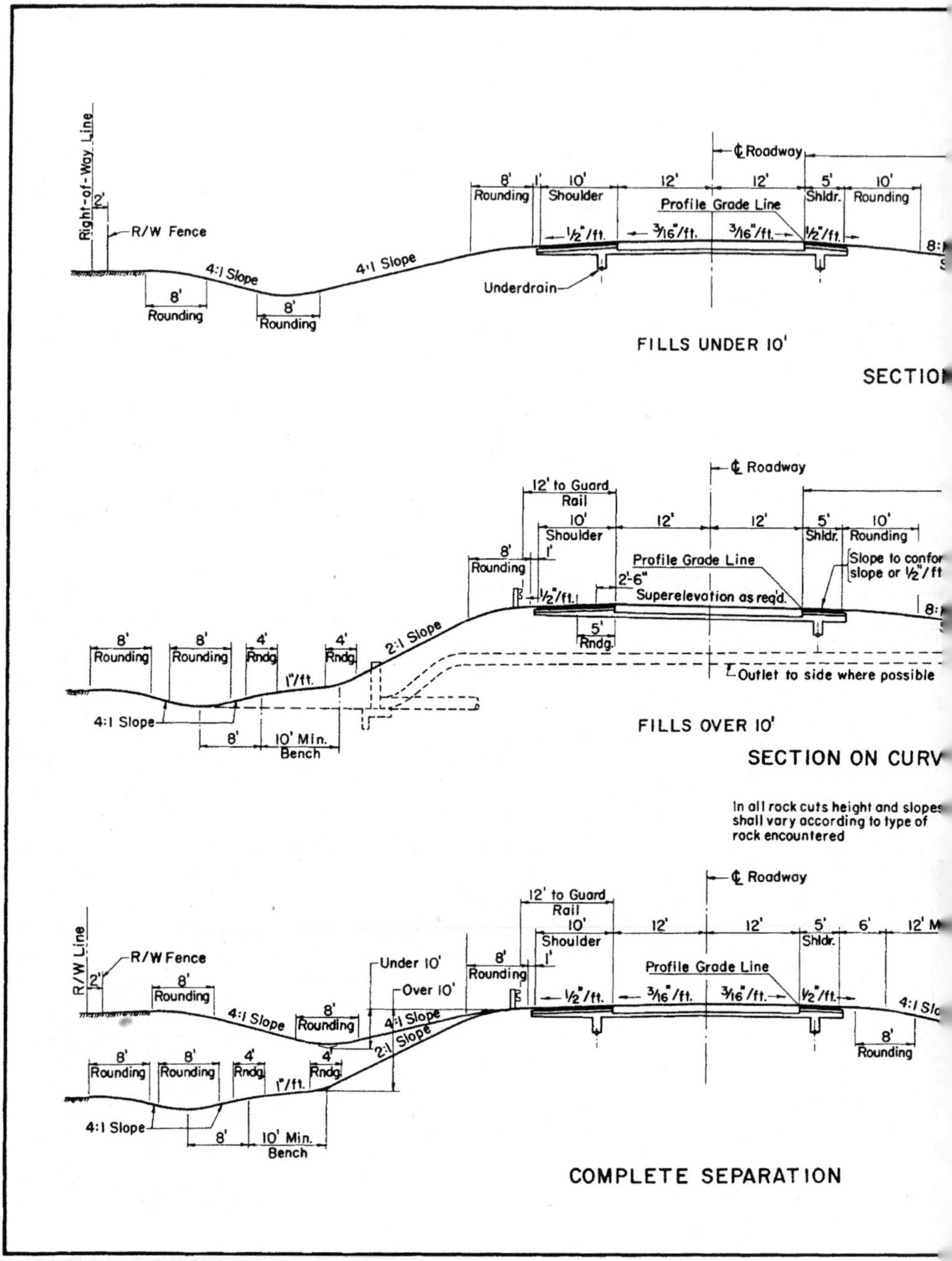

FIGURE 6.3 Interstate road typical-section *(The Ohio Department of Transportation. Interstate 77.)*

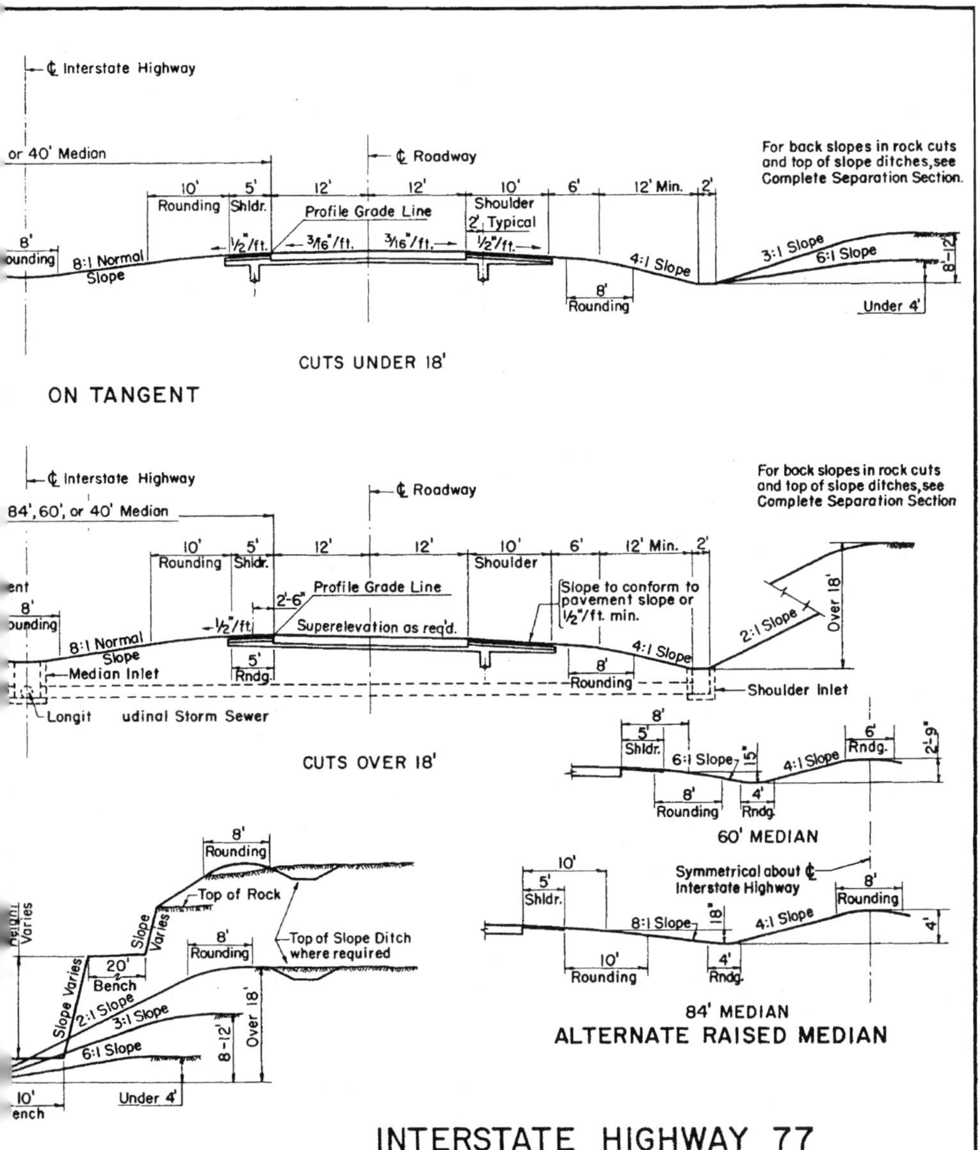

INTERSTATE HIGHWAY 77

TYPICAL ROADWAY SECTIONS

up-station (ahead) are measured from this common point using station 110 + 00.00 as the control. Therefore, if station 111 + 00.00 is to be set in the field, and it is located up-station (ahead) of the station equation, the surveyor measures 100 ft. If, on the other hand, station 115 + 22.36 is to be located, and the point is down-station (back) of the station equation, the surveyor will measure 122.64 ft and set the point. All points, back or ahead, are related to the common station equation, whether in the office or field, and adjusted accordingly.

The above examples are computed in this manner:

111 + 00.00 (ahead) station to be set

110 + 00.00 (ahead) station equation
1 + 00.00, or 100 ft

116 + 45.00 (back) station equation

115 + 22.36 (back) station to be set
1 + 22.64, or 122.64 ft

Refer to Figure 6.1 as we continue with the explanation of typical sections. This figure represents a less complicated roadway and therefore is easier to explain. The width of the pavement is 20.00 ft, it has no median (which is usually representative of secondary cross-roads), and each side of the typical section represents a different construction problem. The left side of the typical (letter A) shows the slope requirements when either side is in cut material; the right side of the typical (letter D) shows the slope requirements when either side is in fill material.

Obviously, not all roadway improvements will have the cut on the left side of centerline or the fill on the right side of centerline, but in order to save space and eliminate extra details, this simplification is used most often.

The typical section is extremely necessary for the contractor, negotiator, and surveyor. The surveyor uses the typical to develop elevations for staking purposes, the highway department negotiator uses it to apprise the property owner of the proposed construction during negotiation, and the contractor uses it to prepare preliminary cost estimates and, later, to construct the final project. Many roadway items are shown in Figure 6.1, such as type and depth of pavement surface, base and subbase materials under the pavement, shoulder treatment, rock aggregate underdrains, guardrail, and earthen side slopes. Letters and numbers are placed beside each construction item and are defined below:

A = The rate of side-slope cut depends on the height of the cut.
B = The width of the pavement is 20 ft.
C = The profile grade of the pavement is the centerline of survey and centerline of construction. Occasionally the two centerlines are different.
D = The ratio of side slopes, when in fill.
E = The limiting stations: In this example, there are two roads with the same typical.

3 = 4 in of subbase material, item 310.
6 = 6 in of aggregate base, item 304.
9 = Guard rail, item 606, modified.
10 = Seeding and mulching, item 659.
12 = 1¼ in asphalt concrete, item 404.
13 = 1¼ in asphalt concrete, item 403.
14 = Bituminous prime coat, item 408.
15 = Aggregate drains, item 605.
18 = Seal coat, item 409.

CONSTRUCTION AND MATERIAL SPECIFICATION MANUAL

The item numbers (such as 310 and 304) shown above are defined in each state's Department of Transportation *Construction and Material Specifications Manual*, published every few years. All contractors refer to the specification guide for highway construction when preparing estimates for highway bidding. The manual includes written material that discusses the contractor's bidding procedures, document preparation; national agencies that govern materials and methods, such as the American Iron & Steel Institute (AISI), the American Association of Nurserymen (AAN), the American Society of Civil Engineers (ASCE), and many others. The manual sets forth methods of construction and defines each referred-to item. The example below shows the details covered by the *Ohio Highway Construction and Specification Manual* and is comprehensive, demanding that certain construction procedures be followed, as well as recommending certain quality materials be set forth as a standard for the construction of each roadway item.

ITEM 310 (SUBBASE)

ITEM 310.01 (description)

This work shall consist of furnishing, placing and compacting subbase, including furnishing and incorporating water required for compaction, in reasonably close conformity with the lines, grades and cross-sections shown on the plans or established by the Engineer. Type I subbase shall meet one of the graduations listed in 310.02 at the time of incorporation into the work.

Type II subbase shall meet grading A, 310.02, after all operations of placing and compaction have been completed.

ITEM 310.02 (materials)

Materials furnished under this item shall be gravel, crushed slag, crushed stone, sand, granulated slag, a mixture of crushed and granulated slags, slacker aggregate or other types of suitable material meeting the requirements of this item and having the approval of the Director.

The excerpt above continues for several additional pages covering such topics as:

1. Construction methods for item 310
2. Method of measurement for item 310
3. Basis of payment for item 310

PROFILE GRADE

The *profile grade* is generally coincident with the centerline survey. There are exceptions, so beware. Profile grades are mentioned in several other chapters, but especially in Chapter 9. However, a brief explanation is included here for purposes of computing elevations on pavement cross slopes, culvert pipe slopes, and earthen side slopes, where each is related to a profile grade elevation. "*Profile*"can refer to an object, a person, or in this situation, a highway centerline, where one can stand aside and view the feature from a 90° perspective. "*Grade*"refers to the ascending or descending elevations of that profile. A grade can refer to an existing surface, such as the earth's contours, or it can refer to a proposed surface, such as the elevations of a proposed highway. The elevations of highways are universally related to mean sea level, where, as early as 1929, the USGS established vertical control throughout the contiguous 48 states. In 1988, the North American Vertical Datum (NAVD) was completed, which, as the name implies, established elevations for the entire North American continent (including Canada).

PROFILE AND CROSS-SLOPE ELEVATIONS

One might anticipate finding profile elevation tables on the typical section drawings. However, each sheet in a set of highway plans has its own special purpose.

The purpose of the typical section is to identify the position of the profile grade (whether centerline or otherwise), to identify the composition of the roadway, and to show the station limitations and other engineering specifics, including the pavement cross slopes, such as $\frac{3}{16}$ in/ft, as shown in Figure 6.1; the shoulder rate, such as 1in/ ft; and the side-slope earth ratios, such as 2:1 and 3:1.

Profile grade elevations can always be found on the plan and profile sheets (see Chapter 9), and on the cross-section sheets (see Chapter 8).

Convert Fractions to Decimal Parts of a Foot

Whenever the surveyor uses pavement cross slopes in computations, the fraction, such as the $\frac{3}{16}$ in/ft, should be reduced to decimal parts of a foot. This is done by reducing the fraction ($\frac{3}{16}$) to a decimal part of an inch by dividing the denominator into the numerator, or 16 divided into 3 equals 0.1875. The decimal parts of the inch now must be divided by 12 (the number of inches per foot) to obtain the decimal part of 1ft: 0.1875 divided by 12 equals 0.015625 ft/ft.

This slope indicator can now be used quite easily to obtain elevations for other types of highway problems, such as sloping culvert pipes, centerline or side ditches, sloping underdrains, sloping bridge approach slabs, or any other elevations on sloping objects can be obtained in a similar manner.

Example:

Assume the profile grade in Figure 6.1 is elevation 600.00, and you are required to compute the edge of pavement elevation at this station. The typical section shows a total pavement width of 20 ft, with a distance of 10 ft from the crown to the edge of pavement. Using the decimal part of a foot, computed above, multiply the number (which represents an incline or decline in the pavement, per each foot) by the 10 ft width. The result is:

$$0.0156 \times 10 = 0.156 \text{ ft}$$

The elevation at the edge of pavement, both sides, after subtracting the 0.16 ft (600.00 minus 0.16) is 599.84 ft above mean sea level. The cross-slope digit in the hundreds column is rounded to the next whole number to obtain the final answer. This is a common practice in surveying, as in most elementary mathematics, when the digit in the thousands column is 5 or above.

Example:

Compute the edge of shoulder for this same profile grade. Figure 6.1 shows the shoulder cross slope at 1in/ft, with a shoulder width of 8 ft when in fill condition, and a 6-ft shoulder width when in cut condition.

The rate of slope (1 in/ft) is already reduced to parts of an inch, and therefore, the final reduction must be in decimals of a foot:

$$1 \div 12 = 0.083333 \text{ ft/ft}$$

The cross-slope elevation difference for the 8-ft-wide shoulder is:

$$0.08333 \times 8 = 0.67 \text{ ft}$$

and subtracting this amount from the edge of pavement elevation results in the edge of shoulder finished grade to be elevation 599.17 (599.84 minus 0.67 equals 599.17) above mean sea level.

The cross-slope elevation difference for the 6-ft-wide shoulder is:

$$0.08333 \times 6 = 0.50 \text{ ft}$$

and subtracting this amount from the edge of pavement elevation results in the edge of shoulder to be elevation 599.34 (599.84 minus 0.50 equals 599.34) above mean sea level.

Tabulations

Computations similar to the above are generally tabulated on computer printout sheets. If printout sheets are not available, the surveyor prepares these tabulations, which include the station, profile grade elevation, edge of shoulder finished elevation, edge of shoulder subgrade elevation, ditch elevation, and other parts of the typical section requiring grading.

EDGE OF SHOULDER SUBGRADE

The *edge of shoulder subgrade* is probably the most important elevation used by the surveyor to direct the contractor's grading personnel in bringing the project to dirt grade. The subgrade of any project is the elevation of the ground located directly below the materials described in the typical section. Figure 6.1 identifies the types of material to be used in constructing this particular highway, and each type of material is labeled with a circled number that identifies the item.

Example:

The top course, number 12, is 1¼ in of asphalt concrete; the second course, number 13, is 1¼ in of asphalt concrete; number 14 is a prime coat (no thickness); number 6 is an aggregate base, 6-in thick; number 3 is a subbase, 4-in thick. These materials comprise the cross section of this pavement, and when their depths are added together and the total subtracted from the profile grade, you have developed the elevation of the subgrade.

The sum of the various items, listed above, equals 12½ in or 1.04 ft and subtracting this depth from the profile grade elevation of 600.00 leaves a subgrade elevation (earthen grade) on the centerline of 598.96 ft.

To compute the elevation of the subgrade at the edge of the *shoulder*, one might consider subtracting the 1.04 ft from the outside edge of shoulder elevation to obtain the subgrade at this point. This would reduce the outer limits (horizontal distance from edge to edge) of the highway shoulders, and would not conform with the engineer's design.

EARTHEN SIDE SLOPE

Notice on Figure 6.1 where the subgrade (bottom of the typical) intersects with the earthen fill slope or cut slope.

This intersection falls somewhere (distance unknown) along the earth side slopes. In fill, the earth slope is identified as 2:1 (pronounced 2 to 1), and in cut, the earth slope is identified as 3:1. The side-slope ratios relate to horizontal and vertical distances, where the first number in the ratio refers to the horizontal distance, and the second number in the ratio refers to a vertical distance. Using the fill side slope (2:1) as an example, the first number represents 2 horizontal feet, and the second number represents 1 vertical foot. Using the cut side slope (3:1) as an example, the first number represents 3 horizontal feet, and the second number represents 1 vertical foot.

Again, refer to Figure 6.1 and notice that the subgrade (bottom of the typical) does not end directly under the edge of shoulder finished grade but continues on, intersecting further out at the side slope. A feeble attempt to establish this edge of shoulder subgrade elevation can be made by applying a triangle calculation from the shoulder, outward, toward the earthen side slope. To use the triangle approach, as suggested above, subtract 1.04 ft from the edge of shoulder finished grade on the fill side, arriving at a subgrade elevation of 598.13 (599.17 minus 1.04 equals 598.13) and then use the earthen side slope ratio of 2:1 to form the long leg of the triangle and the subgrade descent rate of 1"/ft to form the "nearly-horizontal" leg of the triangle. This way you can compute an approximate elevation of 597.97 with a distance from centerline to the subgrade point of >20ft. This is calculated by using the 10-ft pavement width, the 8-ft shoulder width, and knowing the 2:1 side-slope ratio, with approximately 1 vertical foot of material and adding 2 horizontal ft, making a total of >20ft (10 + 8 + 2 = 20).

The only appropriate way to compute the edge of shoulder subgrade is to use computer programmable software that intersects the subgrade information (as found on the typical section) with the side slope. The data that make this intersection possible include the horizontal distances shown on the typical section, the rates of slope (pavement and shoulder), the profile grade elevations, and the earthen side-slope ratio (such as 2:1, 3:1, etc.) for the *top* of the typical section. You then input the *bottom* of the typical section, such as the horizontal distances shown on the typical section, the subgrade elevation at the centerline, the cross slopes of the pavement, any unusual vertical changes at the bottom of the typical, and the final rates of slope for the shoulder. When the data are entered into the computer, a type of bearing-bearing intersect takes place, and the earthen side slopes are intersected with the subgrade. The result is a calculated offset distance from the centerline to the edge of the shoulder subgrade and a calculated elevation. All major roadway slope staking (see Chapter 10) and preliminary dirt grading are based on this computed print out or a hand-prepared substitute.

Although the triangle approach does consider that the subgrade under the shoulder is sloping at 1"/ft (Figure 6.1), in this particular typical, at a point directly under the edge of pavement the subgrade drops vertically an additional 3 inches. You will be surprised when comparing computer subgrade computations with the triangular method (discussed above) that the distance from the centerline to the edge of shoulder subgrade has increased significantly due to the slope (1-in/ft) of the subgrade under the shoulder in combination with the 2:1 slide slope. This merging of two sloping surfaces changes the centerline offset distance as well as the elevation of the point. Obviously, offset distances increase (whether hand or computer calculated) from the centerline to the edge of the shoulder subgrade as the ratio of earthen side slope becomes flatter. Four to one (4:1), six to one (6:1) and eight to one (8:1) slopes increase the horizontal distance off centerline to 4, 6, and 8 extra feet, respectively, when the vertical composition of the roadway is approximately 1 ft thick.

The majority of all slope-staking (see Chapter 10) is related to the edge of the shoulder subgrade (in fill) or the outside toe (bottom) of a ditch (in cut), and therefore, when the distance from the centerline to the subgrade intersection is estimated short (as will occur with the triangle method), so will the outer limits of construction be short, and the roadway will not have sufficient width at the subgrade (earthen fill) level to support the pavement, a guard rail, and a shoulder. This can become very critical to the contractor when the state highway officials request additional fill material to be placed on the side slopes, thus completing the full width of the roadway as designed.

SUPERELEVATED SECTION

We have briefly discussed, the superelevated portion of a highway and will now cover the grading and preliminary design features used by engineers when developing superelevation.

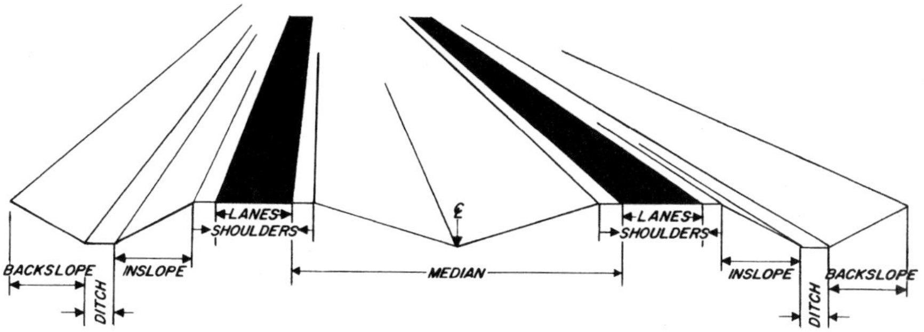

FIGURE 6.4 Two-lane roadway with median *(The Ohio Department of Transportation.* Plan Reading Course. *Central Office. Columbus, Ohio.)*

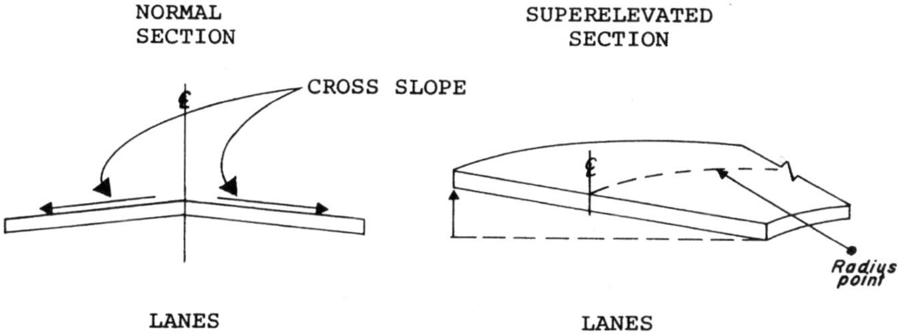

FIGURE 6.5 Crown versus superelevated sections *(The Ohio Department of Transportation.* Plan Reading Course. *Central Office. Columbus, Ohio.)*

Figure 6.4 illustrates a normal roadway with median, and Figure 6.5 illustrates a normal roadway in crown position and also shows the roadway when the crown has been removed and the cross slope reaches full superelevation.

This material will provide the surveyor with background that can be beneficial when elevations are required on construction projects that have insufficient mathematical data from which to finish the layout, and the contractor is eager to proceed.

The easy explanation of a superelevated highway is that one edge of the roadway is raised higher than the other edge to compensate for centrifugal force (the outward force exerted as vehicles are traversing around a curved alignment). This simple explanation is true, but many other engineering specifications must be met before the superelevation is established. Several of the design features are:

1. Vehicle speed, in miles per hour
2. Tire and roadway friction
3. Climate, critical when in freezing localities
4. Safe minimum curve radius
5. Practical maximum superelevation rate
6. Length of transition from crown to full superelevation

We will briefly discuss the various situations that affect the design of superelevation. For more detailed information, consult your local state highway design engineer, engineering consultants in private practice, and/or highway and civil engineering handbooks from your local library.

Vehicle Speed

Vehicle speeds vary in many ways and are subject to many conditions. In urban areas, the speed of automobiles is subject to the number of cross-street intersections, the stop-light controls, the population density in the area, the types of vehicles using the roadways (trucks, buses, and passenger cars), the width of the roadway, existing grades, and driving skills and age of the population. In rural areas the speed of vehicles is generally dependent on terrain. The flatter the topography, the more likely the increase in speed; conversely, the more hilly the area, where grades are more severe, the slower the speed. (Barnett, 1940).

Minimum Safe Radius

Vehicles traveling around curves where the radius is constant will always exert a force toward the outside of the circle, and the highway engineer provides a design, through superelevation, that minimizes this outside movement. The design increases the pavement cross slope at a calculated rate that keeps the vehicle tracking around a curve without moving from the vehicle's lane into another lane in order to maintain that curve (Barnett, 1940). The minimum safe radius for certain velocities is computed based on vehicle speeds, in miles per hour.

The original equation (Barnett, 1940) used to determine the minimum safe radius of a curve is shown below and is computed based on the weight (W) of the vehicle, the velocity, the acceleration of gravity (in feet per second per second), the radius of the curve, and the friction of the tires and roadway (Highway Research Board, 1936). The minimum safe radius (original equation) is:

$$\left(\frac{Wv^2}{gR}\right) = \mathrm{WS} + \mathrm{W} \tag{6.1}$$

Where:

- W = The weight of the vehicle
- v = The velocity, in feet per second
- g = Acceleration of gravity, in feet per second per second
- R = The radius of the curve, in feet
- S = The superelevation slope
- F = The friction factor

This original equation was later reduced by eliminating the weight of the vehicle (W), changing the velocity from feet per second to miles per hour, and substituting 32.16 for gravity. The final equation (Barnett, 1940) assumes a superelevation rate of 0.10 and a maximum safe factor for frictional resistance of 0.16 and is as follows (Highway Research Board, 1936):

$$\text{Minimum safe radius} = 0.258\, V^2 \tag{6.2}$$

Where:

- R = radius
- V = velocity, in miles per hour

Therefore, for speeds of 30, 40, 50, and 60 miles per hour (mi/ h) the safe radius is 232, 412, 644, and 928 ft, respectively (Barnett, 1940). This means that no radius of lesser length should be used for these design speeds. Longer radii, which translate to flatter curves, are always welcome.

The minimum safe radius for a design speed of 70 mi/ hr is computed using a slightly different factor for friction, which is 0.14, and the equation is as follows (Barnett, 1940):

$$\text{Minimum safe } R = 0.279\ V^2 \tag{6.3}$$

Therefore, for speeds of 70 mi/h, the safe radius is approximately 1370 ft, or the equivalent of a 4.2° curve (4° 12' 00"). See Chapter 11 for horizontal curve computations.

Tire and Roadway Friction

Engineers have traditionally refrained from designing a superelevation rate that, by itself, equalizes the centrifugal force around a known curve, due to the possibility the vehicle may slide off the pavement and move toward the inside of the radius because of the steep incline. To compensate for this overdesign, the engineer relies on the friction of tire surfaces and roadway surfaces in combination with superelevation to equalize this effect. The effects of friction and the method of computing friction are beyond the scope of this book, but many professors of engineering, physicists, and scientists have worked on this problem and have concluded, based on driving tests, that a factor of 0.16 represents an amount of resistance to transverse sliding for speeds of 30 to 60 mi/h, and a factor of 0.14 for speeds of 70 mi/h (Highway Research Board, 1936).

Climate

Obviously, the climate in the local area plays a significant role in superelevation (including friction) due to transverse sliding on freezing pavement. Tests have been made where the recommended superelevation rate should not exceed 0.10 ft/ft, or about 1¼ in/ft of pavement cross slope where freezing occurs. It has also been suggested that rates of 1½ in/ft can be used in design when the roadway projects are in nonfreezing areas (Barnett, 1940). However, I have seldom seen the superelevation cross-slope rate exceed 0.08 ft/ft.

Minimum Curve Radius

This has been discussed earlier in the chapter.

Maximum Superelevation Rate

This has been discussed earlier in the chapter.

Length of Spiral Versus Superelevation Transition

Figure 6.6 shows methods of obtaining superelevation. Tables are prepared by engineering designers which list the station, the centerline profile grade, the edge of pavement elevations throughout the transition (from normal crown to the full superelevation), full superelevations throughout the curve, and finally, the transition elevations that return the pavement to the normal crown position. The length of the total transition is not an arbitrarily chosen number but represents a computation that first solves for the length of spiral and, then solves for the run-out length of the transition. The results of these two computations are combined and the transition distance is established. Both computations are shown below:

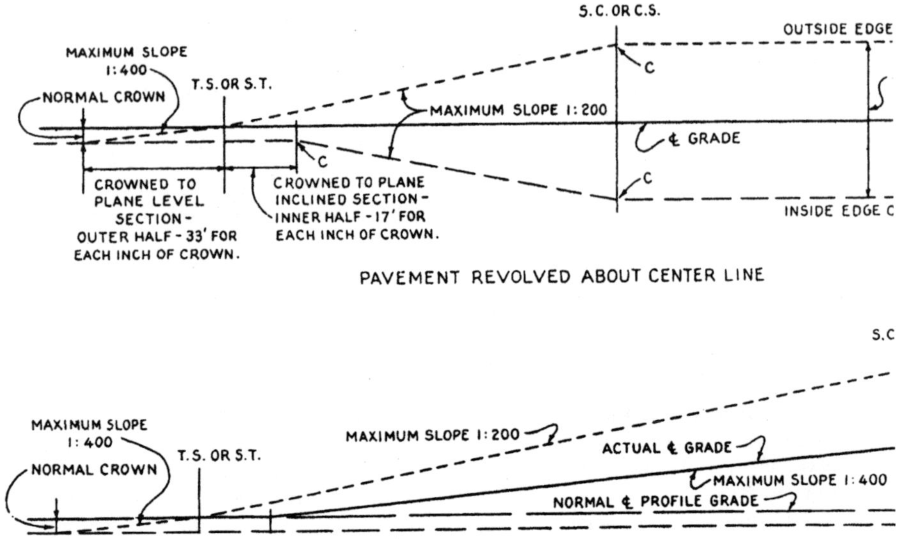

FIGURE 6.6 Geometric diagram showing methods of attaining superelevation. Where necessary, the maximum relative slope of edge of pavement may be increased to 1:150 and 1:175 for speeds of 30 and 40 mph respectively. Short vertical curves may be inserted, if desired, at points marked C. They may be inserted by eye adjustment of grade stakes or forms in the field (Barnett, 1940).

1. Spiral length (L_s). The equation for the spiral length is based on the *radius* of the impending curve, the design *speed* of the roadway, and a constant factor relating to centifugal force. The equation is as follows (Barnett, 1940):

$$L_s = 1.6\,\frac{V^3}{R} \tag{6.4}$$

Where:

L_s = length of spiral
V^3 = velocity, in miles per hour
R = radius of curve
1.6 = a constant based on centrifugal force

Assume:

D_c = 4° = 1432.395 ft radius
V = 70 mi/h

Therefore:

L_s = $1.6\ V^3\backslash R$
= $1.6\ (70)^3\backslash 1432.395$
= 1.6×239.459
= $\pm$ 383 ft

Note: Designers usually round this spiral length to the nearest even increment, such as 400 ft.

The transition length, as defined in this solution, is the total amount of feet required to gradually alter the grade from the edge of pavement while in a normal crown position to the full superelevation of the adjoining curve. This total distance not only includes the length of spiral (as computed above) but also includes a distance called the "tangent run-out." The tangent run-out (Figure 6.6) is a distance computed based on a maximum straight-line slope of 1:400 ratio (not to exceed 1 ft vertically in 400 ft horizontally). Notice in Figure 6.6 that this straight-line slope commences at a point that is represented by the edge of pavement in a normal crown position and travels to the beginning part of the spiral (*TS* or *ST*). The maximum slope ratio then changes to a rate of 1:200 until the full superelevation is reached at the adjacent curve.

2. Tangent run-out solution. Assume you are to compute the tangent run-out distance in the "crowned to plane-level section" (Figure 6.6) and the edge of pavement is 3-in below the normal pavement crown. If the maximum ratio is to be maintained (1:400), 3-in are equal to ¼ ft, and one-fourth of 400 ft is 100 ft, or approximately 33 ft of horizontal distance is required along the outside edge of pavement for each inch of crown. Therefore, with a 3-inch drop from crown to pavement edge, the additional run-out distance must be 3 times 33, or 99 ft (call it 100 ft). Combining this distance with the length of the spiral, which was rounded to 400 ft, gives a total transition length of 500 ft.

The inside edge of pavement requires a slightly different commencing point for the crown to plane-level ratio, and according to Joseph Barnett (1940) and shown in Figure 6.6, the tangent run-out distance is 17 ft/ in of cross slope, and in this example, the run-out distance must be 17 times 3, or 51 ft (call it 50 ft).

In summation, the outer edge of pavement commences its superelevation transition 100 ft either back (on tangent) or ahead (on tangent) of the flat part of the spiral, while the inside edge of pavement commences its superelevation transition 50 ft into the flat part of the spiral. For more information regarding horizontal spirals and their applications, see Chapters 12 and 13.

SUPERELEVATION CASE STUDY

For another example of superelevation, its computations, and the representative drawings found in a set of highway construction plans, we will use an actual case study from roadway project APD 483 (28), West Virginia Department of Highways.

Superelevation drawings are usually in the form of "tabulations" and are called "superelevation tables". These tables are prepared by the engineer and used by the surveyor to establish individual finished paving grades, or the surveyor will set guideline string elevations for the paving machine to operate along. These same finished grade elevations (top of pavement) are used to establish the *subgrade* at the edge of pavement by subtracting the depth of the pavement and base material composition from the finished grade. However, when dirt grade is established, two methods are used most often: (1) locate the station and drive grade hubs on each side of the pavement to match the subgrade elevation (wooden hubs 1½ in^2 by various lengths, usually 6 to 18 in long) and (2) set lath (wooden guard stakes 4 ft high) 2, 3, or 5 ft outside the edge of pavement, and prorate the pavement cross-slope grade to match these offsets.

The following represents the curve data from which we will develop the superelevation from an actual case study, the cross-slope rates, the minimum length of spiral, and the minimum length of run-out and, through diagramming, will show how the finished grade edge of pavement elevations are computed within the transition. Remember, the transition length and the spiral length are not the same.

Table 6.1, below, shows the actual elevation data included in the test project, being an example of a portion of the superelevation table from Project APD 483 (28)

Table 6.1 represents one-half of the computations found on the engineer's drawing. The second half of the table is not shown but is computed in a similar manner.

MAINLINE SUPERELEVATION DATA

P.I. Station 553 + 52.39

T.S. = station 545 + 61.61	L_c = 764.38 ft
S.C. = station 549 + 61.61	Run-out = 150 ft
C.S. = station 557 + 25.98	D_c = 2° 00' 00" (Right)
S.T. = station 561 + 25.98	S.E. = 0.065 ft/ft
	L_s = 400 ft (set)

TABLE 6.1 Example of a Portion of the Superelevation from Project APD 483 (28).

Station	44 ft left	Profile grade	44 ft right
543 + 00	1155.46	1155.84	1155.46
+ 25	1154.52	1154.89	1154.51
+ 50	1153.56	1153.94	1153.56
+ 75	1152.62	1152.99	1152.61
544 + 00	1151.68	1152.04	1151.66
+ 25	1150.78	1151.09	1150.71
+ 50	1149.89	1150.14	1149.76
+ 75	1149.00	1149.19	1148.81
545 + 00	1148.12	1148.24	1147.86
+ 25	1147.25	1147.31	1146.93
+ 50	1146.43	1146.42	1146.04
T.S. + 61.61	Begin	Spiral	
+ 75	1145.66	1145.56	1145.19
546 + 00	1144.94	1144.75	1144.37
+ 25	1144.26	1143.97	1143.58
+ 50	1143.62	1143.23	1142.80
+ 75	1143.02	1142.53	1142.04
547 + 00	1142.55	1141.87	1141.28
+ 25	1141.93	1141.25	1140.56
+ 50	1141.44	1140.66	1139.88
+ 75	1140.99	1140.12	1139.24
548 + 00	1140.58	1139.61	1138.63
+ 25	1140.21	1139.14	1138.07
548 + 50	1139.88	1138.71	1137.54
+ 75	1139.58	1138.31	1137.05
549 + 00	1139.32	1137.96	1136.60
+ 25	1139.08	1137.63	1136.18
+ 50	1138.80	1137.29	1135.78
S.C. + 61.61	Begin	Curve	
+ 75	1138.50	1136.96	1135.41
550 + 00	1137.18	1136.62	1135.06
+ 25	1137.84	1136.29	1134.72
+ 50	1137.51*	1135.95*	1134.39*

* Full superelevation throughout the remainder of the curve.

Source The West Virginia Department of Highways. *Project No. APD 483 (28).* Union District. Monongalia County.

Computing the Minimum Length of Spiral

The plan data are as follows:

$$D_c = 2° \ 00' \ 00''$$
$$R = 2864.79'$$
$$\text{Velocity} = 70 \text{ mi/ h}$$

Using Eq. 6.4, we have the following:

$$L_s = 1.6 \frac{V^3}{R} \tag{6.4}$$

$$L_s = 1.6 \frac{(70)^3}{2864.79 \text{ ft}}$$

$$L_s = 191.56 \text{ ft (call it 200 ft)}$$

Obviously, by comparing this computation with the length of spiral shown on the plan, there is a considerable difference (400 ft versus 200 ft). Remember, the result of Eq. 6.4 is in terms of a *minimum* spiral length, where the adjacent curve is 2° and where the velocity of the roadway is 70 mi/ h. It is not unusual that longer spiral lengths are chosen by the designers since this permits a more gradual transition and improves the roadway safety.

COMPUTING THE MINIMUM LENGTH OF SPIRAL RUN-OUT

Before continuing with the run-out computation, review the typical section for the actual case study to determine the elevation difference from the crown (or profile grade, when pavements are separated) to the edge of pavement. This elevation difference is used to compute the run-out length, as was shown earlier in the chapter. Notice the limiting stations of this curve in Figure 6.7.

station 543 + 75 to station 550 + 00 (transition)
station 550 + 00 to station 556 + 75
station 556 + 75 to station 563 + 00 (transition)

The normal typical, with $\frac{3}{16}$ in slopes breaking down from the profile grade, is shown at the top of the figure, and the superelevation typical, with a cross-slope rate of 0.065 ft/ft, is shown at the bottom of the figure.

As previously discussed, the run-out computation is subject to the elevation difference in the normal cross slope of the pavement, and in this example, the pavement has a cross slope of $\frac{3}{16}$ in/ft (or 0.0156 ft/ft). Multiply the cross slope rate by the 24-ft pavement width which gives $0.0156 \times 24 = 0.3744$ ft (rounded to 0.37 ft). The minimum horizontal run-out distance is, therefore, 33 ft per each inch of elevation difference ($4\frac{1}{2}$ in are the equivalent to 0.37 ft), or $4.5 \times 33 = 148.50$ ft (call it 150 ft).

To determine where the run-out commences, review Table 6.2. The text information, in our case study example, indicates that the transition, or the beginning of the run-out, commences

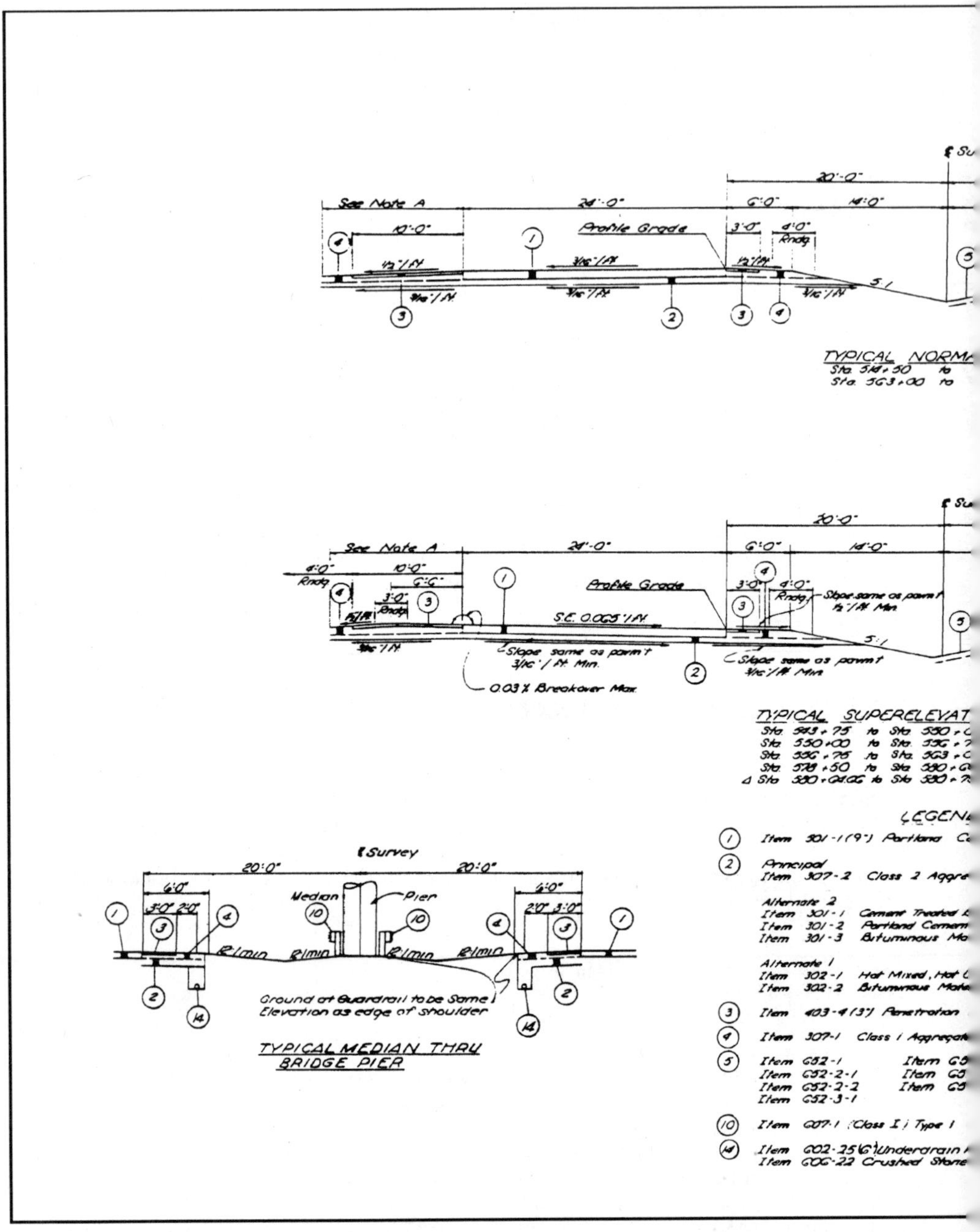

FIGURE 6.7 Case study of Interstate project APD 483 (28) *(The West Virginia Department of Highways. Union District.)*

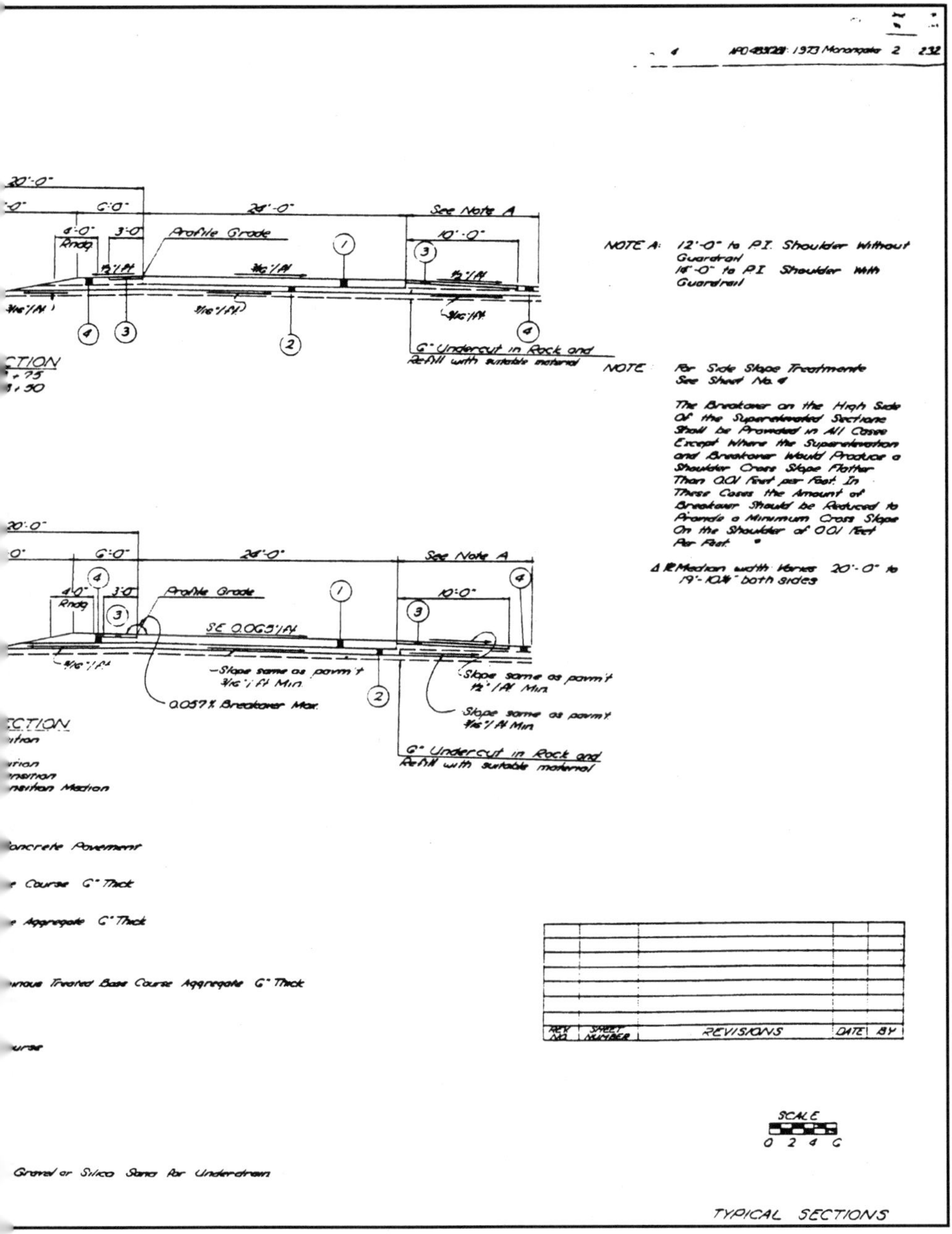
1973 Monongalia 2 232
See Note A
Profile Grade
NOTE A: 12'-0" to P.I. Shoulder Without Guardrail
16'-0" to P.I. Shoulder With Guardrail
6" Undercut in Rock and Refill with suitable material
NOTE: For Side Slope Treatments See Sheet No. 4
The Breakover on the High Side Of the Superelevated Sections Shall be Provided in All Cases Except Where the Superelevation and Breakover Would Produce a Shoulder Cross Slope Flatter Than 0.01 Feet per Foot. In These Cases the Amount of Breakover Should be Reduced to Provide a Minimum Cross Slope On the Shoulder of 0.01 Feet Per Foot.
SE 0.065'/Ft
Slope same as pavm't 3/16"/Ft Min.
0.057% Breakover Max.
Slope same as pavm't 1/2"/Ft Min.
Slope same as pavm't 3/16"/Ft Min.
Concrete Pavement
Course 6" Thick
Aggregate 6" Thick
Treated Base Course Aggregate 6" Thick
Gravel or Silica Sand for Underdrain
REV. NO. | SHEET NUMBER | REVISIONS | DATE | BY
SCALE
0 2 4 6
TYPICAL SECTIONS

TABLE 6.2 Example of a Portion of the Superelevation Table from Project No. APD 483 (28) Showing the Cross-Slope Rate for each 25 ft Station Interval.

Cross-slope	Station	44 ft left	Profile grade	44 ft right	Cross-slope
–0.0158	543 + 00	1155.46	1155.84	1155.46	–0.0158
–0.0154	+ 25	1154.52	1154.89	1154.51	–0.0154
–0.0158	+ 50	1153.56	1153.94	1153.56	–0.0158
–0.0154	+ 75	1152.62	1152.99	1152.61	–0.0154
–0.0150	544 + 00	1151.68	1152.04	1151.66	–0.0150
–0.0129	+ 25	1150.78	1151.09	1150.71	–0.0158
–0.0104	+ 50	1149.89	1150.14	1149.76	–0.0158
–0.0079	+ 75	1149.00	1149.19	1148.81	–0.0158
–0.0050	545 + 00	1148.12	1148.24	1147.86	–0.0158
–0.0025	+ 25	1147.25	1147.31	1146.93	–0.0158
0.0004 level	+ 50	1146.43	1146.42	1146.04	–0.0158
	T.S. + 61.61	Begin	Spiral		
+0.0041	+ 75	1145.66	1145.56	1145.19	–0.0154
+0.0079	546 + 00	1144.94	1144.75	1144.37	–0.0158
+0.0121	+ 25	1144.26	1143.97	1143.58	–0.0162
+0.0162	+ 50	1143.62	1143.23	1142.80	–0.0179
+0.0204	+ 75	1143.02	1142.53	1142.04	–0.0204
+0.0283 ?	547 + 00	1142.55	1141.87	1141.28	–0.0245
+0.0283	+ 25	1141.93	1141.25	1140.56	–0.0287
+0.0325	+ 50	1141.44	1140.66	1139.88	–0.0325
+0.0362	+ 75	1140.99	1140.12	1139.24	–0.0366
+0.0404	548 + 00	1140.58	1139.61	1138.63	–0.0408
+0.0445	+ 25	1140.21	1139.14	1138.07	–0.0445
+0.0487	+ 50	1139.88	1138.71	1137.54	–0.0487
+0.0529	+ 75	1139.58	1138.31	1137.05	–0.0525
+0.0566	549 + 00	1139.32	1137.96	1136.60	–0.0566
+0.0604	+ 25	1139.08	1137.63	1136.18	–0.0604
+0.0629	+ 50	1138.80	1137.29	1135.78	–0.0629
	S.C. + 61.61	Begin	Curve		
+0.0641	+ 75	1138.50	1136.96	1135.41	–0.0645
+0.0233 ?	550 + 00	1137.18	1136.62	1135.06	–0.0650
+0.0645	+ 25	1137.84	1136.29	1134.72	–0.0650
+0.065	+ 50	1137.51	1135.95	1134.39	–0.0650

at station 543 + 75 and the transition ends at station 550 + 00, making a total of 625 ft. To get the beginning and ending of the run-out distance from the superelevation table compute each station for its individual cross slope and compare one with another. This way the surveyor becomes thoroughly familiar with the entire transition and will find drafting errors that must be brought to the attention of the contractor or the governmental agency and corrected.

Superelevation Table Review

Table 6.2 is an extension of the original plan data, with cross slope ratios computed for both left and right sides of the pavement. These computations are made by subtracting the elevation of each pavement edge from the profile grade, noting whether the pavement edge is above or below the profile grade and dividing this difference by 24 (the width of pavement, in feet). You are probably wondering why the superelevation table shows distances of 44 ft left and right for the edge of pavement, while we divided by 24 ft to obtain the cross slope at each station. Figure 6.7 answers the question, by showing a 20 ft median width (both sides of centerline), and therefore, the pavement width of 24 ft plus the median width of 20 ft results in a 44-ft distance from centerline. Chapter 9 explains in greater detail that the profile grade of a highway is computed on centerline and then shifted to the inside edge of pavement for layout purposes; therefore, the 24-ft pavement is used to compute each cross slope.

Review the cross slope shown in Table 6.2, remembering that the curve is to the right, and therefore, the outside edge of pavement (left side) is gradually ascending to a level position equal to the profile grade. It then continues to elevate above the profile grade until it reaches full superelevation, while the inside edge of pavement (right side) gradually descends until it reaches full superelevation.

An analysis of the cross-slope rates, on the left pavement edge, shows a degree of consistency commencing at station 544 + 00 and continuing to station 547 + 00, where the cross-slope is identified as +0.0283 ft/ ft and the following station 547 + 25 has the same rate of slope (+0.0283). Unless the pavement cross slope has reached its maximum, this duplication of rates indicates that an error exists and the elevations must be checked before continuing.

The mathematical information in these cells is probably the result of drafting errors that stand out when the cross slope ratios are computed and placed in their appropriate positions, as in Table 6.2.

Determining the Slope-Rate Change

The difference in slope-rate change (why one rate is 0.0162 and the adjacent rate is 0.0204) can be computed in the following manner (refer to Figure 6.8): The diagram, in the tangent run-out portion, shows a full cross slope of 3⁄16 in/ ft (0.0156 ft/ ft) for the 24-ft pavement width at station 544 + 00, with the pavement edge ascending and becoming level with the profile grade at station 545 + 50. The rate change for each successive 25-ft station follows. The ratio is:

$$\frac{0.0156}{150} = \frac{x}{25}$$

$$x = \frac{0.39}{150}$$

$$= 0.0026$$

Where:

0.0156 ft/ ft	=	the normal cross slope
150 ft	=	the tangent run-out distance
25 ft	=	pavement interval to be graded
x	=	slope-rate change per each 25 ft

Therefore, the rate change at station 544 + 25 and throughout the run-out is:

$$\begin{array}{l} 0.0156 \text{ at station } 544 + 00 \\ \underline{-.0026} \\ 0.0130 \text{ at station } 544 + 25 \end{array}$$

$$\begin{array}{r}-.0026\\ \hline 0.0104\end{array}\text{ at station }544 + 50$$

$$\begin{array}{r}-.0026\\ \hline 0.0078\end{array}\text{ at station }544 + 75$$

$$\begin{array}{r}-.0026\\ \hline 0.0052\end{array}\text{ at station }545 + 00$$

$$\begin{array}{r}-.0026\\ \hline 0.0026\end{array}\text{ at station }545 + 25$$

$$\begin{array}{r}-.0026\\ \hline 0.0000\end{array}\text{ at station }545 + 50$$

There are some small differences when the above slope rates are compared with the case-study plan rates in Table 6.2. These differences are usually caused by the rounding of numbers and attempting to keep similar pavement elevations on each side of the roadway. These small variations in cross-slope rates become insignificant when elevations are computed, but the mathematical principal used by the engineer can be duplicated, as we have done here, to check elevations or to add new grading points when necessary. This ability often becomes a necessity when the surveyor is required to grade the corners of bridge approach slabs and the bridge is in horizontal curves or spirals.

Continuing with the case-study diagram (Figure 6.8), we have the following slope-rate changes for the remainder of the transition from station 545 + 50 to S.C. station 549 + 61.61 and will show how these computations duplicate the original plan data. You will note that the cross-slope rates change somewhat uniformly throughout this area but commence at station 549 + 50 to taper off until the full superelevation is reached at station 550 + 50.

The rate-change ratio for each 25 foot station interval is:

$$\frac{0.065}{411.61} = \frac{x}{25}$$

$$x = \frac{1.625}{411.61}$$

$$= 0.0039\text{ ft/ft (called 0.0041 on the case study)}$$

Where:

0.065 ft/ ft	=	the full superelevation cross slope
411.61 ft	=	the spiral run-out distance
25 ft	=	pavement interval to be graded
x	=	slope rate change per each 25 ft

Therefore, the rate change at station 545 + 50 to and including spiral station 549 + 61.61 is:

$$\begin{array}{r}0.0000\\ +0.0041\\ \hline 0.0041\end{array}\begin{array}{l}\text{ at station }545 + 50\text{ (level pavement)}\\ \\ \text{ at station }545 + 75\end{array}$$

$$\begin{array}{r}+0.0041\\ \hline 0.0082\end{array}\text{ at station }546 + 00$$

$$\begin{array}{r}+0.0041\\ \hline 0.0123\end{array}\text{ at station }546 + 25$$

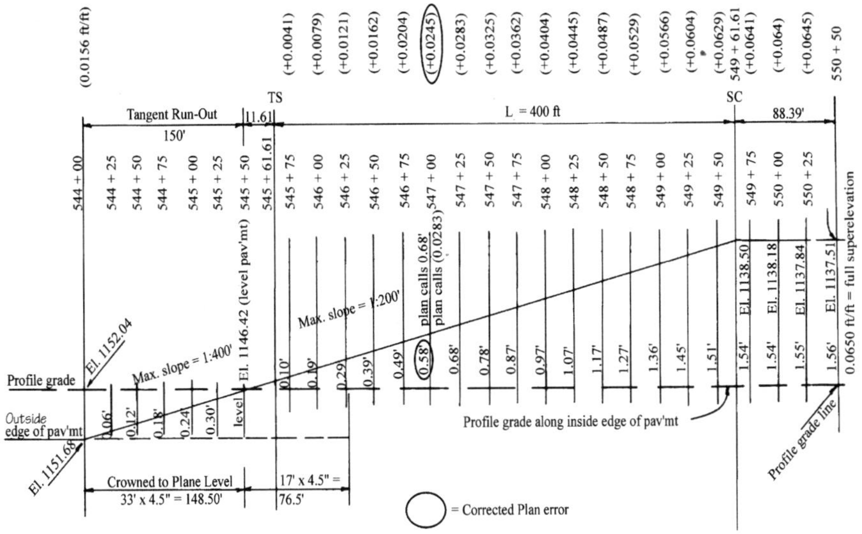

Procedure

1. Establish the profile grade for each 25 ft interval and check against plan-profile sheets.
2. Compute the tangent run-out based on the per inch slope from crown to pavement edge.
3. Compute the slope-rate change from the outside edge of pavement to a level position.
4. Compute the slope-rate change from level (end of run-out) to fulll superelevation.

FIGURE 6.8 Case study of superelevation showing cross slopes

+0.0041
0.0164 at station 546 + 50

+0.0041
0.0205 at station 546 + 75

+0.0041
0.0246 at station 547 + 00 (compare case study elevations)

+0.0041
0.0287 at station 547 + 25

+0.0041
0.0328 at station 547 + 50

+ 0.0041
0.0369 at station 547 + 75

+ 0.0041
0.0410 at station 548 + 00

+ 0.0041
0.0451 at station 548 + 25

+ 0.0041
0.0492 at station 548 + 50

$$\frac{+\,0.0041}{0.0533}$$ at station 548 + 75

$$\frac{+\,0.0041}{0.0574}$$ at station 549 + 00

$$\frac{+\,0.0041}{0.0615}$$ at station 549 + 25

$$\frac{+\,0.0024}{0.0629}$$ (adjusted at the discretion of the engineer) at station 549 + 50

$$\frac{+\,0.0012}{0.0641}$$ (adjusted at the discretion of the engineer) at station 549 + 75

$$\frac{+\,0.0000}{0.0641}$$ (adjusted at the discretion of the engineer) at station 550 + 00 (compare case study elevations)

$$\frac{+\,0.0004}{0.0645}$$ (adjusted at the discretion of the engineer) at station 550 + 25

$$\frac{+\,0.0005}{0.0650}$$ (adjusted at the discretion of the engineer) at station 550 + 50 (full superelevation)

If you compared the rate changes in the listing preceding Figure 6.8 with the plan rate changes in Table 6.2, you should have noticed small differences. These differences, as mentioned before, are usually caused by the rounding of numbers, a difference in the choice of stations used to commence the rate change, the engineer's discretion when adjusting the last few stations in the transition, and an attempt to keep equal elevations at the edge of pavement (both sides) when in crown condition or equal cross-slope rates on each pavement when separated by a median.

Superelevation Table Errors

The discovery of errors in the highway case-study superelevation table is not unusual. The practicing surveyor must always be keenly aware that all numbers in a set of highway plans are suspect and that the professional must search for these errors, correct them quickly, and let the contractor client move on with construction.

The superelevation table analysis, just completed, located two problems: (1) the cross-slope rate (shaded) at station 547 + 00 and (2) the edge of pavement elevation problem (shaded) at station 550 + 00. Each problem would have caused some delay in the construction. The elevation that was listed incorrectly for station 547 + 00, although only 0.10 ft (1142.55 versus 1142.45), was in the high direction, and would have most certainly been discovered during concrete paving operations. The error would probably have been discovered earlier while laying of base material but is less likely to have been noticed in the subgrade preparation.

The second of the two problems, the 1-ft elevation discrepancy at station 550 + 00 (1137.18 versus 1138.18), would have been discovered during all three grading procedures (sub-grade, base material, and paving). The error is of such a magnitude that either 25 or 50 ft on each side of the problem would require regrading if the problem was still in its subgrade stage. If the error, by some chance, reached top of stone stage, many cubic yards of excess material would have to be removed, and the area surrounding the problem, either 25 or 50 ft, would require regrading. In both of these instances, the contractor would have to spend many dollars on this remedial work. If the problem ever got to the concrete paving stage, the entire concrete paving operation would stall, as many as 10 to 80 pavers would be halted, the survey crew would be recalled to locate the problem, corrections would be made, the contractor's grade lines adjusted and, finally, the operation would continue.

We have taken a considerable amount of time to discuss the grades of highways, both in normal crown position and in superelevation. The mathematics behind these engineering characteristics are absolutely necessary to the surveyor. The surveyor is the last individual to have contact with the engineer's design, and this places the surveyor in a unique position of plan interpretation and also plan error prevention.

CHAPTER 7
TOPOGRAPHIC MAPS

INTRODUCTION

Topographic maps are representations of portions of the earth's surface and are intended to accurately show the horizontal position of the various natural and artificial features, and to depict the variation in elevation of the ground surface. Topographic maps can be planimetric (defined as horizontal features), contour (defined as the vertical relief of the ground), or a combination of both planimetric and contour.

The five groups below represent a variety of items that are present on a topographic map.

1. Artificial structures, such as highways, bridges, streets, railroads, residential, industrial and commercial buildings, and subdivisions.
2. Elevations, including the ground surface (hills, valleys, mountains and plains) and such features as excavation, embankments, and landfills.
3. Vegetation, including such features as cultivated fields, residential and rural grass areas, trees, wilderness, and orchards.
4. Drainage, including such items as streams, rivers, lakes, ponds, and borders of large bodies of water.
5. Cities, towns, civil subdivisions, and property lines.

These types of features are usually included on most topographic maps (if applicable) and are easier to recognize when conventional symbols are used. Symbols lessen the confusion between features of common origin and eliminate the explanations that would otherwise become necessary. A number of symbols have become standard practice, and such graphics can be obtained through the state highway departments, the Architectural Society, the Federal Highway Administration, and many commercial publications. Graphic standards (or conventional signs) are included in most highway construction plans.

While certain objects can be described on architectural, engineering, and surveying drawings using standard symbols, the vertical relief (elevations) of the existing ground is presented in the form of contour lines, and occasionally by artistic shading.

Artistic shading will not be covered in this chapter; however, the contour map will be reviewed in detail.

THE CONTOUR MAP

Contour Lines

Contour lines, generally called "contours," represent ground positions of equal elevation, such as a line weaving, curving, and connecting equal elevations within the perimeter of a specified area. Connecting common elevations produces the curves (straight lines are rarely found in nature) that are associated with contour mapping. These elevations are obtained by normal field surveying methods or by aerial photographic means.

Aerial Photography

Aerial photography, in highway engineering, is known as "photogrammetric mapping" and involves a series of controlled photographs, exposed at specified altitudes that, when developed, are viewed stereoscopically through analytical plotters, and topographic maps are developed. In this chapter, we will concentrate on the field survey method and how this information is gathered. Included in this portion of the chapter are graphic sketches that depict certain conditions of the earth's surface and how these conditions are interpreted as contour lines.

Preparation

A surveyor is commissioned to prepare contour maps of a given land mass for various reasons; they include:

1. The development of subdivisions
2. The development of shopping malls
3. A proposal for constructing new buildings
4. Grading plans for commercial buildings, and other construction where an engineering design precedes the actual improvement

The surveyor reviews the site (reconnaissance) and chooses the most practical horizontal layout of it from which elevations (relief) and topographic features can be obtained. Although the property can be irregular in shape, the horizontal layout usually consists of a series of equally spaced parallel lines throughout the property, with stopping points located along the parallel lines, forming a series of small squares similar to a checkerboard in Figure 7.1.

The checkerboard system is usually surveyed with the major controlling line being parallel with some existing feature, such as a curb line, property line, or other physical attribute of the site, and from this initial control, intersecting lines at 90° to the baseline or centerline are surveyed. The distance between intersecting lines is usually 50 ft but can be any length if unusual ground changes occur frequently. Choosing a ground control system that corresponds exactly with every ground change cannot be expected, and therefore, ground changes that do not match the checkerboard system are located as a plus or minus, such as shown in Figure 7.1.

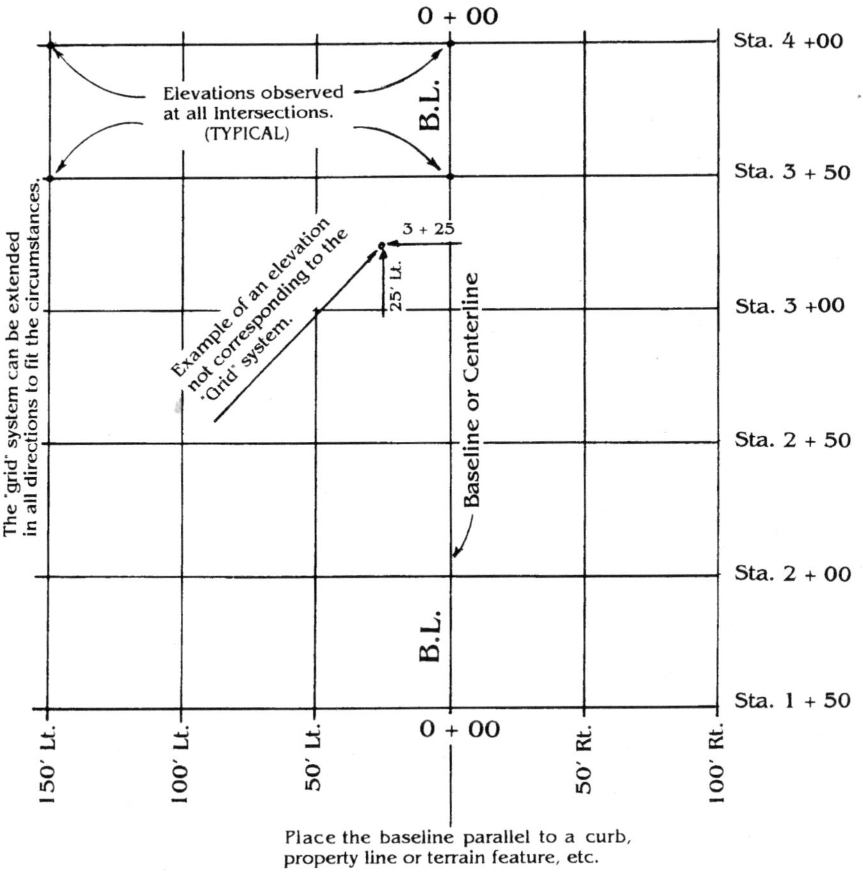

FIGURE 7.1 Example of topographic map grid patterns

As the grid system (checkerboard) develops in the field, a numbering or identification system must be established for each line, such as station 1 + 50 (Figure 7.1) and then succeeding parallel lines identified (in this example) as 2 + 00, 2 + 50, 3 + 00, etc. As the perpendicular lines (at 90° to the commencing station) are surveyed, they must also be identified or numbered, such as 50 ft, 100 ft, etc., left or right, of the zero (0 + 00) or baseline perpendicular. As in Figure 7.1, the horizontal lines are identified with stations, while the vertical lines are identified with distances, left or right, of the zero baseline. Many such systems can be devised, and a sketch of each system is prepared and included at the beginning of the field data. Without this sketch for orientation purposes, office plotting of the topographic and contour features can become difficult, if not impossible. When sketching the grid system in your field notes, orient the drawing so that the north point is toward the top of the page.

The sample "grid" elevation control system, Figure 7.1, represents one method among many that the surveyor establishes in the field before commencing elevation observations.

Horizontal controls (the grid system) are field-located using certain types of surveying equipment, such as transits, electronic measuring equipment, and tapes, that do not relate to elevation gathering. Therefore, it is generally recommended that the surveyor complete one aspect of the survey (grid) before commencing another (elevations). However, under certain conditions, and where large field crews can be divided into separate working groups, both types of observations can be completed simultaneously.

Elevation Data

While the contour map project is in its planning stage, a basis for elevation is being determined. All state and federal highway projects, as well as other significant improvements, are referenced in elevation to mean sea level. Many projects, however, regardless of size, still use assumed elevations as their data. When an assumed datum is chosen, one physical object within the site is chosen, such as the top of a concrete curb, top of an iron pin, top of a fire hydrant, and so forth, to represent the commencing elevation. The surveyor chooses (assumes) the elevation of the object, generally in 100 ft increments, such as 600, 800, or 1000 ft. The assumed datum should certainly represent an elevation where the lowest of the land relief will not lie in negative numbers. Surveyors are also aware of the confusion that occurs when an assumed elevation mirrors an exact elevation. As a means of prohibiting this situation, assumed elevations are purposely chosen 100, 200, or any number of feet different from the (true) elevation range of the area being mapped. As a general statement, assumed elevations are chosen for the purpose of diminishing the cost of the mapping.

U.S. Geological Survey

Mean sea level elevations are those points where the U.S. Geological Survey, U.S. Coast and Geodetic Survey, state and county highway departments, and other organizations have established exact vertical ground control for thousands of objects (monuments) throughout the United States.

Some of the earliest elevations established by these federal agencies date to the year 1929. One of the latest elevation criteria, that of the North American Survey Datum, was in progress as late as 1988.

Therefore, during the planning stage of any topographic contour survey, at least one elevation point must be found or set from which to commence the leveling.

States and/or the federal government have printed material describing where all elevation and horizontal controls are located in each state and throughout the country. It would be an unusual area that did not have access to elevation controls, but if this situation occurs, the surveyor usually locates the nearest control and surveys a closed elevation loop into the project.

Elevation Level Loop

Figure 7.2 is a simplified version of a closed elevation (called level) loop, showing the horizontal location of the automatic level, the rod readings, both backsight (BS) and foresight (FS), the various turning points (TPs), and the original bench mark (BM # 8) from which the surveyor commenced. Table 7.1 shows the method of field notation for running elevation controls into a proposed construction site and a method of checking the accuracy when the level loop is returned to the place of beginning.

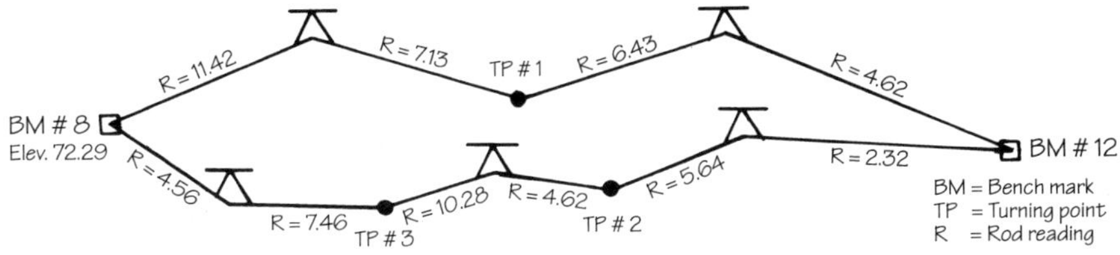

FIGURE 7.2 Diagram of a hypothetical level-loop

TABLE 7.1 Typical Elevation Level Loop in Field Note Form.

Station	BS (+)	HI	FS (–)	Elevation	Remarks
BM # 8	11.42	83.71		72.29	USGS bench
TP # 1			7.13	76.58	
	6.43	83.01			
BM # 12			4.62	78.39	USGS bench
	2.32	80.71			
TP # 2			5.64	75.07	
	4.62	79.69			
TP # 3			10.28	69.41	
	7.46	76.87			
Tie to BM # 8			4.56	72.31	BM = 72.29
	32.25		32.23		

Figure 7.3 is a schematic of the field notes in Table 7.1 and presents in profile view the various level rod readings, with an approximation of the ground profile when the level loop was run.

Preparing the Contour Map

After the elevation control is brought to the site, the ground elevation is obtained at each grid intersection point and field notes are recorded and at the end of each working day are brought to the office for plotting. Intermediate elevations (between grid intersections) and other ground irregularities, such as swales, ground depressions, water courses, and other unusual ground features, require specialized plotting by the office personnel. Not only is the grid system drawn on the topographic map, but each intermediate station and offset is plotted. If extra baselines are located to detail unusual breaks in the ground surface or water courses, these must also be shown horizontally before the elevations can be entered on the drawing. When all elevations are assembled and analyzed, a contour or topographic map can be produced, which indicates, by semi-concentric or curved lines, the elevation relief of the area.

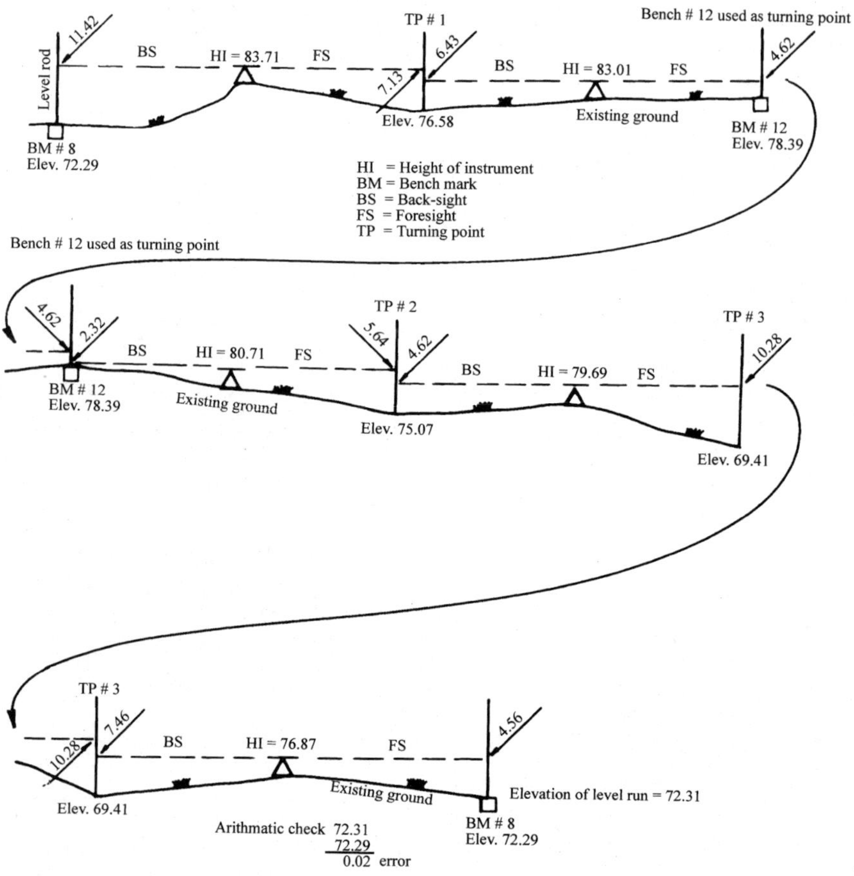

FIGURE 7.3 Level-loop schematic

Interpreting the Contour Map

Figure 7.4 is a plywood representation of a hillside with multiple swales leading to a small stream. The top of the photograph shows a three-dimensional oblique view of the site, and at the bottom of the photograph is a two-dimensional plan view representing the contours of the site. The plywood cutouts are layered on top of each other for instructional purposes, but in nature, the spaces between the plywood would be filled in and would form the slopes we associate with hills and valleys.

Prerequisites to Contouring

Several preliminary features must be addressed and completed before final contour lines can be drawn. These stages are itemized below:

1. Choose a drawing scale that is adequate for the size of the project.
2. Commence the skeleton drawing showing the survey traverse lines from which the topography will be located.
3. Graphically draw the checkerboard grid system located within the perimeter of the site, so as to verify total coverage.
4. Graphically locate other baselines from which unusual topography and elevations are obtained.
5. Reduce all field notes and pencil the elevations in their corresponding locations.
6. Determine the contour interval desired, whether 2 ft, 5 ft, 10 ft, or more.
7. Analyze the corner elevations of each grid intersection, as well as all intermediate points, and commence sketching the contour line that represents common elevations.

Before the lines can be drawn for each contour interval, the approximate location of that elevation between grid intersections must be computed. Some programmable computers are equipped with software that computes and plots this information directly on the preliminary drawing, while other software programs simply compute a distance from a known grid intersection to the contour elevation in question. Figure 7.5 represents a property plat showing the terrain of a proposed site development, with the surveyed grid pattern graphically located and the field elevations reduced and plotted.

Computing Contour Elevations

The site survey in Figure 7.5 is representative of the preliminary field work required before a topographic contour map can be prepared. The property lines are located, the grid system is chosen and surveyed, assumed elevations or governmental controls are established, and the elevation at each grid intersection and intermediate point is observed. Elevations are graphically plotted at each grid point so that the elevation decimal point is coincident with the intersection. Where no intersection occurs, the decimal point represents the location of the elevation, or an arrow is drawn pointing to the designated elevation.

Figure 7.6 represents the development of five contour lines; however, before the lines can be drawn, each grid square must be reviewed to determine the range of elevations between corners. This method assumes that all ground breaks have been located and that between ground breaks the remaining terrain has the same rate of slope. Remember, whenever significant slope changes occur, the station, offset, and elevation must show on the preliminary map. Swales and other water courses, as well as ravines, excavated, or fill areas, will be made easier to contour if the top of the bank (breakover) and the toe of slope are horizontally and vertically located separately and plotted before contours commence.

All unusual elevation changes in the ground that obviously will contribute to a more accurate presentation must be located. Bodies of water, rivers, streams, buildings, or other objects must be located horizontally, where they often disrupt the continuation of the contour line, and where the contour line commences anew on the other side of the object. Having plotted all this information on the preliminary map, you are ready to compute where each contour interval intersects the imaginary straight-line distance between two known elevations.

Establishing the Ratio

Computing elevations between grid intersections without the use of programmable computers can be completed in a manner similar to commercial software programs by using a simple ratio computation.

FIGURE 7.4 Contours in a plywood media *(The Ohio Department of Highways.* Preliminary Training Course. *Division 10. Marietta, Ohio.)*

Example:

(referring to Figure 7.6) The grade at the baseline (0 + 00) intersection with station 1 + 40 is elevation 86.3, and the next northerly adjacent 10-ft intersection (station 1 + 50) is elevation 62.9. The elevation difference is 23.4 ft, and the horizontal distance is 10 ft.

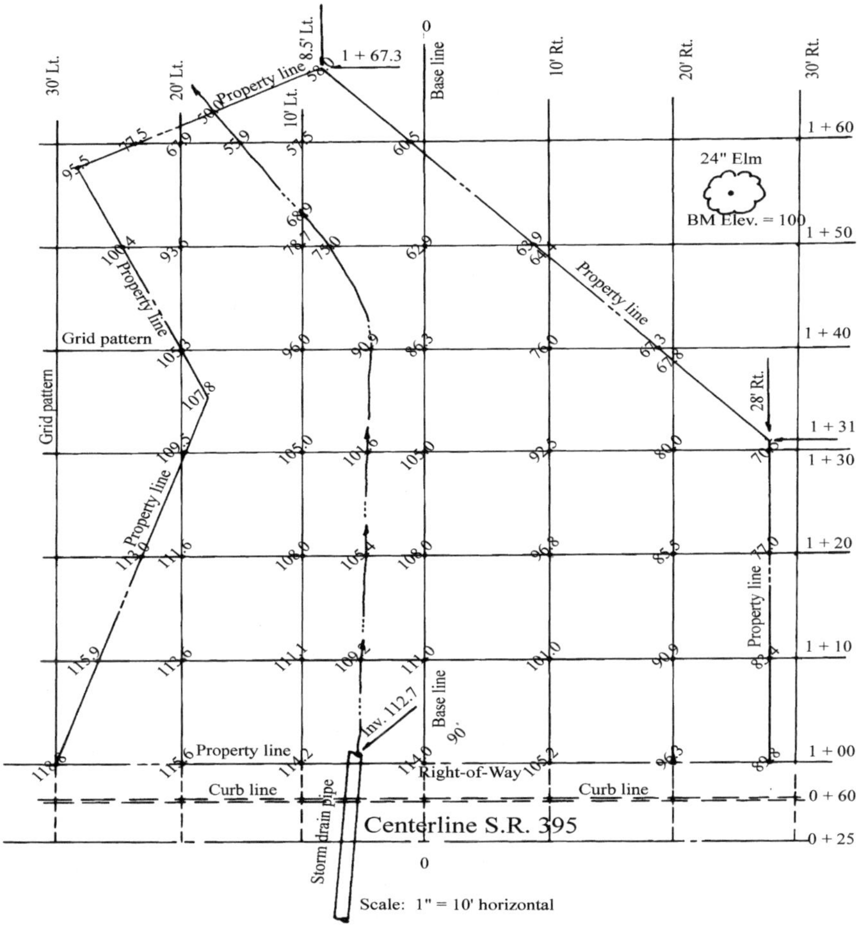

FIGURE 7.5 Commencing the topographic map)

Assume:

The contours (elevations) are to be drawn at 10-ft intervals, and elevation 70 is required. A simple ratio is set up as follows:

Where:

The elevation difference is 23.4 ft.

The distance between elevations is 10 ft.

x = the unknown distance to the contour.

y = the difference in elevation between the existing ground and the contour.

Refer to Figure 7.6, where the ratio to locate contour elevation 70 is as follows:

$$\frac{23.4}{10} = \frac{y}{x}$$

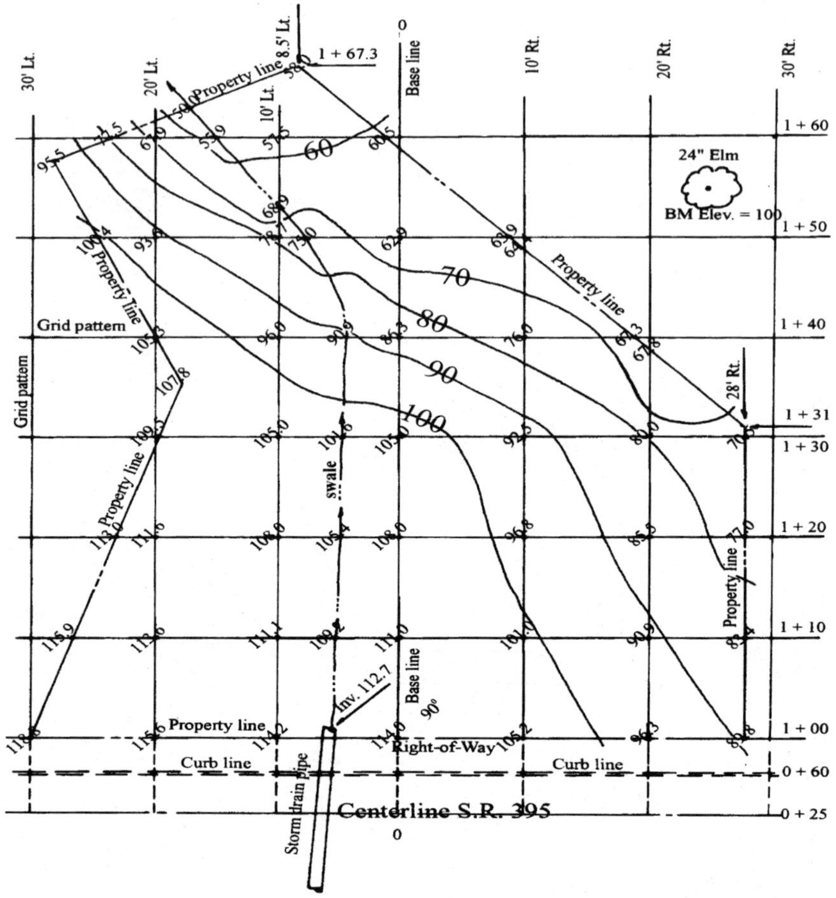

Scale: 1 in = 10 ft horizontal
10 ft = contour interval

FIGURE 7.6 Commencing the contours

Substituting:

$$\frac{23.4}{10} = \frac{16.3}{x}$$

$$x = \frac{10 \times 16.3}{23.4}$$

$$x = 6.96 \text{ ft (call it 7.00 ft)}$$

Contour elevation 70 is, therefore, located 7.00 ft beyond station 1 + 40 due to the 16.3-ft difference in elevation being used in the proportional ratio. The horizontal distance is always measured from the point of known elevation toward the contour interval being computed. Obviously, elevation 86.3 was used in this instance, where subtracting the contour elevation 70 resulted in the difference of 16.3 ft.

The location of each contour interval is penciled on the imaginary line between points and identified, either by color coding, placing the elevation number on the line, such as 70, or placing the elevation so that the decimal point represents the location. Similar computations for other 10-ft intervals (if applicable) are made between each grid intersection and between grid and intermediate ground points. This can become a very time-consuming procedure, but with experience and the use of computer software, the process is accelerated.

Contouring Earth's Different Shapes

Before sketching the contours, you need to have a feeling for different earth shapes and how contour lines are affected by these shapes. In Figure 7.7 we have a simple earth shape with three concentric contours, shown in plan view at the top of the illustration and shown in two-dimensional view at the bottom of the illustration. The top illustration represents the plan view (horizontal), and the bottom illustration represents the profile view (vertical).

The contour lines, represented by three concentric circles (left side of Figure 7.7 show the contours equally spaced, and when equally spaced contours are projected to the profile view, they form a smooth ground line.

The earth shape on the right side of Figure 7.7 also shows three contours, but they are unequally spaced. As these contour lines are projected down to the profile, it becomes clear that the ground line is not smooth and, further, that the closer any two contour lines are to each other, the steeper the ground slope becomes.

An extreme example of close contour lines can be seen as one sketches contours where the ground contains retaining walls, vertical cuts, or caves. In each of these examples, when each contour approaches the obstruction, coming closer and closer together (depending on the height of the obstruction and, of course, the elevation difference between

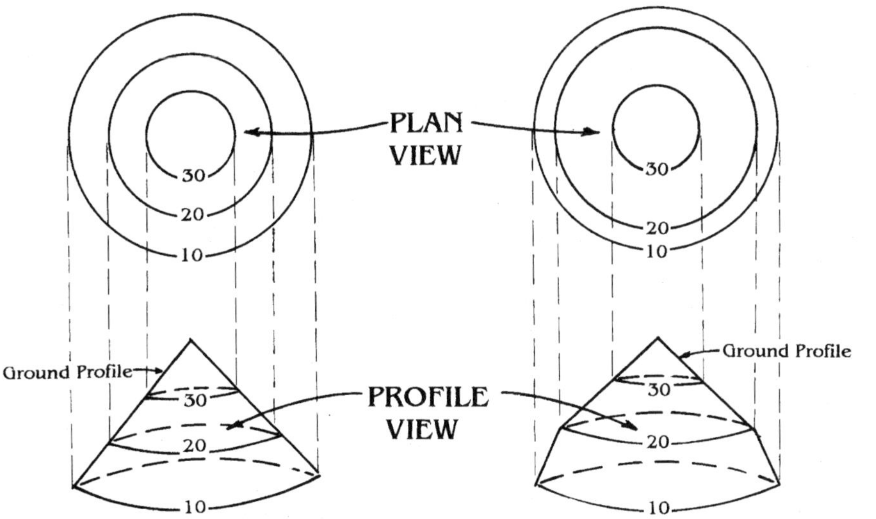

FIGURE 7.7 Concentric and non-concentric contours *(Courtesy: Linda McDougal. From originals prepared by Clifton O. Carey, B.S., C.E., formerly Associate Professor, Geodesy and Surveying, University of Michigan.)*

contour lines), the contour stops at the obstruction and actually moves horizontally across the vertical face of the object.

The following concepts are basic to the development of contour lines:

1. A contour line represents points on the ground of equal elevation.
2. Unless natural or artificial objects interfere, the contour line will close on itself. Due to the limited size of some drawings, not all contours will close within the frame of that particular drawing, but when the contour is continued on the ground, the elevation will find repose.
3. Contours are drawn to represent a line at 90° to the steepest grade in the slope, whether excavation or embankment.
4. Contours of equal elevation do not cross over their own line except where caves or overhanging terrain are situated and then the contour lines becomes coincident.
5. Horizontal distances between contour lines indicate the ground rate of slope. The closer the contours are to each other, the steeper the slope.
6. When a contour line approaches the bank of streams, rivers, or creeks, the contour at the top of bank abruptly changes direction, working diagonally across the bank of the water course, in an up-stream direction, until the contour meets its own elevation in the thread of the water course. The contour then crosses the water course at right angles and continues diagonally across the bank, in a down-stream direction, until the contour meets an equal elevation at the top of the bank.

The following five earth shapes, Figures 7.8 through 7.12, illustrate various contour lines at the bottom of each page and an artist's interpretation of the corresponding ground at the top of the page.

FIELD SURVEY PROCEDURES

To this point, we have concentrated on the mechanics of the contour map, and the characteristics of contours. Some knowledge of the survey process used in obtaining this data is essential. There are two common methods of field surveying used in gathering elevations and physical topography for contouring. They are: (1) the random-shot method, and (2) the grid-system method (previously discussed). The method of elevation gathering when contouring is not necessary is the cross-section method (see Chapter 8).

Random-Shot Method

This method is generally preferred when a comparatively congested area is to be surveyed. The procedure consists of field-locating a closed traverse near the perimeter of the project, using the corners of the traverse and other intermediate survey control points to locate all physical details, such as buildings, roads, bridges, fences, rivers, ponds, lake boundaries, electric poles and lines, telephone poles and lines, cable, high voltage transmission, and subsurface utilities (water lines, cisterns, gas lines, sanitary sewers, septic tanks, storm sewers, etc.).

FIGURE 7.8 Artist's rendition of earth mounds *(Courtesy: Linda McDougal. From originals prepared by Clifton O. Carey, B.S., C.E., formerly Associate Professor, Geodesy and Surveying, University of Michigan.)*

The survey instruments of today include "total stations" (angle and distance measurements observed from one housing) as well as individual "laser" equipment (distance measurements from a separate housing than the transit) with which the location of topographic features can quite easily be obtained. Elevations, for this same area of land, are located by using the checkerboard method previously discussed. However, when the property has an abundance of physical objects or foliage, making the checkerboard control lines inaccessible, a secondary approach is often used. This involves locating smaller grid intersections, which can be rotated to fit the circumstances. A large checkerboard system of intersection controls, set up to satisfy the whole property, will seldom represent the best condition for locating elevations in the congested or irregular parts of the property. Therefore, most surveyors use the smaller grid pattern, occasionally rotated from the larger grid, to provide adequate cover for the site.

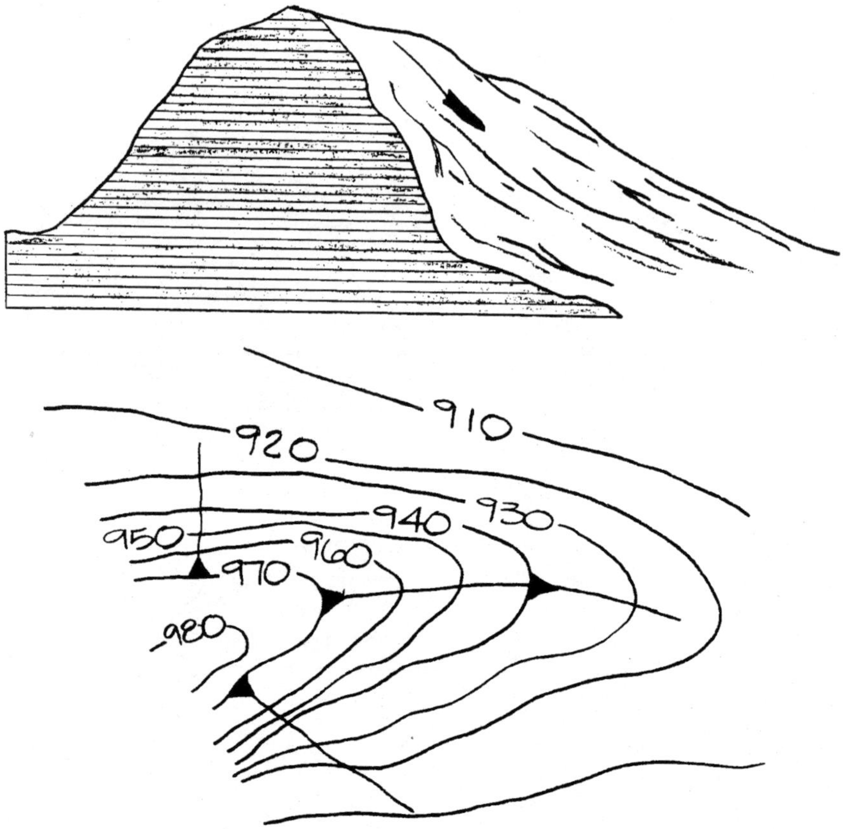

FIGURE 7.9 Artist's rendition of hills *(Courtesy: Linda McDougal. From originals prepared by Clifton O. Carey, B.S., C.E., formerly Associate Professor, Geodesy and Surveying, University of Michigan.)*

When properties are extremely large, the surveyor will probably recommend aerial photography. This type of procedure is seldom appropriate during the summer months when the foliage is heavy, nor is it appropriate when the winter snows cover the ground to depths of several inches (a dusting of snow is ideal for aerial photography). The best months for aerial photography are between November and April; however, projects are flown and mapped throughout the year. When the summer foliage is intense, the surveyor will often need to supplement the ground control by running secondary traverse lines through the wooded areas and mathematically establishing horizontal coordinates and elevations, which are forwarded to the photogrammetist, along with the normal horizontal and vertical control. This provides adequate elevations from which to complete the mapping.

CROSS-SECTION METHOD

This method is generally preferred where the area to be mapped does not require contouring, such as highway design, and where the terrain has great differences in elevation, but

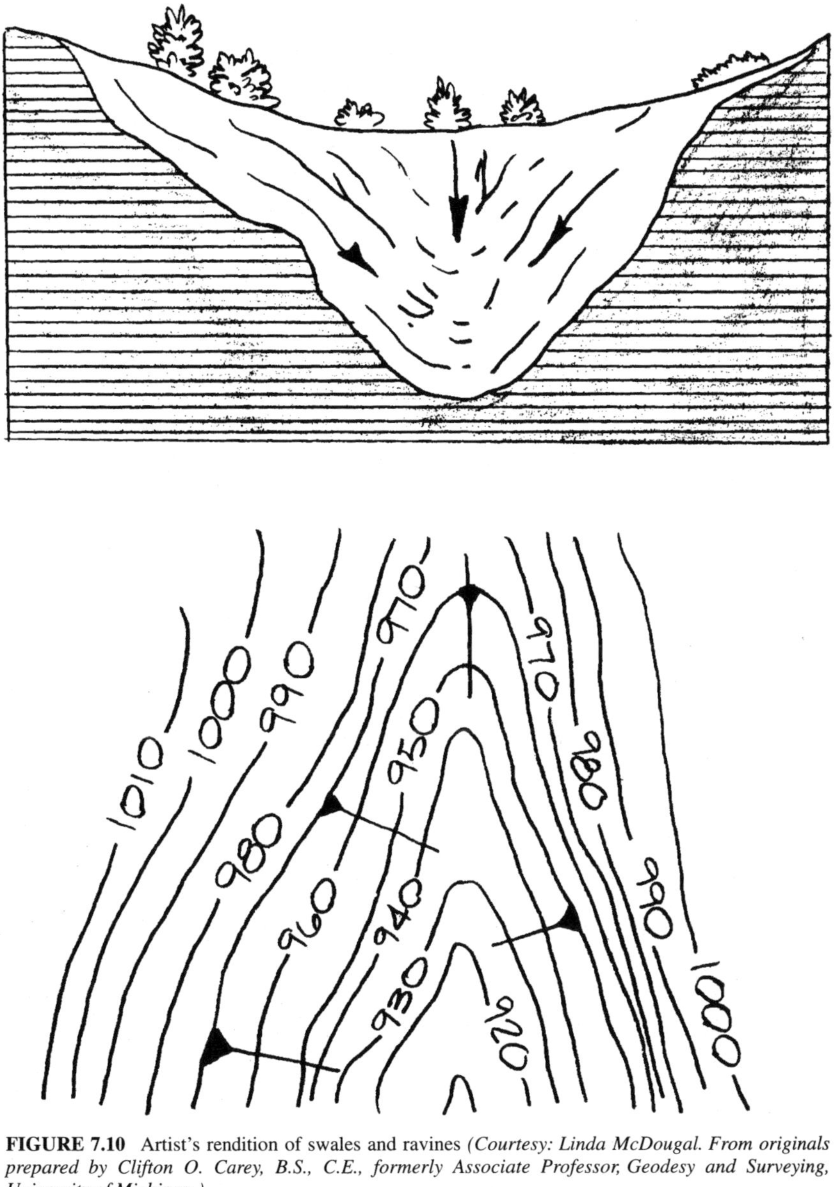

FIGURE 7.10 Artist's rendition of swales and ravines *(Courtesy: Linda McDougal. From originals prepared by Clifton O. Carey, B.S., C.E., formerly Associate Professor, Geodesy and Surveying, University of Michigan.)*

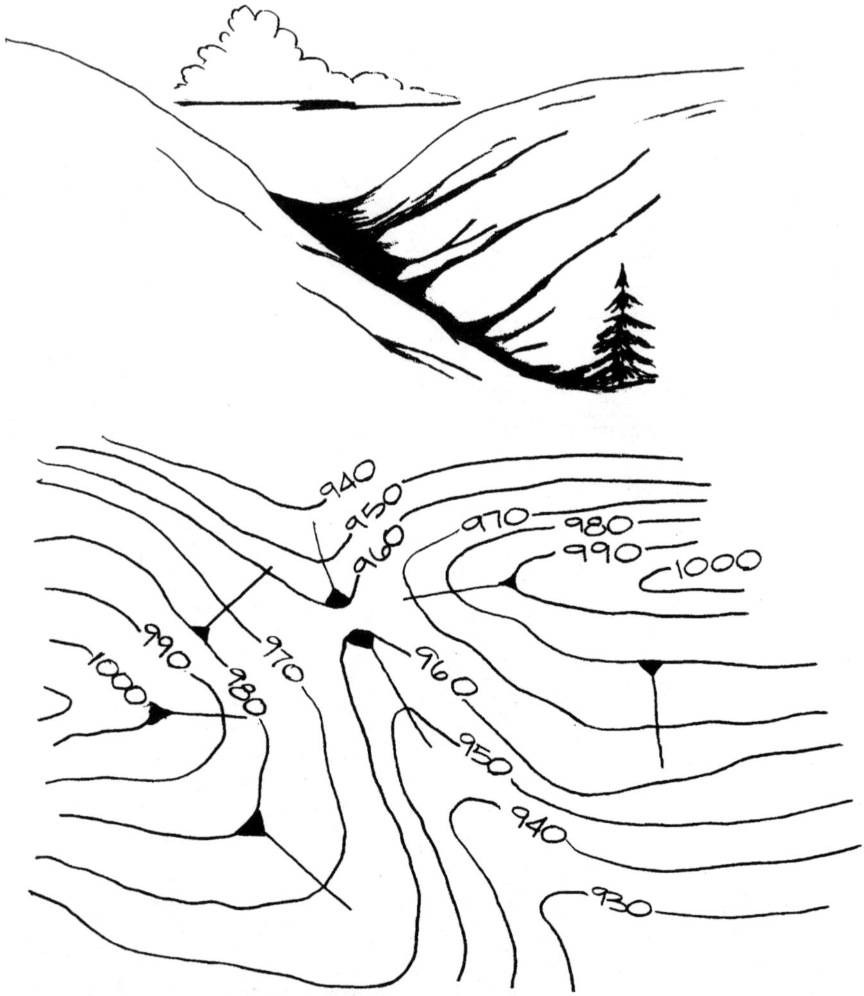

FIGURE 7.11 Artist's rendition of valleys *(Courtesy: Linda McDougal. From originals prepared by Clifton O. Carey, B.S., C.E., formerly Associate Professor, Geodesy and Surveying, University of Michigan.)*

limited physical topography, such as buildings, etc. If the area to be surveyed is long and narrow, as in highway projects, the centerline of the highway is chosen as the control traverse line and each cross-section is observed longitudinally at a maximum of 50 ft intervals.

Elevations are established at each centerline station by running a closed level loop, as discussed earlier in the chapter. Cross-sections are defined as a profile of the ground observed at ninety 90° to a baseline, or centerline. The surveyor visits each highway station individually, and depending on the type of equipment being used, will adjust the work load accordingly. Three of these methods are discussed below:

Automatic Level

If the automatic level is used for elevations, the temporary bench mark used to measure the height of the instrument (HI), as in Table 7.1, will be the elevation located at each 50-ft station during the preliminary level loop, or the surveyor might elect to carry elevations (through the instrument readings) from one of the original bench mark (BM) controls, stopping at the end of a unit of work to verify the results by checking-in to another known elevation. The surveyor will continue to move the level, up or down the terrain, reading the numbers on the level rod (Figure 7.3), adding or subtracting numbers to compute the elevation at each ground break, measuring the horizontal distance, and finally recording the data in the field book.

RDS Level

If the RDS (internal self-reading instrument) is to be used for cross-sectioning, the surveyor sets the instrument on the centerline point, measures the height of the instrument from the ground to the scope, vertically paralleling the ground by setting the low cross-hair at the same height as the instrument. This is the equivalent of reading ground to ground, and no vertical angle adjustments are necessary. The internal mechanism of the RDS instrument is so calibrated that the elevation difference can be calculated using simple mathematics while the notes are being taken, and the distance to the object is read automatically using the internal cross-hairs.

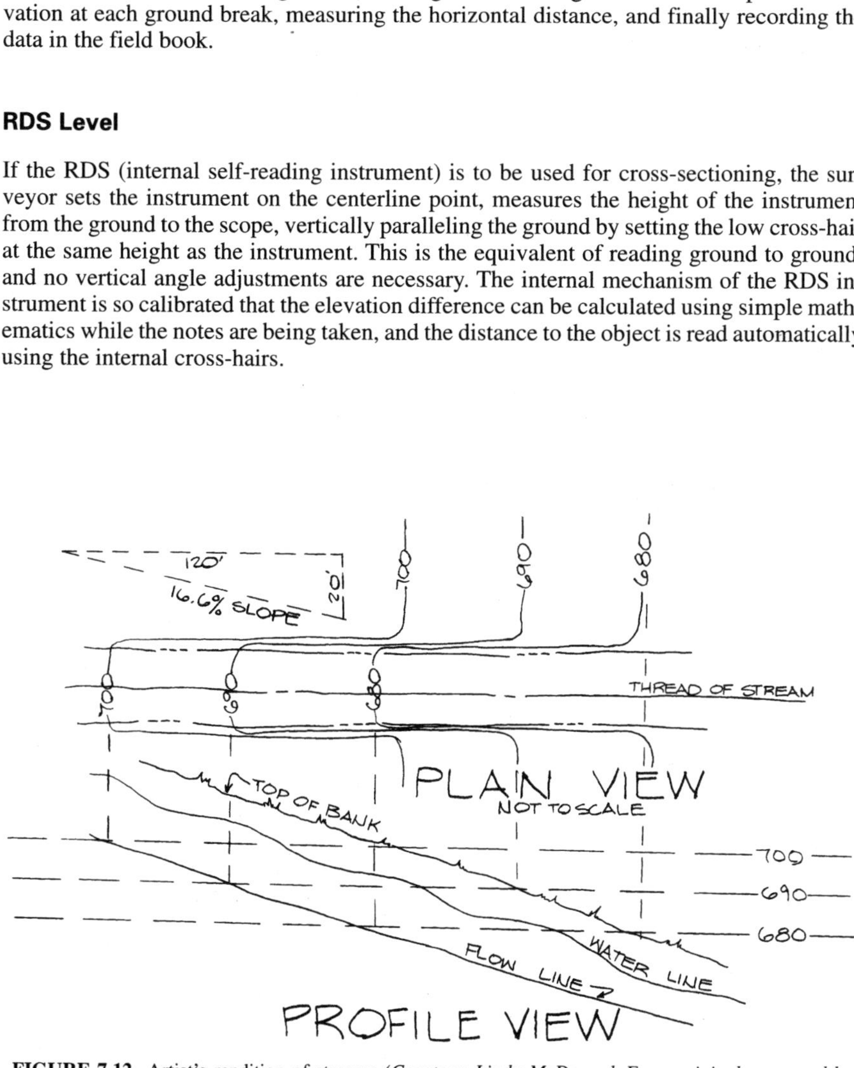

FIGURE 7.12 Artist's rendition of streams *(Courtesy: Linda McDougal. From originals prepared by Clifton O. Carey, B.S., C.E., formerly Associate Professor, Geodesy and Surveying, University of Michigan.)*

Lock-Level

If the lock-level (small, hand-held level) is used to establish the elevation at each ground break, the surveyor or grade foreman steadies the lock-level against a metal, wood, or plastic pole, thus adding support for the lock-level, and therefore, duplicating the process when automatic levels are used, but with considerably less carrying weight and greater maneuverability. The pole support also permits you to rotate 360° without losing the elevation of the lock-level and the operator reads the rod graduations as you would with a surveyor's automatic level. The rod readings are then added or subtracted as previously discussed in Figure 7.2, Table 7.1, and 7.3. The results are very inconclusive and should never be used for multiple elevation gathering, however, this method is often used by contractor's grade-checkers to determine whether the slope of highway cuts, or fills, are progressing according to the engineer's plan.

CHAPTER 8
CROSS SECTIONS

INTRODUCTION

Cross sections are one of several important elevation tools used by engineers in the design of highways and commercial improvements. Most major preengineered plans commence with the cross section. Architects, on the other hand, use the elevation grid or checkerboard system (as discussed in Chapter 7) to develop their interpretation of site elevations. The contour map is the elevation tool used by the architect and upon which design proposals are completed. In architect's language, this is called the "grading plan."

The engineer's cross section, by definition, is a profile of the earth's surface where elevations are observed perpendicular (at right angles) or radial to a known centerline or baseline. This description simply means that ground elevations are taken at each significant ground break and measured, by distance, to the control centerline or baseline and are then graphically plotted using appropriate scales to represent both the horizontal and vertical measurements.

Figures 8.1 and 8.2, show the cross section view at the bottom of each figure and an artist's rendition of the before and after ground line in three-dimensional view at the top of each figure. Other than for this presentation, the artist's rendition does not appear on the engineer's cross-section drawings. Figure 8.1 represents a cut section, and Figure 8.2 represents a fill section. Cut and fill lines can easily be seen due to the standard graphics used in these presentations, where the existing terrain is always shown as a dashed line and the proposed features are always shown as a solid line. Cross sections often extend great horizontal distances from the centerline or baseline, sometimes exceeding 800 ft or more, both left and right of centerline, or as small as 5 to 10 ft both sides of centerline for the location of creeks and swales.

Cross-section drawings are scaled representations of the engineer's design and become a valuable tool for the surveyor during construction layout. This is one of several plan sheets that require postengineering computations before the initial layout can commence.

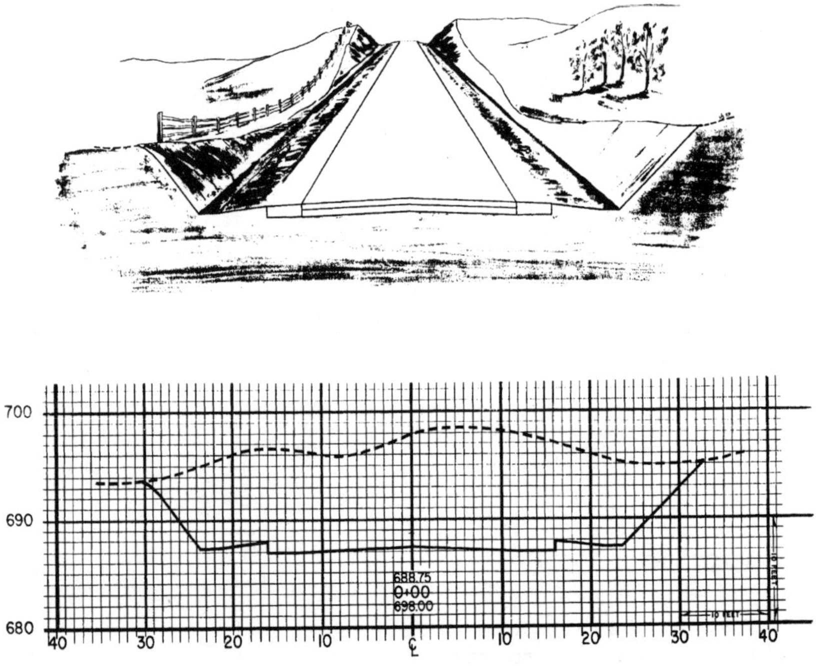

FIGURE 8.1 Cross-section, in cut *(The Ohio Department of Transportation)*

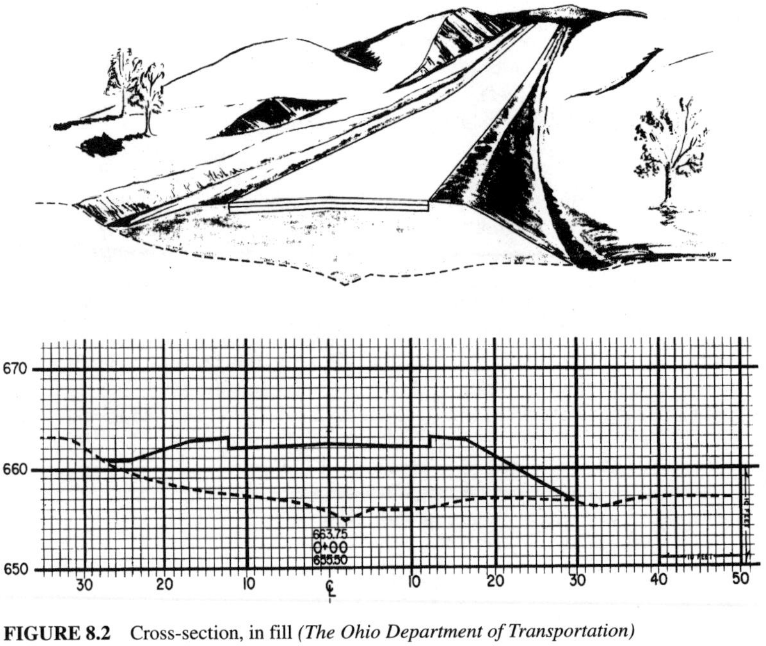

FIGURE 8.2 Cross-section, in fill *(The Ohio Department of Transportation)*

FIELD INFORMATION GATHERING

"Cross sectioning" means the establishing of elevations and distances left and right of centerline (or baseline) that represent the changes in the existing ground elevations. Small ground changes, such as 1 or 2 in in elevation, are not usually observed, but depending on the significance of the design, the surveyor has the discretion to locate all breaks. This is especially true where concrete objects are involved and a new concrete design must be made to match the existing one. Larger ground breaks are always field-located, and their importance cannot be over emphasized. Cross sections form the basis for earthen computations, and therefore, the contractor's payment schedule is subject to the accuracy of these field observations.

The field elevations and offset distances observed during the "taking" of cross sections are used to prepare the drawings on which the proposed highway design will be shown. A cross-section field note example is shown in Figure 8.3 and represents one of many different styles of cross-section note keeping.

The cross section in Figure 8.3 graphically shows the centerline, centered vertically over the highway station, the station of the cross section (station 474 + 50), and whether the information is left or right of centerline. With each elevation recorded, a corresponding distance is noted from the point to the centerline. This type of field information, later plotted as in Figures 8.1 and 8.2, forms the basis for all highway preliminary and final design.

MULTIPLE CROSS SECTIONS

Multiple cross sections, observed throughout a typical project, becomes the tools used by the engineer to make calculations of earthwork volumes. When an earthwork imbalance is computed, the engineer will often adjust the preliminary grade line, either raising or lowering the grade, to obtain a better balance between cuts and fills, and thereby reducing the cost of the project. Preliminary grades (the proposed elevation of a roadway) are established during early stages of design, and during this phase, the preliminary grade (per each 50-ft station) is superimposed on its corresponding cross section of the existing ground elevations.

A typical roadway template is prepared (to eliminate repetitive plotting), cut from thin plastic or cardboard to represent the design requirements of the pavement and shoulder

Left | Cross Section | Right

			1064.17 / 157'	1065.50 / 160.5'			
	1051.29 / 74.7'	1049.21 / 103.7'	1064.13 / 138.7'	1064.97 / 142'	1063.79 / 124.9'	1052.31 / 96.8'	1053.92 / 67.8'
1050.67 / 38'	1050.32 / 34'	1049.21 / 23'	1057.36 / 2.5'	1057.36 / 2.5'	1054.03 / 12.3'	1055.36 / 26.1'	1055.33 / 30.5'

Station 474 + 50

FIGURE 8.3 Field note gathering

portions of the highway. The template is placed on each cross section at the elevation chosen for the profile grade. Cut or fill side slopes are connected to this template at the outside edge of shoulder, based on the engineer's design, and continued until the side slope meets the existing ground. Usually, a typical template will consist of the following:

1. Grass or concrete median widths
2. Typical pavement and shoulder widths
3. Ditches
4. Proposed cut or fill side slopes

When the proposed side slopes are connected to the existing ground, they form a closed figure, and therefore, cut and fill computations can be performed, using programmable computers or hand calculations to determine the quantity of earth translated into an amount of cut material versus fill material, calculated in cubic yards. These computations assume that each two adjacent cross sections will be combined sequentially throughout the project to develop the calculation. For purposes of making a sample volume computation, a second cross section is shown in Figure 8.4.

Before proceeding with the volume computation, review the two cross sections, stations 474 + 50 and 474 + 75, noticing the fractional entries: the numerator representing the elevation of the point and the denominator representing the distance from centerline or baseline to the point. The field information nearer to the centerline is shown at the bottom of the cross section, while the distances that are farther away from centerline are shown moving toward the outside of the page and then wrapping around on the next entry above. This suggested style of cross-section note-keeping signifies that the cross-section stations are increasing up-page, rather than down-page and that the field data on the left of the page represents elevations left of centerline, and the field data on the right of the page represents elevations right of centerline.

COMPUTING EARTHWORK VOLUMES

The following equation is used to compute earthwork volumes (in cubic yards) employing any two adjacent cross sections and is identified as the "average end area." This simply means that each cross section (no depth) is computed for its area (in square feet), each set

Left				Cross Section			Right
				$\frac{1064.54}{137.3'}$	$\frac{1065.52}{161.7'}$		
$\frac{1051.26}{75.7'}$	$\frac{1049.07}{104.7'}$	$\frac{1064.29}{140.7'}$	$\frac{1064.17}{158}$	$\frac{1064.01}{126.2'}$	$\frac{1052.41}{97.9'}$	$\frac{1054.95}{67.8'}$	
$\frac{1050.52}{39'}$	$\frac{1050.25}{35'}$	$\frac{1048.99}{24'}$	$\frac{1057.42}{2.5'}$	$\frac{1057.42}{2.5'}$	$\frac{1054.02}{12.5'}$	$\frac{1055.45}{26.1'}$	$\frac{1055.46}{30.5'}$

Station 474 + 75

FIGURE 8.4 Field note gathering

of two adjacent cross sections is evaluated for its *average* area (in square feet), and finally the square foot average is multiplied by the length between the two cross sections and divided by the number of cubic feet (27) in a cubic yard to produce a volume (in cubic yards).

The average end area equation below takes the average area, in square feet, of two adjacent cross sections (shown as the numerator in the equation) divided by the number of square feet in a cubic yard (shown as the denominator in the equation) and multiplies the result by the distance between cross sections.

$$V = \frac{\frac{(A_1 + A_2)}{2}}{27} \times L \qquad (8.1)$$

Where:

V = volume, in cubic yards
A_1 = area of the first cross section, in square feet
A_2 = area of the second cross section, in square feet
L = length between cross sections

Simplifying the Equation

$$V = \frac{(A_1 + A_2)}{54} \times L \qquad (8.2)$$

A simplified example of adjacent cross sections (Figure 8.5) portrays an embankment (fill) consisting of two stations, 1 + 00 and 1 + 50. When the proposed roadway template is superimposed on each ground cross section, as referred to earlier in the chapter, the result of the graphics will show either cut or fill material but more often will include a combination of cut and fill material. Much caution should be exercised when separating the cut from the fill areas due to the importance of identifying where the separation occurs. A thorough knowledge of the site is obviously an important factor as one chooses the position where a cut or fill begins and ends. Most programmable computers will ask for a taper distance between the estimated cut and fill separation and the next full cross section. If there were tapers in Figure 8.5, these distances would be included in the computer calculation, and in the hand calculation.

When tapers are required in the computation, the commencing point of each taper is considered to contain a 0 ft^2 area, and therefore the average end area from taper to the next full cross section is equal to one-half of the full section.

For the cubic yards computation in Figure 8.5, we will exclude the tapers and continue as follows.

Solving the volume (in cubic yards) for the two adjacent cross sections, Figure 8.5, can be completed by using one or more of the following methods:

1. Programmable computers
2. Plotting the cross section to an accurate scale and estimating the square feet, such as on graph paper (not recommended)
3. Plotting the cross section to an accurate scale and using a planimeter[1] to calculate square feet

1 The planimeter is a mechanical devise used to measure an object's perimeter by applying horizontal and vertical factors to the reading of the instrument and ultimately computing the square feet within the object.

4. Using mathematical data, such as located in the cross-section sample, and solving for rectangles and triangles

Volume (in square feet)

$$\text{Area} = A_1 = 14 \times 40 = 560 \text{ ft}^2$$

$$= A_1 = \frac{14 \times 56}{2} = 392 \text{ ft}^2$$

$$= A_1 = \frac{14 \times 28}{2} = 196 \text{ ft}^2$$

$$\text{Total} = 1148 \text{ ft}^2$$

(Cross sections A_1 and A_2 are equal in area.)

Volume (in cubic yards)

$$\text{Volume} = \frac{(A_1 + A_2)}{54} \times \text{length}$$

$$= \frac{2296}{54} \times 50 \text{ ft}$$

$$= 2125.9 \text{ yd}^3$$

(using the simplified equation)

Volumes by Computer

The programmable computer (method 1) is used universally to compute volumes. See Figure 8.6 for this example, noticing that the cross sections are plotted to a scale of 1 in equals 20 ft horizontally, and 1 in equals 10 ft vertically. Figure. 8.7 shows the actual computer output.

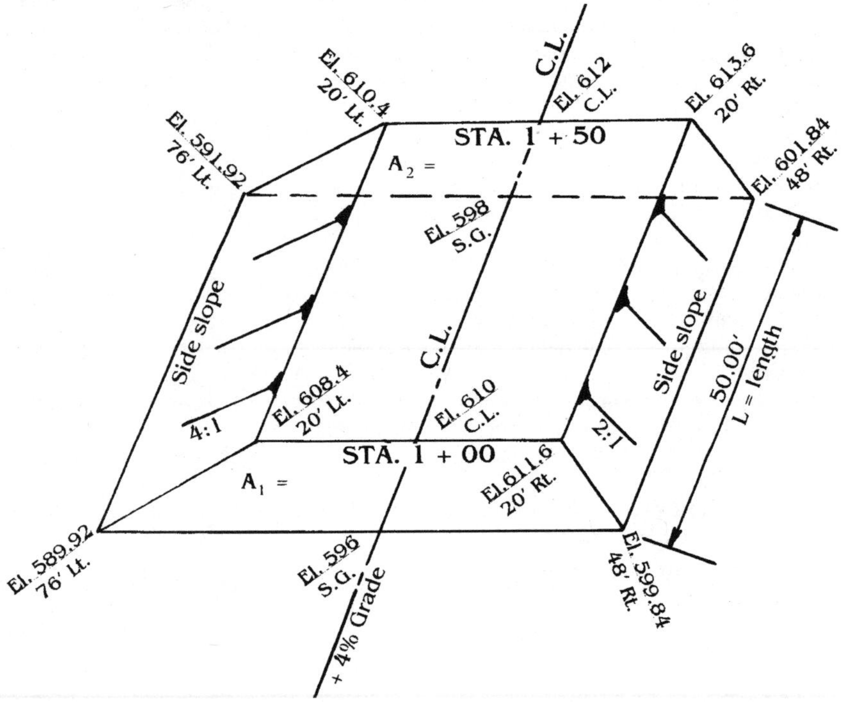

FIGURE 8.5 Average end area

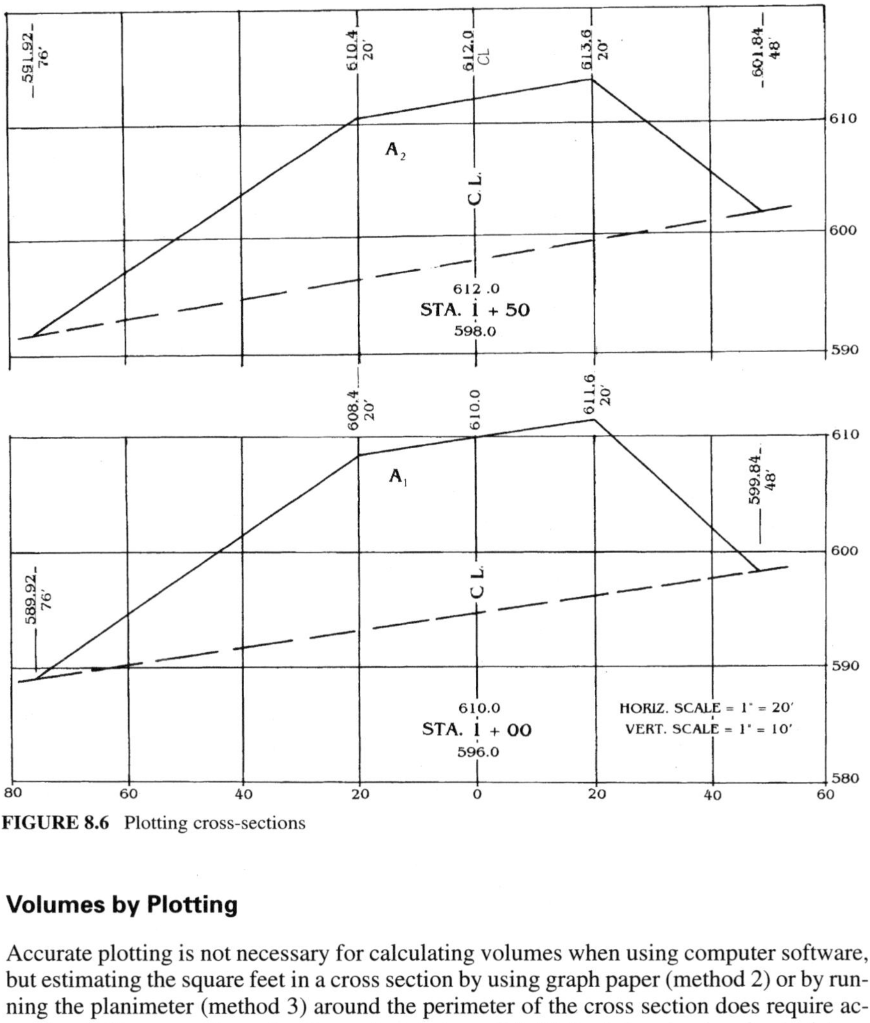

FIGURE 8.6 Plotting cross-sections

Volumes by Plotting

Accurate plotting is not necessary for calculating volumes when using computer software, but estimating the square feet in a cross section by using graph paper (method 2) or by running the planimeter (method 3) around the perimeter of the cross section does require accurate plotting procedures for the system to be valid. Method 4, using hand calculators, does not require accurate plotting since the distances required for the volume computations are obtained by multiplying, dividing, adding, or subtracting numbers.

Volumes for Commodities

Cross section plotting, as shown in Figure 8.6, becomes a valuable tool in determining volumes for such commodities as earthwork (just discussed), coal, rock, and certain stockpiles of minerals, for which the value in dollars is appreciable. Many chemical, coal, power, steel, and other companies, who stockpile hundreds or thousands of cubic yards of their specific commodity, use standard survey methods of cross sectioning or aerial photography to compute these volumes. The commodities, called "stores," represent large sums of dollars in assets and, of course, are subject to federal and/or state taxes; therefore, accuracy is imperative. When companies merge, as in arbitrage consolidations, an accurate accounting of these stores (assets) is demanded and the dollar value of the commodity becomes one of many entities leading to the final consolidation.

Programmable Computer Graphics

The cross-section example (Figure 8.6), represents two centerline stations of a typical highway project. The graphics are simplified in this example; however, the normal cross section consists of multiple existing ground elevation points as well as numerous proposed pavement, shoulder, ditch, and side slope elevations, all leading to a difficult earth-volume computation, if not processed through programmable computers. Figure 8.7 represents typical computer input data.

Listing of earthwork volume per cross section.
SECTION 1, STATION 1+00

OFFSET	ELEVATION	LINE NUMBER
1+00 (STA.)	1.00 (Code)	0
–76.00"	589.92	1
–20.00"	608.40	2
+20.00'	611.60	3
+48.00'	599.84	4

SECTION 2, STATION 1+50

OFFSET	ELEVATION	LINE NUMBER
1+50 (STA.)	1.00 (Code)	0
–76.00'	591.92	1
–20.00'	610.40	2
+20.00'	613.60	3
+48.00'	601.84	4

The field data, shown above, represent the proposed top elevations.
Code 1 identifies that the cross section has no tapers.
File name for the top (proposed) of the cross section is A-1T.
Negative numbers represent points left of centerline.

SECTION 1, STATION 1+00

OFFSET	ELEVATION	LINE NUMBER
1+00 (STA.)	1.00 (Code)	0
–76.00	589.92	1
+48.00	599.84	2

SECTION 2, STATION 1+50

OFFSET	ELEVATION	LINE NUMBER
1+50 (STA.)	1.00 (Code)	0
–76.00	591.92	1
+48.00	601.84	2

The field data, shown above, represent the existing bottom elevations.
Code 1 identifies that the cross section has no tapers.
File name for the bottom (existing) of the cross section is A-1B
Negative numbers represent points left of centerline.

FIGURE 8.7 Typical cross section computer input

Volumes When Tapers Are Applied

Earlier in the chapter we discussed the location of earthen tapers and their significance in the total volume process. If the two cross sections in Figure 8.6 had a 10-ft taper leading into station 1 + 00 and a 25-ft taper leading away from station 1 + 50, the computations would be as follows:

Where

Begin taper at station 0 + 90: area = 0.00 ft^2
Full cross section at station 1 + 00: area = 1148 ft^2
Full cross section at station 1 + 50: area = 1148 ft^2
End taper at station 1 + 75: area = 0.00 ft^2

Therefore, the cubic yard volume between stations 0 + 90 and 1 + 00 is:

$$V = \frac{A_1 + A_2}{54} \times L \tag{8.2}$$

$$= \frac{0 + 1148}{54} \times 10$$

$$= 212.6 \text{ yd}^3$$

Therefore, the cubic yard volume between stations 1 + 00 and 1 + 50 is:

$$V = \frac{A_1 + A_2}{54} \times \text{length} \tag{8.2}$$

$$= \frac{1148 + 1148}{54} \times 50$$

$$= 2125.9 \text{ yd}^3$$

Therefore, the cubic yard volume between stations 1 + 50 and 1 + 75 is:

$$V = \frac{1148 + 0}{54} \times 25 \tag{8.2}$$

$$= 51.35 \text{ yd}^3$$

Total cubic yards = 2870.0

This analysis is overly simplified, but when computing the volumes of cut and fill in highway projects or surveying quantities of stockpile material for large industrial plants, the cut and fill analysis, especially the location of the tapers, becomes very critical.

Programmable Computer Ouput

When the computer data, shown in Figure 8.7 has been input, a command is issued to compute the earthwork volume. The computer searches for the files (A-1T and A-1B) in which the data are stored, matching station numbers (1 + 00 and 1 + 50), comparing the top entry data with the bottom entry data for each station, computing the number of square feet

TYPICAL COMPUTER OUTPUT DATA
Earthwork volumes for files: A-1T and A-1B

STATION	Cut, in ft^2	Fill, in ft^2	Cut, in yd^3	Fill, in yd^3	Cut Accume, in yd^3
1 + 00	0.00	1148.00	0.00	0.00	
1 + 50	0.00	1148.00	0.00	2125.9*	0.00

*Amount of excess fill material

FIGURE 8.8 Typical cross section computer output

in each section, and finally reducing this number to cubic yards. As the computer is progressing sequentially through each successive pair of cross sections, excavation (cut) and embankment (fill) material is analyzed and shown in its respective columns. A sample computer printout is shown in Figure 8.8.

Compaction Factor

The computer operator also can specify whether the fill material is to be computed using a compaction factor. Compaction is that process where the soil is closely and firmly packed, filling voids in otherwise loose, airy, and less-compacted soil. Earthwork compaction is a requirement for all embankment (fill) projects constructed under county, state, and federal highway standards, as well as all engineered and architectural designed projects whose construction specifications refer to federal and/or state heavy construction codes. Compaction factors are based on actual tested conditions where the results indicate that the compacted fill requires additional material ranging between 15 and 20 percent, but most frequently is computed at 15 percent.

In the example in Figure 8.8, no compaction factor was initiated, and therefore the fill, in cubic yards, does not represent added material. If the recommended 15 percent compaction factor is used in this example, an additional 318.9 yd^3 would have been computed for a total of 2444.8 yd^3.

Computed earthwork volumes, in which compaction factors are applied, universally increase the cost of a project. Many highway improvements of old were designed without carefully considering compaction, forcing the contractor to complete the improvement without additional compensation. This could not happen in today's architecture, engineering, and construction climate.

P · A · R · T · 3

MATHEMATICS

CHAPTER 9
VERTICAL CURVES

INTRODUCTION

Occasionally there will be projects where you will be asked to establish an elevation ("set grade") at locations where the engineers during their design process have not created that elevation. As a slang expression, this is frequently called "setting grade," although a grade is considered more than establishing one elevation and primarily identifies a change in slope, such as you see in Figures 9.1 to 9.3. The word "grade," of course, is used interchangeably with "slope," so don't be confused by the two usages. The context should make clear the meaning.

To set an elevation (at the inside edge of pavement, for example), you must first calculate the elevation of the point on the centerline survey that corresponds to that edge of pavement station. As we will be discussing later in this chapter, the calculated centerline grades, regardless of the median width, are transferred to the inside edge of pavement, whether in a horizontal tangent or horizontal curve. To set grade on the outside edge of pavement (away from centerline), you will need to follow the cross slopes shown on the typical section and superelevated section to determine the mathematical change from one edge of pavement to the other. Chapter 6 explains pavement cross slopes.

The centerline elevation is not difficult to calculate for locations on straight-line vertical tangents. The design plans will contain a profile view of the centerline (as in Figure 9.1), showing the proposed elevation changes along the centerline of the highway, expressed as a percentage of grade. If the required centerline elevation is located in a vertical curve (as in Figure 9.3), the elevation calculation gets a bit more complicated. In this chapter, you'll learn how to do these types of calculations and most other mathematical manipulations needed to cope with vertical curves.

The chapter also includes some of the design standards used in setting the vertical profile of a highway; however, most often this information will be furnished by the design engineers as the plan is being developed for construction. Surveyors do have the responsibility of coping with last-minute plan changes, and this chapter may help you cope with these changes by providing some elementary knowledge of the design process.

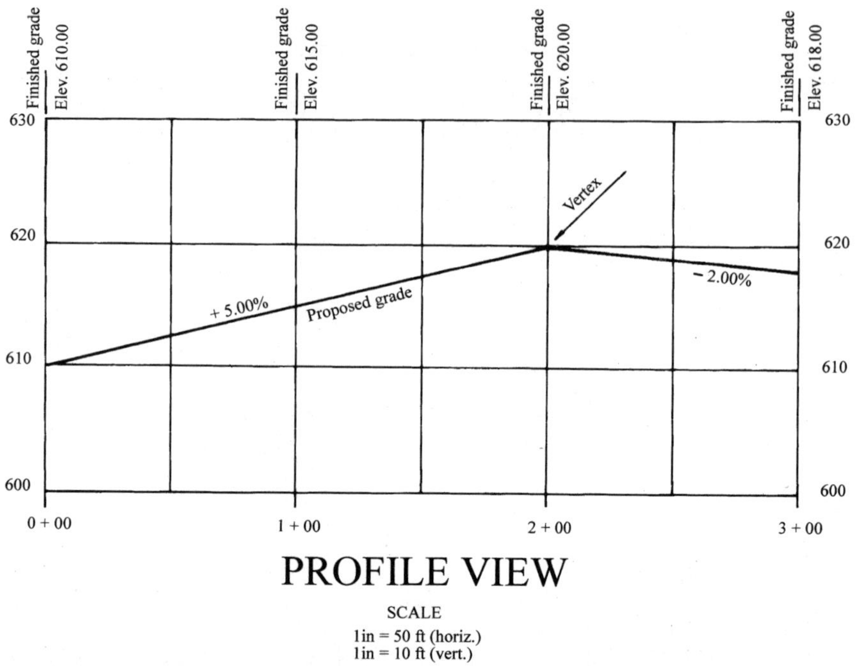

FIGURE 9.1 Preliminary profile grid for vertical curves

VERTICAL CURVES VERSUS HORIZONTAL CURVES

The curve used in a design to make a smooth transition between vertical tangents is the parabolic curve. Parabolic curves are ideal for vertical designs in that they are easy to compute, both in the field and office, and where the curve lies an equal distance from its tangent line, as when similar distances are measured from either end of the curve. This will be covered in more detail as we proceed in the chapter, but this mathematical phenomenon allows parabolic curves to be fit to any of the four possible types of vertical tangent intersections, such as both slopes ascending, both slopes descending, or the two slopes approaching in opposite directions (see Figure 9.3). The other examples of grades, both in the same direction and either descending or ascending, are outlined later in this chapter under the heading "Determining the Algebraic Difference."

You will be introduced to two additional curves when you get to the chapter that describes horizontal aspects of the centerline. The curves used to make smooth horizontal transitions between horizontal tangents are circular curves or spiral and circular curve combinations. Neither of these horizontal curves are parabolas, so you will eventually learn the mathematics for all three different types of curves used in highway design. Before starting the vertical computations, we will review some background information relevant to the design of roadway grades.

BACKGROUND INFORMATION

Choosing the Profile

As the engineer considers the profile of a new improvement, he or she carefully reviews the existing topography, such as existing roadways, proposed roadways, railroads, rivers and streams, and other natural or artificial obstructions which might affect the design. This information is gathered during the preliminary stages and quite often constrains the new grades. Some topographic features, such as railroads or waterways, must remain in their vertical position with only minimal changes made. Intersecting existing city streets will probably remain at their present elevation with minimal changes. Therefore, many critical existing features remain stationary, and the new profile is designed to merge with these features.

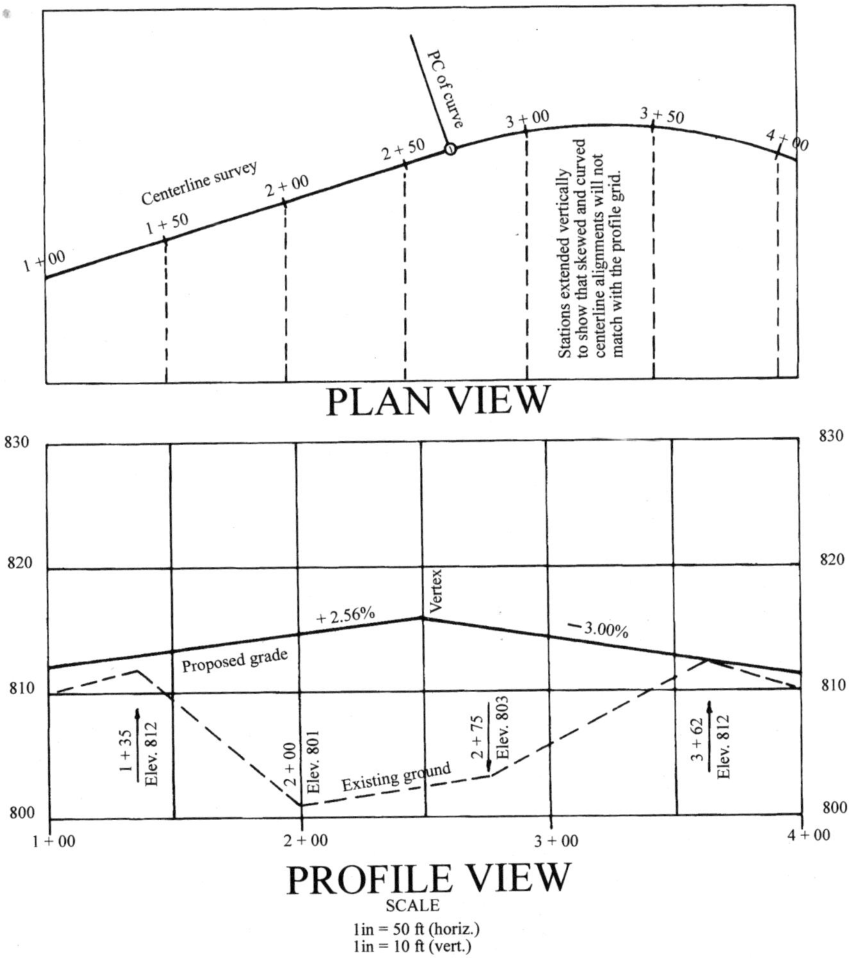

FIGURE 9.2 Plan and profile drawing

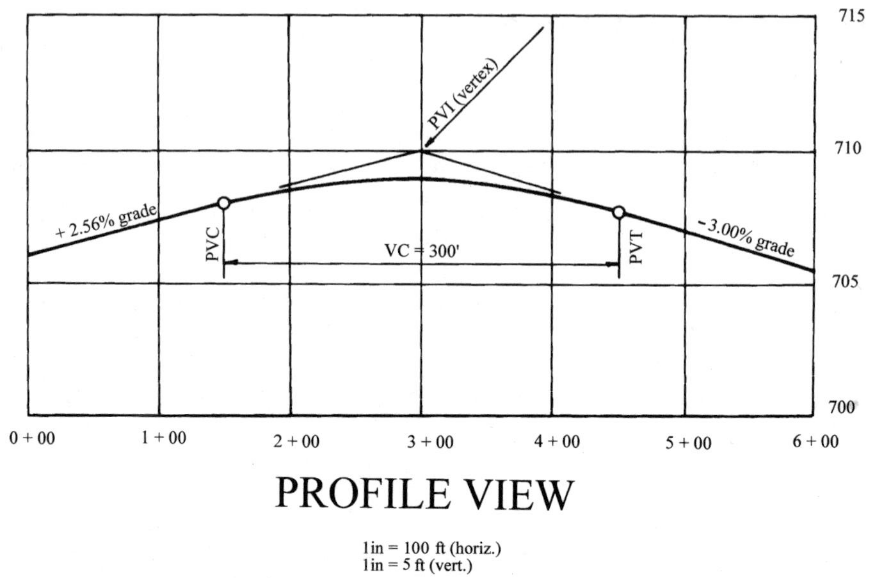

FIGURE 9.3 Typical vertical curve

Earthwork Balance

Proposed grades for a centerline are designed so that the existing ground line is replaced by smooth slopes and smooth transitions but closely following the ascending or descending terrain of the existing ground, except where cuts of various magnitudes are made through hills or where deep ravines are traversed by bridges.

Determining the grades of a new highway is very subjective, and a designer could easily set any number of entirely different grades to fit the same terrain. Where possible, a design is chosen that will balance the amount of earth excavated (the cuts) with the amount of earth needed for fill, while meeting all state and federal requirements. Occasionally, earthwork balancing cannot be done, often resulting in an increase in construction costs due to the need to purchase borrow material or to pay for the distribution of excess excavation.

Grades Have a Sign Convention

The proposed grades are expressed as a percentage (feet of vertical elevation change per 100 horizontal feet). The grades also have a plus or minus sign attached to them. The convention is that the slope is a plus grade if the elevations along that tangent are increasing (ascending) as the stationing increases. A minus grade has decreasing elevations as the stationing increases.

For example, a plus 2 percent grade (+2.00%) indicates an ascending slope of the highway rising 2.00 ft vertically in each 100 ft of horizontal distance. The basis for determining plus or minus grades relates only to the direction of increasing stations, not the direction a motorist might be traveling. These signs need to be included in the computations you'll see later in the chapter.

Maximum and Minimum Grades

Grades on State or Interstate highways fall within a narrow range for safety reasons. A +5 percent on the mainline, and a +7 percent on entrance and exit ramps are considered maximum grades for most automobile and truck traffic during inclement weather. Minus slopes are not as critical unless they have the potential of causing harm to the breaking systems of the vehicles during extended down-hill travel. Minimum grades are generally limited to 0.30 percent to provide for adequate drainage.

Vertical Clearance

When a new roadway is made to intersect existing topography, and grade separations occur, certain vertical clearances must be respected. On roadways built before World War II, vertical clearances of 14.00 to 15.00 ft were acceptable. Today, vertical clearances are in excess of 17.00 ft with changes periodically. "Vertical clearance," by definition, is the vertical distance from the highest traveled point on the roadway to the lowest point on the structural steel or concrete member of the intersecting feature.

Graphics for the Profile View

Plan and profile views are generally drafted on commercially prepared sheets with the plan view at the top and the profile view at the bottom. The bottom half of the drawing is divided into 1 in^2 grid lines. On this grid, the profile view shows the slopes along the vertical tangents of the proposed centerline from the side view, as if one were to severe the highway down the centerline and look at it from the right side, standing at a position 90° perpendicular to the cut. The axis from left to right on the profile view shows horizontal distances, with stationing always increasing from left to right across the sheet. The axis from top to bottom on the profile view shows vertical distances. The elevations that existed before any construction had begun are shown as the existing ground profile (see Figure 9.2), conventionally drawn as a dashed line. The proposed slopes along the centerline are also shown, as a solid line.

Proposed elevations are usually calculated and shown at all intersections (PIs) of vertical tangents and at all even 25-ft or 50-ft stations along the centerline. The elevations shown will be feet above mean sea level, established based on the vertical database of 1929 or, for later elevation control, the North American Vertical database of 1988.

The drafting scale for either view will be indicated on the drawing. In general, the most widely used scales in highway planning are 1 in equals 50.00 ft for the horizontal scale and 1 in equals 10.00 ft for the vertical scale. This ratio of 1:5 is used primarily to exaggerate the profile view for clarity.

The Drafting Process

The horizontal centerline is positioned on the top half of the plan sheet, with the commencing station (whether 0 + 00 or any other station) near the left border edge of the drawing and in line with one of the prepared grid lines in the lower half of the drawing. This establishes the commencing point for the profile view also. If commercial sheets are not available, a grid consisting of 1 in increments is drawn with horizontal and vertical lines across the lower half of the drawing. Representative even-foot profile stations are located using the prepared grid system rather than dropping perpendicular lines from the stations located on the plan view in the upper half of the drawing. Perpendiculars dropped in such a manner would not necessarily have distances equal to 50.00 ft, depending on the orientation of the centerline tangents or curves shown on the plan view at the top of the plan sheet. Figure 9.2 shows a typical plan and profile view.

Figure 9.2 shows the standard graphic location of the horizontal and vertical alignment for most highways, which includes:

1. Centerline stations at the bottom of the grid
2. Vertical reference elevations on either side of the grid
3. A dashed line to portray the existing ground
4. The elevation noted for each major existing ground break point near the bottom of the grid
5. The proposed profile as a solid line

Centerline Versus Edge of Pavement Grades

Profile grades are designed and computed to be located along the centerline of survey, and all engineering drawings reflect this procedure. However, for construction layout purposes, the inside edge of pavement is considered the profile grade line, and this is identified on the typical section drawings (see Chapter 6). For highways that have pavement separations, such as a median, the profile grade is considered to be located on the inside edge of the pavement for each of the two directions of traffic. Shifting the centerline grade to the pavement edge offers no mathematical problems when in a horizontal tangent condition. However, some concern has surfaced concerning the centerline profile grade being shifted to the inside edge of pavement when the highway is in a horizontal curve. Such a procedure causes elevation problems due to the difference in concentric curve lengths.

One hundred foot centerline station increments are decreased or increased according to the median width, and thus a different rate of grade is computed for each 100 ft. These elevation problems are considered minimal, but the surveyor must be aware of this situation, and occasionally will need to compute intermediate grades between stations, based on the new rate of slope. At the present time, no methods are being introduced to reconcile this situation at the design stage.

Another place to be cautious is at entrance or exit ramps. The profile grade along the mainline is the grade along the centerline of survey, but the profiles that are shown at entrance or exit ramps are along one of the edges of pavement, not the centerline, and usually the outside edge of pavement. Careful observation of the plan is necessary to determine exact location.

Bifurcated Highways

When pavements are specifically designed to be separated (either horizontally, vertically, or both), they are referred to as bifurcated, and will probably have two separate profile grade sheets. A careful review of the design drawings is essential.

Design Standards

Several design standards must be satisfied before choosing an acceptable length of vertical curve. These standards are set by state and federal agencies and include:

1. Minimum and/or maximum grade requirements
2. Minimum curve lengths, based on design speed and critical-sight-distance

Minimum Allowable Vertical Curve

The minimum length of a vertical curve cannot be less than the algebraic difference in grades, divided by the maximum allowable change in grade per each 100 ft (Davis, Foote, & Kelley, 1968). In Figure 9.3, the length of the vertical curve is set at 300 ft, but assuming that the tangent grades meet federal guide-lines, the minimum length of vertical curve is computed as follows:

First example: Assume the allowable change in grade per each 100 ft = 4%

Plus grade = +2.56%
Minus grade = –3.00%

$$+2.56 - (-3)$$
$$+2.56 + 3 = 5.56$$
$$5.56 \div 4 = 1.39 \text{ stations}$$
or 139 ft

Second example: Assume the known data in Figure 9.4:

$$+5.00 - (-3.25) = 8.25 = \text{algebraic difference of gradients}$$

Assume: a 5.00% grade is the maximum allowable

$$\text{Therefore: } \frac{8.25}{5} = 1.65 \text{ stations, or } 165 \text{ ft}$$

Note that the sign of the grade will affect the subtraction. A reminder on how to handle algebraic subtractions is given later in the chapter, under "Determining the Algebraic Difference."

Choosing the Length of the Vertical Curve

In Figure 9.4, the vertical curve is set at 300 ft, far exceeding the minimum length, so it is a permissible curve. Longer lengths of vertical curves are not only set to match stations or portions thereof but are quite often set at greater than minimum distances to decrease the high point of the curve and thereby clear certain obstacles that otherwise could cause design concern. The engineer may also choose curve lengths that are designed to begin and end on even stations, half stations, or quarter stations, such as 4 + 00, 4 + 25, 4 + 50, or 4 + 75.

Nomenclature

The vertical curve, like the horizontal curve, has certain designations that identify the parts of the curve, and for purposes of keeping the two types of curves separate, the nomenclature for vertical curves always includes the letter V in the various acronyms. By convention, the stationing determines direction. Therefore, the beginning of the vertical curve has the smallest station and is known as the "point of vertical curve" or PVC, the end of the vertical curve is known as the "point of vertical tangent" or PVT, and the vertex is known as the "point of vertical intersection" or PVI. The grade on the vertical tangent, as the vertical tangent begins, is known as the "grade in" and the grade on the vertical tangent, as the vertical curve ends, is known as the "grade out." Curves that form a hill are in "crown, "crest, or "summit" condition; curves that form a valley are in "sag" condition.

Mosaics Used in Design

For purposes of a preliminary design, PVI (vertex) elevations are usually set by the engineer or computed from other controls along the tangent, such as intersecting roadways or other topographic features which control the new profile. A proposed profile is temporarily designed, as shown in Figure 9.2, using short ground distances as the basis for the design. Highways, of course, are many miles in length, and therefore, the final proposed profile grades, as well as horizontal alignments, are designed according to an overview of larger portions of the project.

An overview of the project is completed by assembling hundreds, even thousands, of lineal feet of photographic mosaics (aerial photographs) attached to one another, thus permitting a full view of the horizontal and vertical location. Aerial photography includes existing horizontal topographic features, as well as existing contour elevations, from which the ground relief is plotted on profile drawings and studied. During these preliminary studies, more than one horizontal alignment is chosen, the elevations from each study are lifted from the contour mapping and projected onto engineering drawings, and new profile grades established.

Mosaics to Grade Lines

Grade lines are then analyzed for earthwork volumes and, from these studies, one is chosen by the engineer to represent the most cost-efficient proposal, not only in construction costs but also right-of-way costs, and where the proposal satisfies all state and/or federal planning requirements.

Computing the Vertical Curve

With the vertical tangents set, we can now turn to computing elevations in the vertical curves, which brings us back to the parabola.

A parabolic curve is egg-shaped, so it does not have a uniform radius. It does, however, have a slope that changes uniformly throughout its length (Urquhart, 1959). That characteristic, plus the design convention of fitting the parabola into the intersection of the two vertical tangents so that an equal horizontal amount of the curve is located on each side of the tangent intersection, results in a few simple equations for elevation or vertical offsets. Fitting the parabola equally on both sides of the vertex results in the following properties of a vertical curve on a highway:

1. A perpendicular line extended from the vertex (point B on Figure 9.4) passes through the chord AC of the curve at its midpoint (point D).
2. The vertical curve intersects that perpendicular projection at the midpoint (point E) of the line BD.
3. Equal horizontal distances measured from the beginning of the vertical curve or the end of the curve to known points within the curve will have identical vertical offsets from the tangent lines to the curve.

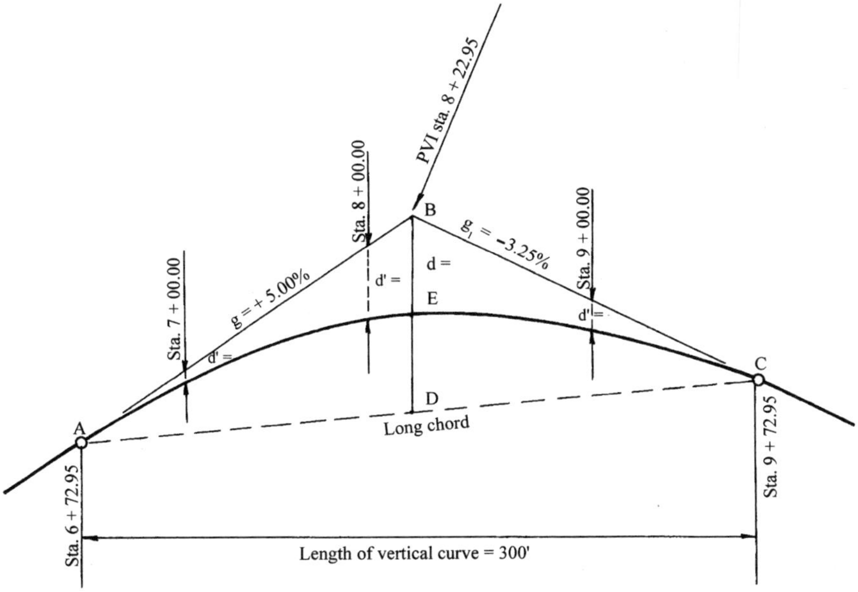

FIGURE 9.4 Vertical curve offset computations

For a vertical curve on a highway, the arc length of the parabola does not need to be known or computed. The only distances that are needed are along straight horizontal and vertical lines, which can be calculated using one of four simple equations, given with examples in the rest of the chapter.

The first equation lets you solve for the perpendicular "offset" distance d from the vertex to the horizontal midpoint of the vertical curve. The second and third equations are two different ways to calculate the perpendicular offset distance d' from any point on a tangent to its corresponding vertical counterpart on the vertical curve.

The final equation combines parts of the other equations so you can directly calculate the elevation at any point on the curve in one step rather than the two steps needed when using the traditional (offset) equations. If you're interested in seeing the derivation of these equations, or other details about parabolas, refer to any geometry textbook or any of the civil engineering textbooks listed in the Bibliography.

Elevation at Midpoint

The first equation you need is the equation that solves for the distance, in feet, between the vertex and the midpoint of the curve, the distance d. The equation is referred to as the traditional method, and it is:

$$d = \frac{L}{800}(g - g_1) \tag{9.1}$$

Where:

L = horizontal length of the entire vertical curve
800 = constant
g = grade-in
g_1 = grade-out
d = distance from vertex to midpoint of curve

Using the known data in Figure 9.5, we have:

$$d = \frac{300}{800}[+5.00 - (-3.25)]$$

$$= 0.375(8.25)$$

$$= 3.09 \text{ ft}$$

Determine the Algebraic Difference

Several of these vertical curve equations use the algebraic difference in grades as part of the computation. Be careful when the grade for g_1 is a negative number. Remember that subtracting a negative number actually involves adding that number. So, $g - g_1$ (Figure 9.5) for example, becomes $g - (-3.25) = g + 3.25$, or a total algebraic expression of $5 + 3.25 = 8.25$.

Figure 9.6 shows all four possible combinations where g and g_1 are either positive or negative grades, as a reminder of how to handle each situation.

COMPUTING TANGENT OFFSETS

The second equation (the "tangent offset distance") can be computed from the following:

$$d' = \left(\frac{g_1 - g}{2L}\right)(l)^2 \qquad (9.2)$$

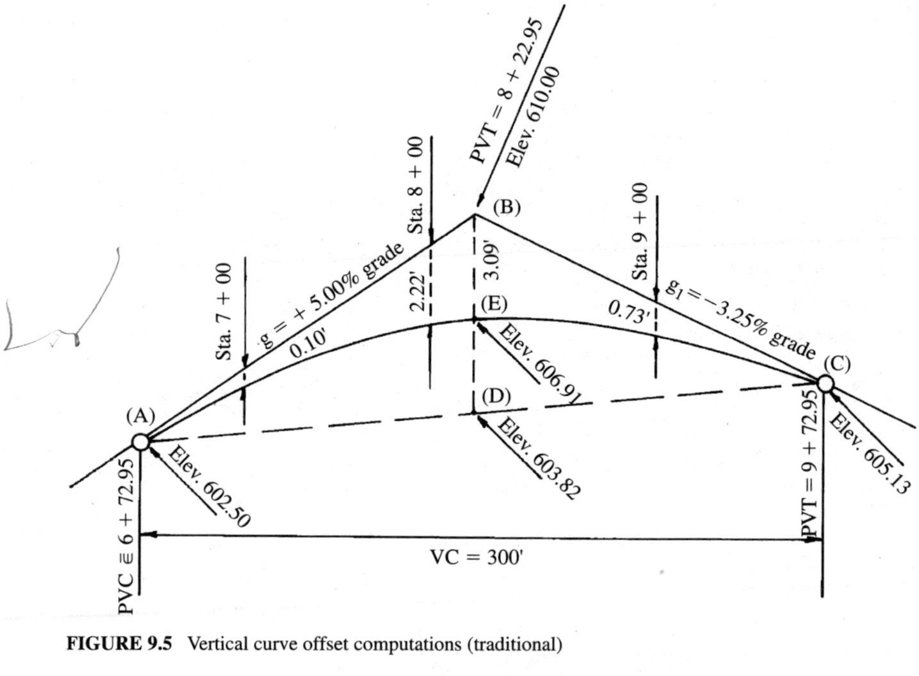

FIGURE 9.5 Vertical curve offset computations (traditional)

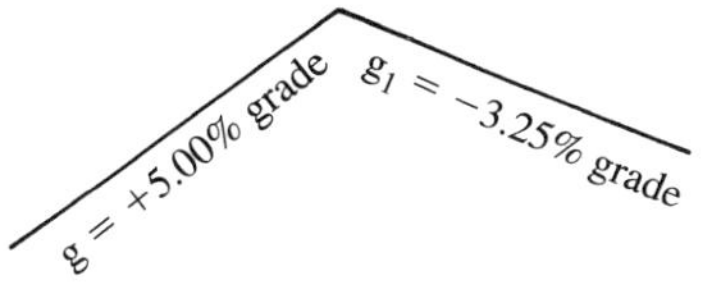

Algebraic expression when the grades are in opposite directions (crest)

$$(g - g_1)$$
$$= 5 - (-3.25)$$
$$= 5 + 3.25 = 8.25$$

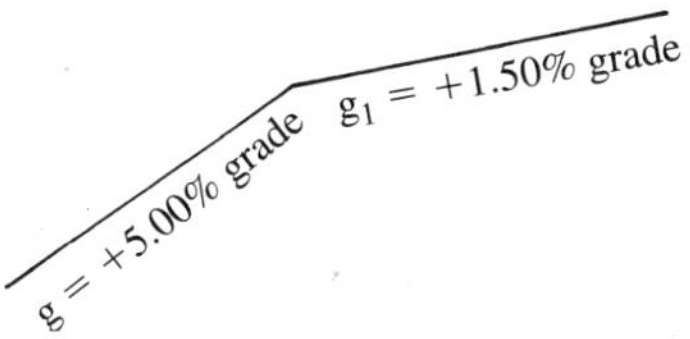

Algebraic expression when the grades are in the same direction and ascending.

$$(g - g_1)$$
$$= 5 - (+1.50)$$
$$= 5 - 1.50 = 3.50$$

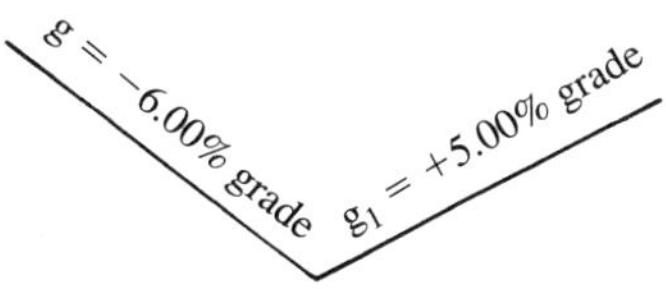

Algebraic expression when the grades are in opposite directions (sag)

$$(g - g_1)$$
$$= -6 - (+5)$$
$$= -6 - 5 = 11$$

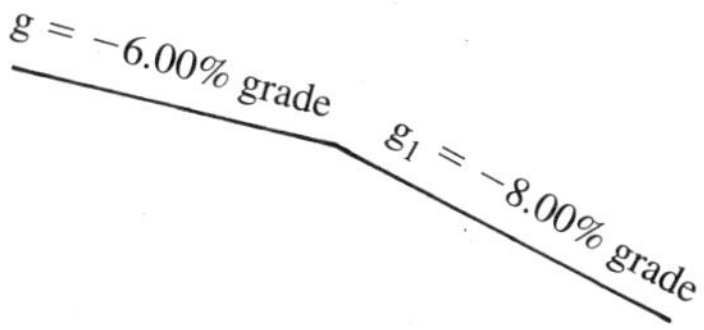

Algebraic expression when the grades are in the same direction and descend ing

$$(g - g_1)$$
$$= -6 - (-8)$$
$$= -6 + 8 = 2$$

FIGURE 9.6 Algebraic expressions reduced

Where:

g_1 = grade-out, in feet per foot
g = grade-in, in feet per foot
L = total horizontal length of vertical curve
l = distance from PVC or PVT to any station, including the midpoint
d' = offset distance from tangent to curve

Using the known data in Figure 9.5, we have:

$$d' = \frac{[-0.0325 - (+0.05)]}{2(300)}(150)^2$$

$$= \frac{-0.0825}{600}(150)^2$$

$$= -3.09 \text{ ft}$$

Note: This equation will give you the vertical offset distance from any point on either tangent to the corresponding point on the vertical curve. Also, a negative value for d' tells you the vertical curve is located at a lower elevation than the vertical tangent line. For the rest of this text that minus sign will be ignored. The graphics in the figures will make it obvious whether the curve is in crest or sag.

Other tangent offsets can be computed by reducing the equation within the brackets to a constant, and thereafter complete the solution by the square of l:

$$= \frac{0.0825}{600} = 0.0001375 \text{ (constant)}$$

Tangent offset at station 7 + 00, Figure 9.5 computed using the constant times the square of l:

$$\text{Station } 7 + 00 - \text{station } 6 + 72.95 = 27.05 \text{ ft}$$

$$d' = 0.0001375 \times (27.05)^2$$

$$= 0.10 \text{ ft}$$

Tangent offset at station 8 + 00, Figure 9.5, computed using the constant times the square of l:

$$\text{Station } 8 + 00 - \text{station } 6 + 72.95 = 127.05 \text{ ft}$$

$$d' = 0.0001375 \times (127.05)^2$$

$$= 2.22 \text{ ft}$$

Tangent offset at station 9 + 00, Figure 9.5, computed using the constant times the square of l. Note that station 9 + 00 is up-station of the midpoint, and therefore the PVT must be used in the computation, not the PVC, so l can never be greater than half of L (the total length of the vertical curve) because you are then in a location on the curve where the other tangent and grade are controlling.

$$\text{Station } 9 + 72.95 - \text{station } 9 + 00 = 72.95 \text{ ft}$$

$$d' = 0.0001375 \times (72.95)^2$$

$$= 0.73 \text{ ft}$$

Computing the Tangent Offsets (T\ditional Method)

Tangent offsets can also be computed using the traditional d' equation, where the solution cannot be completed without first solving for d in Eq. (9.1):

$$d' = \frac{l^2}{\left(\frac{L}{2}\right)^2} \times (d) \qquad (9.3)$$

Where:

d' = tangent to curve distance
L = length of vertical curve
d = midpoint of curve to vertex (computed in Eq. (9.1).]
l = distance to the point from PVC or PVT

Solving for d' at station 7 + 00, in Figure 9.5, we have:

$$d' = \frac{(27.05)^2}{\left(\frac{300}{2}\right)^2} \times 3.09 \text{ ft}$$

$$= \frac{731.7025}{22500} \times 3.09 \text{ ft}$$

$$= 0.10 \text{ ft}$$

Elevations on Tangent Lines

To obtain the actual elevation on the curve requires a second computation, that of establishing the elevation on the tangent line at the point that is perpendicular to the point on the curve and applying the d' computation in the appropriate direction.

Example:

For station 7 + 00, Figure 9.7, compute the tangent elevation by multiplying the distance to the required station by the percentage of grade and then correcting in the appropriate direction, such as:

$$27.05 \times 0.05 = 1.35 \text{ ft}$$

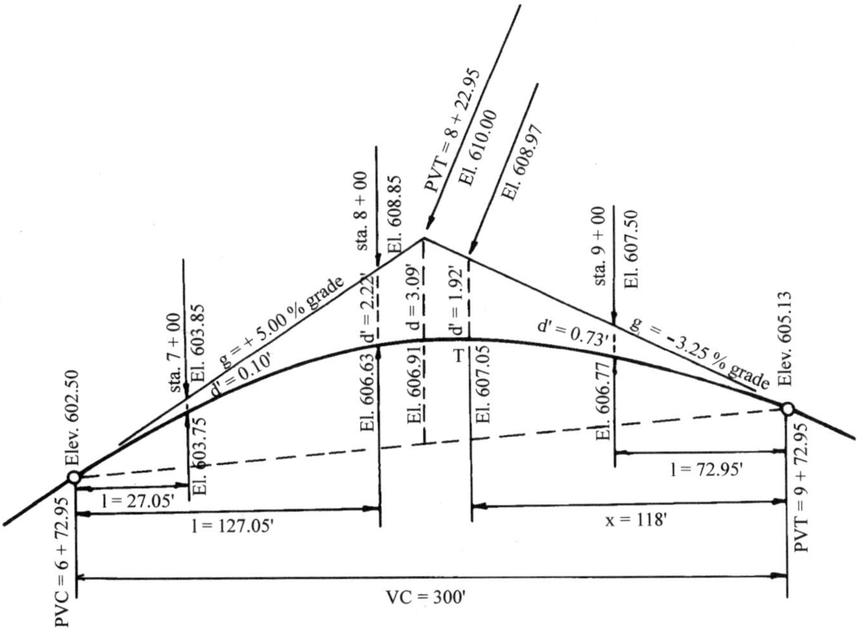

FIGURE 9.7 Computing the highest point

Elevation at PVC station 6 + 72.95 = 602.50
Elevation on the tangent line at station 7 + 00 = 602.50 + 1.35 = 603.85
Elevation on curve at station 7 + 00 = 603.85 – 0.10 = 603.75

Compute Elevations Direct (Crown condition)

Direct vertical curve elevations can be computed using the following equation, which includes two previous solutions, (1) the elevation on tangent and (2) the tangent offset at the same point, and applying this result in the appropriate direction:

$$\text{elev. at } PVC \text{ or } PVT + (g \text{ or } g_1)(l) - \left(\frac{g_1 - g}{2L}\right)(l)^2$$

Where:

g = grade-in
g_1 = grade-out
L = length of vertical curve
l = distance from PVC or PVT to the station on curve

Example:
Station 7 + 00, Figure 9.7, lies within the half-length of the vertical curve where the +5.00 percent grade controls; therefore:

$$\text{elev. at } PVC + (0.05)(27.05) - \left[\frac{-0.0325 - (+0.05)}{2 \times 300}\right](27.05)^2 \qquad (9.4)$$

$$602.50 + 1.3525 - (0.0001375)(731.7025)$$

$$602.50 + 1.3525 - 0.10 = 603.75$$

Example:
Station 8 + 00, Figure 9.7, also lies within the half-length of the vertical curve where the +5.00 percent grade controls; therefore:

$$(0.05)(127.05) - \left[\frac{-0.0325 - (+0.05)}{2 \times 300}\right](127.05)^2 + 602.50 \qquad (9.4)$$

$$6.352 - (0.0001375)(16141.7025) + 602.50$$

$$6.352 - 2.22 + 602.50 = 606.63$$

Example:
Station 8 + 22.95 (vertex), Figure 9.7, is the station of the vertex and/or midpoint of the curve and can be computed using either the grade-in or grade-out. By using the grade-in (+0.05), the computation is as follows:

$$(0.05)(150) - (0.0001375)(150)^2 + 602.50$$

$$7.50 - 3.09 + 602.50 = 606.91$$

Example:
Station 9 + 00, Figure 9.7. *Note:* The forward half of the vertical curve is computed using the grade-out percentage of slope, and the PVT station as the source of vertical control:

Station 9 + 72.95 – station 9 + 00 = 72.95 ft.

The grade-out in Figure 9.7 is a minus (–3.25 percent)

$$(0.0325)(72.95) - (0.0001375)(72.95)^2 + 605.13$$

$$2.37 - 0.73 + 605.13 = 606.77$$

COMPUTING CONTROL ELEVATIONS

In some of the previous examples, we used the elevation of the PVC and the PVT to complete the solution. Before continuing, we should review the process that establishes these control elevations.

Assume: The vertex of the vertical curve is set at elevation 610.00 (Figure 9.7). The grade-in is a plus (+5.00 percent), and the grade-out is a minus (–3.25 percent). Compute the elevation at PVC station 6 + 72.95.

Subtracting the PVC station from the PVI (vertex) station gives a distance of 150 ft. Remember that the vertex point is vertically perpendicular to the midpoint of the vertical curve and that all vertical curves are computed based on their horizontal distances. Using the grade-in percentage to calculate the elevation at the PVC (station 6 + 72.95) simply involves multiplying the grade percentage (+5.00 percent) by the horizontal distance (150 ft), such as:

$$0.05 \times 150 = 7.50 \text{ ft}$$

and adding or subtracting as appropriate. In this example, the PVC is lower than the vertex; therefore, 7.50 ft must be subtracted from elevation 610.00, resulting in the PVC having a grade of 602.50. Likewise, the PVT elevation is established by using the grade-out (–3.25 percent) and multiplying by the horizontal distance, such as:

$$0.0325 \times 150 = 4.875 \text{ ft}$$

and again, in this example, the PVT is lower than the vertex; therefore, the 4.875 ft must be subtracted from the vertex, resulting in the PVT having a grade of 605.13.

Average Elevation Between PVC and PVT

Refer to Figure 9.5 and compute the elevation at point *D*, which is the elevation average (mean) between the PVC (point *A*) and the PVT (point *C*), or 605.13 minus 602.50 equals 2.63 ft, divided by 2 equals 1.315 ft. Either add this distance to the PVC or subtract it from the PVT, as appropriate, to obtain the average elevation at point *D*, such as: 602.50 plus 1.315 equals 603.815 or 605.13 minus 1.315 equals 603.815, which rounded to the nearest one-hundredth of a foot equals 603.82.

COMPUTING THE MIDPOINT OF THE CURVE

Compute the elevation at point *E* using the mean between points *B* and *D*, or 610.00 – 603.82 = 6.18 ÷ 2 = 3.09 ft, and the resulting grade at point *E* is 603.82 + 3.09 = elevation 606.91. This same result was computed using the traditional Eq. (9.1) and (9.3).

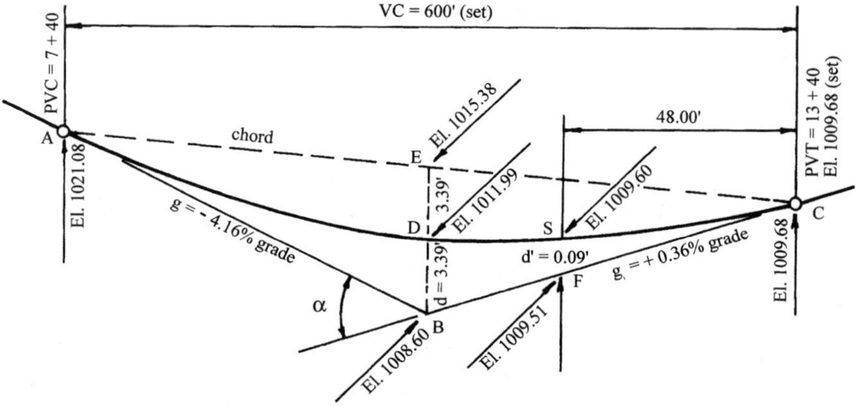

FIGURE 9.8 Computing the lowest point

Location of High and Low Curve Point

Refer to Figure 9.7, where the grade-in, grade-out, vertex elevation, and length of curve are identical to Figure 9.5 except for the computation of the "high-point." The sag point is shown in Figure 9.8. Crown or sag points can be computed from the following equation (Breed et. al., 1958):

$$x = \frac{Lg_1}{g - g_1} \tag{9.5}$$

Note: Grade (g_1) is used in this solution because the high point of the curve (point T) is located in the forward half of the curve, and therefore, we assume the grade-out controls this computation. As you will observe in later computations, this is not true. Computing for the T distance, using data in Figure 9.7, we have:

$$x = \frac{Lg_1}{g - g_1} \tag{9.5}$$

$$= \frac{(300)(-3.25)}{8.25}$$

$$= 118.18 \text{ ft}$$

In the example, "Location of High and Low Curve Points," point T is computed as though it is located in the forward half of the curve, and as noted in that heading, we assumed that the grade-out percentage would be used in the equation. This is not totally incorrect; however, as you will see in Eq. (9.6) the grade-in (g) rate of slope is used to compute this same point, and the result locates the high point in the identical position. Therefore, using either grade will position the crown (or sag) in its correct location, and the distance to the point is either subtracted from the PVT, or the distance is added to the PVC based on your control (grade-in or grade-out) percentage of slope (Breed et. al., 1958):

$$x = \frac{Lg}{g - g_1} \tag{9.6}$$

$$= \frac{(300)(5)}{8.25}$$

$$= 181.82 \text{ ft, (or rounded} = 182 \text{ ft)}$$

Computing the Tangent Offset (Traditional Method)

The tangent offset distance to the crown, point *T* (Figure 9.7) can be computed using the traditional *d'* equation.

$$d' = \frac{(118.18)^2}{\left(\frac{L}{2}\right)^2} \times 3.09 \tag{9.3}$$

$$= \frac{13966.512}{(150)^2} \times 3.09$$

$$= 1.92 \text{ ft}$$

COMPUTING THE CURVE ELEVATION

The elevation on the tangent, opposite point *T*, Figure 9.7 is:

$$118.18 \times 0.0325 + \text{elevation at PVT } 9 + 72.95$$

$$3.84 + 605.13 = 608.97$$

Therefore, adjusting for the offset distance, the elevation at point *T* becomes:

$$608.97 - 1.92 = 607.05$$

Required Sight Distance

The minimum allowable curve length, in a crown situation, is controlled by the design speed and the "critical sight distance," CD, between two vehicles approaching in opposite directions. It is generally assumed that the drivers of both vehicles to have an eye height of 4 to 4.5 ft above the pavement level, and under adverse conditions, predictable distances are traveled while the drivers are reacting to an emergency.

Two equations are used to compute the minimum allowable curve length:

1. Where the critical distance is equal to, or less than, the length of the vertical curve
2. Where the critical distance is greater than the length of the vertical curve, placing one of the vehicles on the vertical tangent, rather than both vehicles traveling on the vertical curve portion of the roadway.

A supplemental solution, Eq. (9.7), is used to make this determination (Figure 9.7).

$$\text{CD} = \frac{(g - g_1)(L)}{8} \tag{9.7}$$

Where:

CD = critical sight distance

$g - g_1$ = algebraic difference in grades

L = length of vertical curve, in stations
8 = constant

This equation is from Breed (1971):

$$CD = \frac{[5.00 - (-3.25)](3)}{8}$$

$$= \frac{(8.25)(3)}{8} = 3.09 \text{ stations, or } CD > L$$

Therefore, the supplemental equation shows 3.09 stations for the critical distance CD versus 3.00 stations for L, and Eq. (9.8) is used to compute the minimum allowable length of curve (Breed, 1958). Conversely, if the critical distance CD is equal to or less than, L, Eq. (9.9) is used to compute the minimum allowable length of curve.

When the critical sight distance CD is greater than the length of curve L, such as computed above, the following equation must be used (Breed, 1958).

$$L = \frac{2(g - g_1)CD - 8h}{g - g_1} \tag{9.8}$$

$$= \frac{2(8.25)(3.09) - 8(4.5)}{8.25}$$

$$= \frac{50.985 - 36}{8.25}$$

$$= 1.82 \text{ stations, or } 182 \text{ ft}$$

When the critical sight distance CD is less than L, the following equation must be used (Breed, 1958).

$$L = \frac{(CD)^2 (g - g_1)}{8h} \tag{9.9}$$

Where:

L = length of vertical curve, in stations
CD = critical sight distance, in stations
$g - g_1$ = algebraic difference in grades
h = eye height above pavement (4 to 4.5 ft)

Assume: $CD = 2.75$ stations (for comparison):

$$L = \frac{(2.75)^2 [5 - (-3.25)]}{8 \times 4.5}$$

$$= \frac{(7.5625)(8.25)}{36}$$

$$= 1.73 \text{ stations, or } 173 \text{ ft}$$

VERTICAL CURVES IN SAG POSITION

Figure 9.8 is an example of the vertical curve in a sag position, which is computed in the same manner as the parabolic curve in a crown position, except the critical sight distance

equations do not have the same significance, due to the constant view experie
vers in down-hill travel while approaching one another in opposite directions. T
the vertical curve, in this example, is set at 600 ft, the PVT station 13 + 40 is c
end of the curve, and tangent grades ($g = -4.16$ percent; $g_1 = +0.36$ percent) are provided.

By applying the various equations outlined on previous pages, you can develop the various parts of the parabolic curve. In this example, actual engineering data are used from design plans prepared for Interstate 79 in West Virginia.

Using Figure 9.8, determine the following: (1) the elevation of point *B*; (2) the elevation of point *A*; (3) the elevation of point *D*; (4) the elevation of point *E*, and (5) the location and elevation of the lowest point on the curve, point *S*.

1. Determine the elevation at point *B*. Point B, the vertex PVI, can be approached from either the PVC or the PVT. Using the more traditional manner, the computation for point *B* would require establishing the grade elevation at point *C*, followed by a computation for point *B* as shown below:

   $0.0036 \times 300 = 1.08$ ft
   Point *C* elevation = 1009.68 (set)
   Point *B* elevation = 1009.68 – 1.08 = 1008.60

2. Determine the elevation at point *A*:

   $0.0416 \times 300 = 12.48$ ft
   Point *B* elevation = 1008.60
   Point *A* elevation = 1008.60 + 12.48 = 1021.08

3. Determine the elevation at point *D*:

$$d = \frac{L}{800}(g - g_1) \tag{9.1}$$

$$= \frac{600}{800}[(-4.16) - (+0.36)]$$

$$= 0.75\,(-4.16 - 0.36)$$

$$= 0.75\,(+4.52)$$

$$= 3.39 \text{ ft}$$

   Point *B* elevation = 1008.60

   Point *D* elevation = 1008.60 + 3.39 = 1011.99

4. Determine the elevation at point *E*:

   Point *E* lies at the mid point between *A* and *C*; therefore,

$$\text{Point } E = \frac{(A) - (C)}{2} = \frac{1021.08 - 1009.68}{2} = 5.70 \text{ ft}$$

   Point *E* elevation = 1021.08 – 5.70 = 1015.38

5. Locate lowest point in the sag position and test the location using the grade-in and grade-out tangents:

$$X = \frac{Lg}{g - g_1} \tag{9.6}$$

Where:

X	=	horizontal distance to lowest sag point
g	=	grade-in (–4.16 percent)
g_1	=	grade-out (+0.36 percent)
$g - g_1$	=	algebraic expression between grades
L	=	length of total curve, in stations

Using the grade-in slope to compute point S, we have:

$$X = \frac{(6)(-4.16)}{-4.16 - (+0.36)}$$

$$= \frac{24.96}{4.52}$$

$$= 5.52 \text{ stations, or } 552 \text{ ft}$$

Therefore, the lowest point in the sag is station 12 + 92.00, or 552.00 ft from PVC station 7 + 40.

Locate Lowest Point in Sag

Compute point S using the grade-out slope, and the reduced equation is:

$$X = \frac{(6)(+0.36)}{4.52} \tag{9.6}$$

$$= 0.4778 \text{ stations, or } 48 \text{ ft}$$

Therefore, the lowest point in the sag is 48.00 ft (rounded) from the PVT station 13 + 40, or station 12 + 92.00, verifying that either of the rates of slope (grade-in or grade-out) will compute the same location, but from different end points.

Compute the Grade (Traditional Method)

Locate the elevation on the vertical tangent, opposite point S:

$$48 \times 0.0036 = 0.17 \text{ ft}$$

$$\text{Point } C \text{ elevation} = 1009.68$$

$$\text{Point } F \text{ elevation} = 1009.68 - 0.17 = 1009.51$$

COMPUTE THE TANGENT OFFSET AT POINT S

$$d' = \frac{l^2}{\left(\frac{L}{2}\right)^2} \times (d) \tag{9.3}$$

$$= \frac{(48)^2}{(300)^2} \times 3.39$$

$$= 0.09 \text{ ft}$$

Therefore, the elevation at point $S = 1009.51 + 0.09 = 1009.60$

Compute Elevations Direct (Sag Condition)

Elevations on vertical curves, in a sag condition, can be computed directly using the following equation:

$$\text{elev. at } PVC \text{ or } PVT - (g \text{ or } g_1)\,(l) - \left(\frac{g_1 - g}{2L}\right)(l)^2 \tag{9.10}$$

Where:

g	=	grade-in
g_1	=	grade-out
L	=	length of vertical curve, in feet
l	=	distance from PVC or PVT to the curve station
$g_1 - g$	=	algebraic expression between gradients

Example:

Compute the elevation at station 12 + 92 (point S, Figure 9.8). Station 12 + 92.00 is upstation from the midpoint, and therefore, the grade-out must be used in the computation. Refer to Figure 9.8.

$$\text{elev. at } PVT - \left[(g_1)(l) - \left(\frac{g_1 - g}{2L}\right)(l)^2\right] \tag{9.10}$$

$$1009.68 - \left[(+0.0036)(48) - \left(\frac{0.0036 - (-0.0416)}{2 \times 600)}\right)(48)^2\right]$$

$$1009.68 - \left[(0.1728) - \left(\frac{0.0452}{1200}\right)(2304)\right]$$

$$1009.68 - [(0.1728) - (0.0867)]$$

$$1009.68 - 0.08 = 1009.60$$

The elevation at the lowest point S, station 12 + 92.00, therefore becomes 1009.60. The signs (+ or –) require much care when solving for elevations direct, and often the easiest and most reliable method is the traditional computation used to solve d and d_1, Eqs. (9.1) and (9.3).

As a second example, compute the elevation direct for the midpoint of the curve, station 10 + 40.00, point D, Figure 9.8.

$$\text{elev. at } PVC - \left[(g)(l) - \left(\frac{g_1 - g}{2L}\right)(l)^2\right] \tag{9.10}$$

$$1021.08 - \left[(-0.0416)(300) - \left(\frac{+0.0036 - (-0.0416)}{2 \times 600}\right)(300)^2\right]$$

$$1021.08 - \left[-12.48 - \left(\frac{0.0452}{1200}\right)(90000)\right]$$

$$1021.08 - [-12.48 - 3.39]$$

$$1021.08 - 9.09 = 1011.99$$

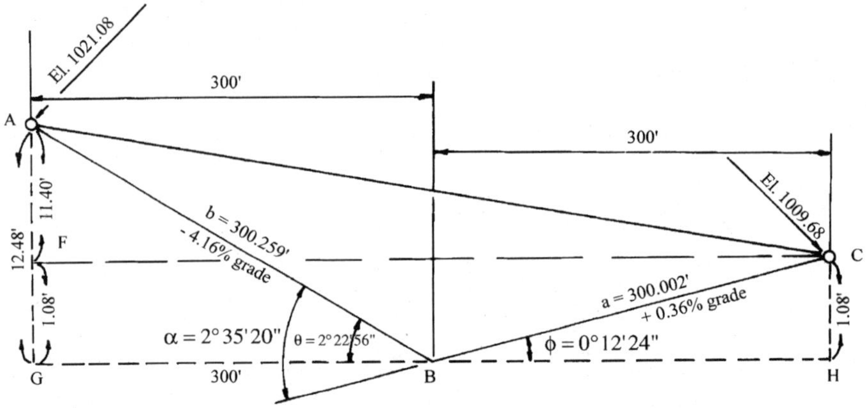

FIGURE 9.9 Alternate methods of computing grades

Two additional computations can be made, Figure (9.9): (1) by computing offsets, using the percentage of grade for each vertical tangent, subtracting one offset from the other, and applying the difference in the appropriate direction, and (2) computing the angles between grades, using the rates of slope to solve the hypotenuse of each triangle.

1. The grade-in slope produces an offset of 12.48 ft (0.0416 × 300 equals 12.48 ft), and the grade-out slope produces an offset of 1.08 ft (0.0036 × 300 equals 1.08 ft). Subtracting one offset from the other, such as 12.48 ft minus 1.08 ft equals 11.40 ft, and applying this difference in the appropriate direction (elevation 1009.68 at point *C* plus 11.40 ft = elevation 1021.08) at point *A*.
2. If the elevation of the PI is known, compute the angles between the two grades, solve the short leg of each triangle *A, G, B* and *B, H, C*, and calculate the elevation accordingly.

SOLVE FOR THE ANGLES (under method 2 above and then solve for the hypotenuse and legs of the triangle)

$$\tan.\theta = \frac{12.48}{300} \qquad \tan.\ \theta = \frac{1.08}{300}$$

$$\tan.\theta = 0.041600 \qquad \tan.\theta = 0.0036$$

$$\tan.\theta = 2.382130921^\circ \qquad \tan.\theta = 0.206263915^\circ$$

$$\theta = 2^\circ\ 22'\ 55.6'' \qquad \theta = 0^\circ\ 12'\ 23''$$

COMBINING THE ANGLES $\alpha = 2^\circ\ 22'\ 55.6'' + 0^\circ\ 12'\ 23'' = 2^\circ\ 35'\ 19''$
SOLVING THE HYPOTENUSE OF EACH TRIANGLE

$$a = \sqrt{300^2 + 1.08^2}$$

$$a = 300.002\ ft$$

$$b = \sqrt{300^2 + 12.48^2}$$

$$b = 300.259\ \text{ft}$$

CHAPTER 10
SLOPE STAKING

INTRODUCTION

Slope staking is the art and mathematics surveyors use to precisely determine the initial point of a cut (excavation) or fill (embankment) before a contractor begins to build a highway or other site grading project. Chapter 8 set forth certain cross-section applications in the development of highway plans. To this list, add the tremendous importance cross sections have on slope staking, in that each proposed pavement, earth slope, and ditch slope, side hill bench slope, and sub-surface fill bench slope is graphically depicted on the cross section, and where each change in elevation is eventually computed. The cross section then acts as the depository of design information required to construct the highway, and the surveyor turns to this drawing for guidance when slope staking.

Proposed highway profile grades (see Chapter 9) seldom follow the existing ground, and earthwork excavation (cuts) and embankment (fills) make up for this difference.

Ground separations that occur between proposed and existing grades are connected by side slopes that are regulated by federal, state and county requirements. These side slopes act as earthen "tie lines," connecting the existing ground with the proposed ground. Slope staking is the art and science of field-locating this exact tie line. Cross section drawings are the tools for approximating this cut or fill point on paper; in this chapter we will learn that slope staking is the mathematical way of finalizing this location.

Figures 10.1 and 10.2 represent proposed highway finished grades, in both cut and fill conditions. The fill section (Figure 10.1) is designed with side slope ratios of 2:1 and 6:1, whereas the combination cut and fill section (Figure 10.2) is designed with a side slope (fill) ratio of 3:1, ditch slopes of 2:1, and a back slope (cut) ratio of 2:1. The "fill bench," as proposed in Figure 10.2, is designed by the engineer to minimize landslides where embankment material is to be placed on sloping hillsides. The field staking of fill benches is as important as embankment or excavation side slopes, and therefore, slope staking is required.

Wherever construction precision is required, the surveyor who has a firm background in slope staking has the advantage over colleagues who have little or no slope staking experience.

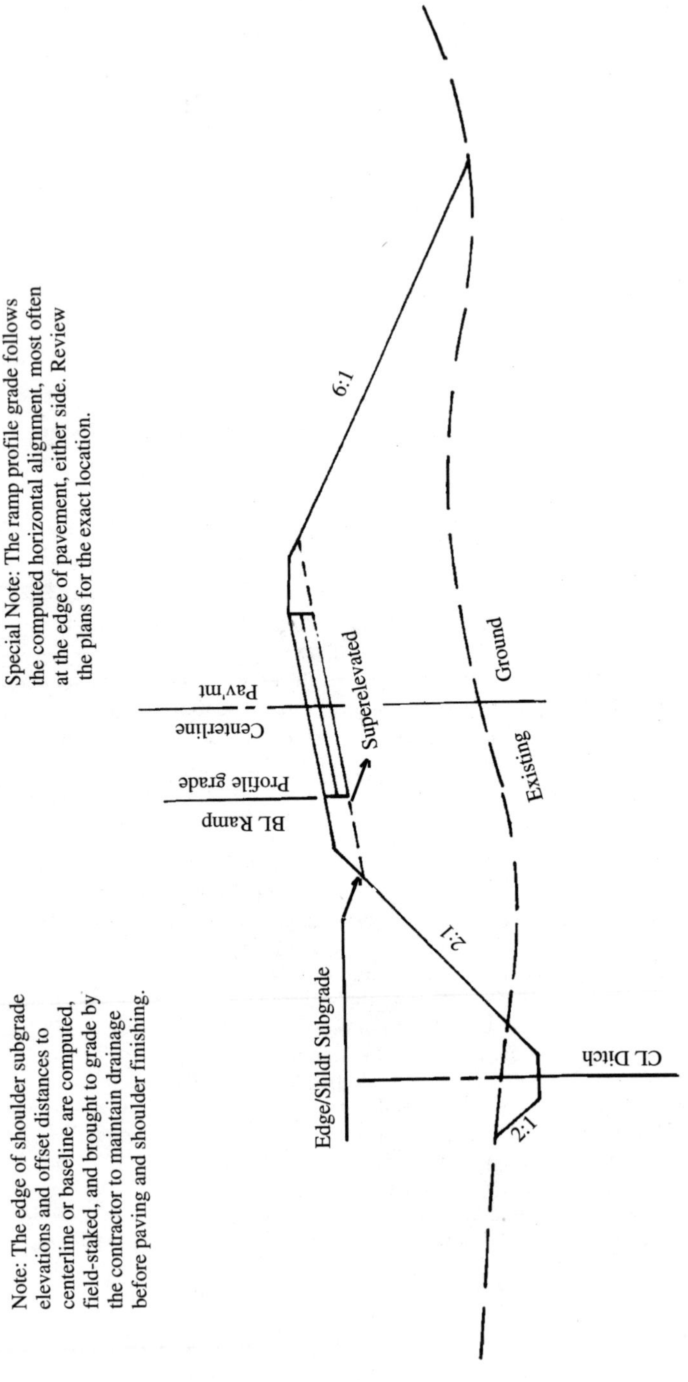

FIGURE 10.1 Slope staking in fill

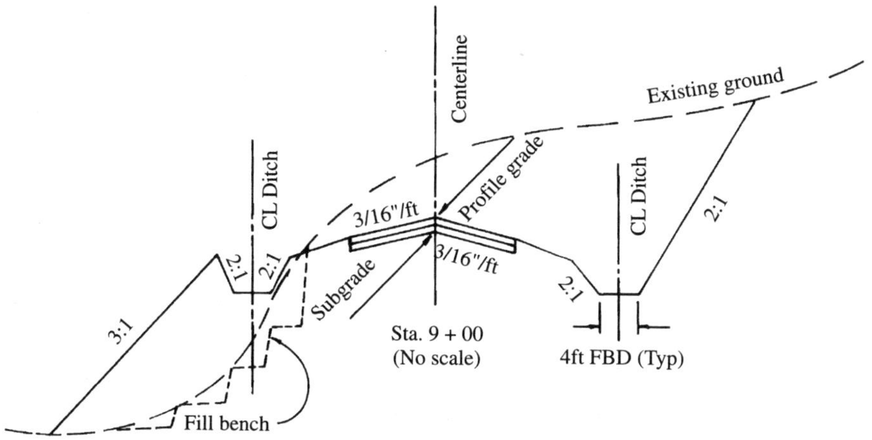

FIGURE 10.2 Slope staking in cut and fill condition

RATES OF SLOPE

Both cut and fill material appear in the examples (Figures 10.1 and 10.2) with certain defined slope ratios. These slope ratios require some additional explanation. Rates of slope are represented on the drawings (cross sections) when the project is being designed. The engineer uses a special cross section sheet which is divided faintly into 1-in squares. The 1-in square pattern on the graph paper expedites the drafting of "slope ratios" by a simple process, rather than computing elevations for each ground break and subsequently drawing the slope based on these computations. Drafting instruments, such as triangles, are used to connect two points, one point representing the horizontal span of a specified number of graph squares (inches), and the second point representing the vertical drop in a specified number of graph squares (inches). The line between these points represents a specific slope ratio.

Example:

Assume that a specified ratio of 6:1 is to be plotted. The first number in the ratio refers to the horizontal distance and the second number refers to the vertical distance. Therefore, six graph squares would be set off horizontally, and one graph square would be set off vertically, and these two points would be connected. This slope represents a rather flat condition for highway construction, but additional rates of slope are identified below.

FRACTIONAL AND NONFRACTIONAL RATES OF SLOPE

Cross-section slopes, such as: $\frac{1}{6}$:1, $\frac{1}{2}$:1, $\frac{3}{4}$:1, 1:1, 2:1, 3:1, 4:1, 6:1, 8:1, 24:1, etc., are drafted using the method described above. Except for the fractional slope ratios, the draftsperson will measure squares equal to the first number in the ratio, then will measure squares equal to the second number in the ratio, and finally will connect the two points diagonally. When a fractional ratio is used in the design, reverse the process and measure the second number as though it were horizontal and the first number as though it were vertical. Therefore, the $\frac{1}{6}$:1 ratio is measured one square horizontally to six squares vertically and the points are connected. This slope represents a steep condition for highway construction and will only be found in rock cuts.

COMPUTER PRINT-OUT OF SLOPE ELEVATIONS

Precise mathematical ground break-point elevations are not required on the plans prepared by consulting engineers; however, subsequent to bid letting, the successful contractor is usually required to furnish and receive approval of these computations before survey layout and construction can commence. These computations are referred to as a "work-book" and are subcontracted to engineering or surveying firms employed by the contractor as their representative in all survey and engineering aspects of the project. These computations are office-generated using programmable computer software specifically designed for this purpose.

The print-out provides all the computed information required by the surveyor to slope stake and includes:

1. Centerline station
2. Offset distance from each proposed break to centerline
3. Rate of slope between proposed breaks
4. Computed elevations for each proposed break point

When all this mathematical information is assembled in a small hand-held booklet, the surveyor no longer needs to carry the full or quarter-size drawings.

FIELD DATA REQUIRED BEFORE SLOPE STAKING

Before slope staking commences, there are certain historical methods used by the surveyor to expedite this process. These are outlined below:

1. Prepare a list of centerline stations and the approximate construction limit for each.[1] Use the cross-section sheets to scale this information, or better yet, use the computed work-book.
2. During the preliminary field survey, locate "work-hubs" within 10 ft or more of these estimated limits.
3. Complete an elevation level-loop using each of the work-hubs as "turning points," noting the elevation, distance from centerline, and station on the back of the work-hub guard stake.[2] The setting of work-hubs and measuring distances in this preliminary

1 Construction limits are considered the outer-most land area where the contractor will operate. Clearing limits are the outer-most vegetation growth that requires removal before construction.

2 Turning points, as referred to in a level loop, are ground points where an elevation is observed as the surveyor enters and exits the turn. Readings observed at turning points are used in the computation to determine mathematical accuracy of the level loop, while side shots are one-time isolated elevations that cannot be included in the mathematical check, and are therefore subject to error without means of verification.

stage eliminates the need to make long measurements from centerline when slope staking begins. All proposed ground breaks shown in the work-book are then related to the work-hub, rather than the centerline.

4. Enter the field with a copy of the construction plans, full, or quarter-size, or if available, use the slope stake work-book.
5. Prepare or purchase copies of slope-stake note taking forms that expedite the process and that later can be used as support data for your computations.[3] Most slope-stake forms (see Figure 10.3) provide space to duplicate the cut and fill information supplied to the contractor. However, if space is not provided on the form, change the form or keep a secondary tabulation of data. Throughout your career in surveying, and especially during slope staking, you will often be required to support your conclusions should disputes arise.

CONSTRUCTION LIMITS AND WORK-HUBS

The sample cut cross section (Figure 10.4) shows the location of the construction limit, both sides of the highway, and the location of the work-hubs set prior to slope staking. The remaining figure typically identifies the centerline-of-survey, profile grade elevation, the subgrade point, existing ground, rates of proposed slope, pavement width, and pavement cross-slope data. Usually, the pavement cross-slope data will be found on the typical and superelevated drawings, previously discussed in Chapter 6. Remember, the work-hub is traditionally set 10 ft or more beyond the construction limit, set perpendicular (90°) to centerline tangents, and set concentric (radial) to centerline curves. The work-hub acts (vertically) as the bench elevation from which the slope stake cut or fills are computed and also acts (horizontally) as the control for measuring distances to each break point, which is shown on the slope stake card for the contractor.

Equipment operators, after construction commences and the initial survey stakes are destroyed, have no centerline from which to reference their movement, and therefore, the work-hub is the source from which all construction is oriented. Work-hubs are traditionally located on both sides of the highway and occasionally are color coded (even stations one color, half stations another color) to prevent the hub on one side from being misaligned with another station on the opposite side while grading is in process. Line of sight, between work-hubs on both sides of the roadway, is used by the grade foreman to help direct the earth-moving equipment as the side slopes are being developed.

CONVERTING INCHES PER FOOT TO FEET PER FOOT

Earlier in the chapter, we reviewed the relation of side slope and ditch ratios (2:1, 3:1, etc.) and the drafting method used to develop these slopes. A careful review of Figure 10.4 shows additional pavement and shoulder cross slopes, designed in inches per foot but converted to feet per foot for easier elevation computations. Notice the pavement cross slope is $\frac{3}{16}$ in/ ft, and the shoulder cross slope is $\frac{1}{2}$ in/ per ft. To reduce this design criteria to more manageable numbers, we usually convert the $\frac{3}{16}$ in to a number representing the slope in 1 ft. This is done in two steps:

3 A sample slope staking form with "heading" is supplied in Part 4, Appendix.

Bench	Mark	Elev. =		Station or B L	"Plan" Data	Trial (4) (slope)	Trial (3) (slope)	Trial (2) (slope)	Trial (1) (slope)	Work-Hub
B - S	H. I.	F - S	Elev.		Center-line distance to plan break	______	______	______	______	Center-line distance to Work-Hub
1)										
2)										
3)										
4)										
Place	paddle	info.	here		Elevation					Elevation

Bench	Mark	Elev. =		Station or B L	"Plan" Data	Trial (4) (slope)	Trial (3) (slope)	Trial (2) (slope)	Trial (1) (slope)	Work-Hub
B - S	H. I.	F - S	Elev.		Center-line distance to plan break	______	______	______	______	Center-line distance to Work-Hub
1)										
2)										
3)										
4)										
Place	paddle	info.	here		Elevation					Elevation

Bench	Mark	Elev. =		Station or B L	"Plan" Data	Trial (4) (slope)	Trial (3) (slope)	Trial (2) (slope)	Trial (1) (slope)	Work Hub
B - S	H. I.	F - S	Elev.		Center-line distance to plan break	______	______	______	______	Center-line distance to Work-Hub
1)										
2)										
3)										
4)										
Place	paddle	info.	here		Elevation					Elevation

B L = Baseline; B - S = Backsight; F - S = Foresight; H. I. = Height of Instrument;

FIGURE 10.3 Slope stake form

1. Dividing the denominator of the fraction (16) into the numerator (3), such as: 3 ÷ 16 = 0.1875
2. Dividing this result by 12 (number of inches in a foot): 0.1875 ÷ 12 = 0.0156 ft/ft. This is referred to as feet per foot, or the pavement is changing grade at a rate of 0.0156 ft per each horizontal foot. Engineers and surveyors use this type of cross-slope mathematics to determine elevations when portions of 1 ft, or many horizontal feet are being graded for the purpose of simplifying fractional rates.

Example:

Compute the elevation at the opposite edge of pavement, using a pavement width of 24 ft, where the cross-slope rate is 3⁄16 in/ft. The result is 0.0156 ¥ 24 = 0.3744 ft. The outside edge of pavement, in this example, is 0.37 ft lower than the profile grade of the pavement in crown position.

Usually, maximum superelevation cross-slope rates are designed at 1-in/ ft (0.0833), and as a means of comparing this maximum slope with the previous cross-slope rate of 3⁄16 in/ ft (0.0156), we need only to multiply the rate (0.0833 ft/ ft) by the pavement width (24 ft). The result is that the edge of pavement is 2.00 ft lower or higher than the profile grade. Contrast this result (2.00 ft) with the previous example, where the edge of pavement was (0.37 ft) lower than the profile grade, and you can recognize the magnitude of full superelevation in a 24-ft wide pavement. Cross-slope rates, whether on pavement, storm drains, or other sloping objects, can be computed quite easily using this conversion method.

Convert inches/foot to feet/foot

3⁄16 in/ft

$16\sqrt{3} = 0.1875$

$12\sqrt{0.1875} = 0.0156$ ft / ft

$0.0156 \times 20 = 0.31$ ft

½ in/ft

$2\sqrt{1} = 0.5$

$12\sqrt{0.5} = 0.0416$ ft / ft

$0.0416 \times 10 = 0.42$ ft

A	=	Elev 600.00	centerline
B	=	Elev 599.69	20 ft Rt.
C	=	Elev 599.27	30 ft Rt.
D	=	Elev 596.27	36 ft Rt.
E	=	Elev 596.27	40 ft Rt.
F	=	Elev 606.27	60 ft Rt.
G	=	Elev 606.77	72 ft Rt.
H	=	Elev 625.00	108 ft Rt.
J	=	Elev 627.75	115 ft Rt.

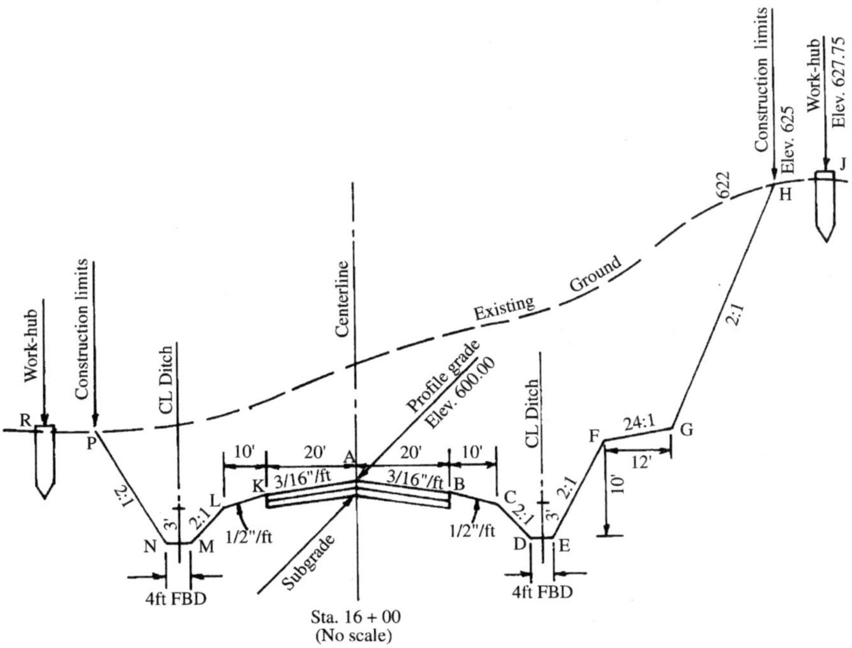

FIGURE 10.4 Slope staking the mainline

NOTE KEEPING

Construction staking, particularly slope staking, generally requires a three-party crew. Two of the members measure distances, hold rods for elevations, and carry wood products and equipment. The surveyor's responsibility is to operate the leveling instrument, carry the cross-section plans or computation work-book, and compute the slope-stake location. A two-party slope-stake crew, unless the project is small, is usually not cost effective, since the surveyor is forced to remain near the vicinity of a single "station" while assisting with the measuring. Slope staking is generally expedited when the leveling instrument is situated so that several observations and slope stakes can be set from a single instrument location.

In the following example, assume an elevation level loop has been completed and the elevation of the work-hub at station 16 + 00 is 627.75. The surveyor sets up the level, observes a backsight (B-S) by placing the rod on the work-hub, and in this example, reads 4.95 ft on the level rod. B-S readings are always added to bench mark elevations to obtain the height of instrument (HI), and therefore, the instrument eye level is elevation 632.70 (627.75 + 4.95 = 632.70). Commence entering the slope-stake data on the form (below) by filling in the H.I. elevation, the B-S reading, the station, and the work-hub elevation and distance.

Bench	Mark	Elev. =		Station or BL	"Plan" Data	Trial (4) (slope)	Trial (3) (slope)	Trial (2) (slope)	Trial (1) (slope)	Work-Hub
B - S	H. I.	F - S	Elev.	16 + 00 CL	Centerline distance to plan break	____	____	____	____	Center-line distance to Work-Hub 115' Rt.
1) **4.95**	**632.70**									
2)										
3)										
4)										
Place	paddle	info.	here		Elevation					Elevation **627.75**

At this point in the development of our slope-stake form, the next entry originates from the cross-section sheet or the work-book where the proposed ground breaks are shown. The slope stake (often called the "catch-point") is being set at the outer limits of construction, and therefore, the next "plan" data to be entered on the form is the proposed ground break closest to the catch-point. Refer to Figure 10.4, where the closest ground break to catch-point H is identified as point *G*. Point *G* is located 72 ft right of the centerline, as noted in the tabulation and which can be computed by adding the various horizontal distances from point *A* to point *G*. The elevation of point *G* is 606.77 (rounded to 606.8) and can be computed by using the data in Figure 10.4, commencing at the profile grade elevation (600.00) on centerline and adjusting as you add or subtract vertical distances toward point *G*.

In the form that follows, we have entered, in the shaded area, the distance and elevation for point *G*.

Point *H*, the catch-point, while not yet computed, is subject to the mathematical relation between the existing field elevations and plan data. You will notice in the following form, that the various trials are divided into three portions, (1), the upper part refers to the trial number and the rate of engineered slope, (2), the middle part refers to the trial distance from centerline and the elevation of the test, and (3), the lower part shows the comparison between the horizontal and vertical distances.

Bench	Mark	Elev. =	627.75	Station or BL	"Plan" Data	Trial (4) (slope)	Trial (3) (slope)	Trial (2) (slope)	Trial (1) (slope)	Work-Hub
B - S	H. I.	F - S	Elev.	16 + 00 C L	Centerline distance to plan break 72' Rt.					Center-line distance to Work-Hub 115' Rt.
1) **4.95**	**632.70**									
2)										
3)										
4)										
Place	paddle	info.	here		Elevation **606.8**					Elevation **627.75**

Bench	Mark	Elev. =	627.75	Station or BL	"Plan" Data	Trial (4) (slope)	Trial (3) (slope)	Trial (2) (slope)	Trial (1) (slope)	Work-Hub
B - S	H. I.	F - S	Elev.	16 + 00 C L	Centerline distance to plan break **72' Rt.**				Part 1	Center-line distance to Work-Hub **115' Rt.**
1) **4.95**	**632.70**									
2)										
3)										
4)										
Place	paddle	info.	here		Elevation **606.8**				Part 2 / Part 3	Elevation **627.75**

Each field test commences by determining the elevation of the estimated point (placing the level rod and reading a foresight), followed by measuring the distance from the work-hub to this test point. The form provides immediate horizontal and vertical comparisons between any two points. Trial number 1 was made by measuring 5 ft toward the centerline, and therefore, the test point is 110 ft from centerline. The rod reading at this ground point is 7.70 ft, and is called a "foresight" (F-S). Foresight readings are always subtracted from the H.I. and the ground at this location is elevation 625.00. When the correct mathematical ratio has been attained, the surveyor flags the point, placing wood lath, noting the station, offset distance, and elevation on the back of the paddle (side away from centerline), and usually attaches blue flagging.[3]

3 Blue flagging generally represents the zero point where construction begins. Offset stakes (5 to 10 ft) are placed behind the zero point, and the contractor's construction information is attached to the offset stake. Yellow flagging is usually attached to offset stakes but might vary according to individual state requirements.

Bench	Mark	Elev. =	627.75	Station or BL	"Plan" Data	Trial (4) (slope)	Trial (3) (slope)	Trial (2) (slope)	Trial (1) (slope)	Work-Hub
B - S	H. I.	F - S	Elev.	16 + 00 C L	Centerline distance to plan break **72' Rt.**	______	______	______	(1) ______ (2:1)	Center-line distance to Work-Hub **115' Rt.**
1) **4.95**	**632.70**	**7.70**								
2)										
3)										
4)										
Place	paddle	info.	here		Elevation **606.8**				**110'** **625.00**	Elevation **627.75**
									38' **18.2'**	

To make the final comparison for trial test number 1, horizontal and vertical differences are placed on the slope-stake form and the slope "ratio" is compared. In addition to the notations on the form above, please note the mathematics below:

Horizontal

110 ft from centerline to test point 1
72 ft from centerline to control point G

Difference = 38 ft

Vertical:

Elevation at test point 1 = 625.00
Elevation at control point G = 606.8 (rounded)

Difference = 18.2 ft

The backslope ratio from points *G* to *H* is designed at a 2:1 ratio; therefore, trial point 1, with a horizontal difference of 38 ft, is greater than twice the vertical difference (18.2 ft × 2 = 36.4 ft).

Therefore, test trial number 1 does not satisfy the engineering requirement of a 2:1 slope and, if staked, would place an unnecessary excavation burden of an extra 1.6 ft of material on the contractor.

As noted on the form below, two additional trials were made. Test trial number 2 shows a horizontal difference between point *G* and the catch-point of 33 ft, with a vertical differ-

Bench	Mark	Elev. =	627.75	Station or BL	"Plan" Data	Trial (4) (slope)	Trial (3) (slope)	Trial (2) (slope)	Trial (1) (slope)	Work-Hub
B - S	H. I.	F - S	Elev.	16 + 00 C L	Centerline distance to plan break **72' Rt.**	______	(3) ______ (2:1)	(3) ______ (2:1)	(1) ______ (2:1)	Center-line distance to Work-Hub **115' Rt.**
1) **4.95**	**632.70**	**7.70**	**625.00**							
2)		**10.70**	**622.00**							
3)		**7.70**	**625.00**							
4)										
Place	paddle	info.	here		Elevation **606.8**		**108'** **625.00**	**105'** **622.00**	**110'** **625.00**	Elevation **627.75**
							36' **18.2'**	**33'** **15.2'**	**38'** **18.2'**	

ence of 15.2 ft. The ratio of 2:1 has not been met under these conditions ($2 \times 15.2 = 30.4$) and therefore, if approved, the contractor would be commencing the cut 2.6 ft too close to the centerline. Test trial number 3 shows a horizontal difference of 36 ft, and a vertical difference of 18.2 ft. This point is acceptable as a catch-point and is shown boxed and shaded in the form below. For field purposes, the correct ratio is usually circled on the form.

A slope stake located within 0.40 ft of the exact calculated ratio is acceptable as a commencing point, however, a more refined location should be attempted when the excavation is in rock.

SLOPE-STAKE INSTRUCTIONS TO THE CONTRACTOR

After locating the slope-stake point for the right side at centerline station 16 + 00, the lath at this catch-point is set (blue flagging attached) with yellow flagging attached to the work-hub. The surveyor now prepares the construction information to be supplied to the contractor. This information (notation) should be noted using a universal method that is understood by the engineers, state highway personnel, and the contractor. The sample paddle (Figure 10.5) represents a universally accepted method of notation, originally conceived during the peak construction of the Interstate Highway System and now regarded as standard practice. (Church, 1956–1970).

The notation stake (paddle) must have sufficient information to horizontally and vertically locate each proposed ground break as designed in the plan. No additional grading information should be necessary until the approximate roadway subgrade elevations are attained. At this time, the contractor will request the centerline to be relocated and hubs (blue tops) set on the edge of pavement at each station.[4]

The subgrade of a proposed highway improvement is the top earthen grade directly under the roadway base course. Chapter 6 covered the information required for certain base and pavement materials, including thickness, width, and other pavement items that remain uniform (typical) throughout the improvement. Finished grades are computed and shown on the engineer's plan, while subgrade elevations are generally the surveyor's responsibility and are calculated on larger projects when the work-book is prepared.

Review the sample-slope stake paddle (Figure 10.5) where the first entry on the paddle is the station and offset distance to centerline. The work-hub elevation, although not visible, is placed on the back of the paddle. The second entry, the left half of the paddle, lists OG (original ground) and the right half of the paddle lists C– (for cut) and a number based on the elevation difference between the work-hub and the ground at the catch-point.

Farther to the right is a number in parentheses (or circled) that denotes the distance from the work-hub to the catch-point. These two items are important; if the ground is altered in elevation through excess construction work, the surveyor has proof that the original slope stake was set correctly.

Each subsequent item, on the left half of the paddle, describes the destination from the work-hub to the next break point, the actual minus – digit (for cut) or plus + digit (for fill), and the distance (in parentheses or circled) to the point from the work-hub. The contractor's grade supervisor will follow the cut or fill progress from break point to break point, not always returning to the work-hub for measurements. If the contractor's supervisor is confident that the location and elevation of the previous point is accurate, the supervisor

4 Blue top hubs (hubs with blue paint or crayon) are set at the edge of pavement or at 2-ft offsets and are graded to the subgrade elevation (top of dirt) for closer plan elevations before placing base material.

16 + 00

115' Rt.

OG (original ground)	C – 2.8 (7)
2:1 to bk. of bench	C – 21 (43)
24:1 to fnt. of bench	C – 21.5 (55)
2:1 to toe of ditch	C – 31.5 (75)
4-ft flat bottom ditch	C – 31.5 (79)
2:1 to edge of shldr	C – 28.5 (85)
1/2"/ ft to E/Pav'mt	C – 28.1 (95)
3/16"/ ft to FG @ C.L.	C – 27.8 (115)

FIGURE 10.5 Slope stake paddle notation

can subtract distances noted in the parentheses (or circled), as well as relate elevations, and proceed directly to the next break point. It would be unwise, however, to construct many break points without returning to the work-hub for verification or calling upon the survey crew for their conformation.

Note: The paddle information (Figure 10.5) in this example is based on the cross section in Figure 10.4 and therefore, as we continue interpreting the information on the paddle, we note the third entry identifies a horizontal distance of 43 ft, a cut (C–) of 21 ft, together representing a coordinate to the back (bk) of the 24:1 bench (point *G*) in Figure 10.4. The 43 ft notation represents the distance from the *work-hub* to the back of the bench, and the cut of 21 ft represents the vertical distance from the *work-hub* to the ground level at the back of the bench.

Caution: The 2:1 earthen side slope cut does *not* commence at the work-hub but follows the instructions in the second entry on the paddle, which identifies that the catch-point was found 7 ft from the work-hub, and that the elevation of the catch-point was 2.8 ft below the work-hub.

As we continue to the fourth entry on the paddle, the distance noted is 55 ft, and the cut is 21.5 ft, together forming the coordinate to the front (fnt) of a 24:1 bench (point *F*), in Figure 10.4.

When this excavation is completed, the elevation of the 24:1 bench is between 21 and 21.5 ft below the work-hub. The grade supervisor can either elect to use point *F* as the horizontal and vertical control for each additional ground break or can call for additional stakes to be set by the survey crew.

If the survey crew is called, they will probably be requested to relocate a new centerline and show rough cuts to the subgrade by markings on the centerline lath.

The fifth and sixth entries on the paddle describe distances of 75 and 79 ft to the back (bk) and front (fnt), respectively, of the 4-ft flat-bottom ditch, with cuts of 31.5 ft. Paddle entry number 7 shows a distance of 85 ft to the outside of a shoulder, with a cut of 28.5 ft. Before this finished edge of shoulder is graded, however, the surveyor will probably be requested to return to the site and regrade the centerline and edge of shoulder subgrade. Paddle entries number 8 and 9 identify elevations and distances to the edge of shoulder and centerline, but are never used for grading purposes by the grade supervisor. The centerline will be located by the survey crew using centerline reference ties, and the edge of shoulder will be measured from this centerline. Remember, the distance to the edge of shoulder subgrade and the elevation will usually be computed as part of the materials supplied in the work-book.

Therefore, the finished shoulder and pavement elevations are constructed using the combined mathematical information found on the typical sections, cross sections, pavement details, superelevation tables, and plan and profile sheets. In other words, all detailed plans and elevations must be reviewed thoroughly before the finished grades are set.

DUPLICATE INFORMATION ON THE SLOPE STAKE FORM

Whenever slope-stake information is attached to the offset paddle, in preparation for excavation or embankment, the surveyor must keep a duplicate of these instructions. Refer to the sample form below, where the shaded area is available for this data. Take advantage of the duplication; it can save you much concern at a later time, when the grades are in question.

Bench Mark Elev. =				Station or BL	"Plan" Data	Trial (4) (slope)	Trial (3) (slope)	Trial (2) (slope)	Trial (1) (slope)	Work-Hub
B - S	H. I.	F - S	Elev.		Centerline distance to plan break					Center-line distance to Work-Hub
1)										
2)										
3)										
4)										
Place paddle info. here					Elevation					Elevation

CHAPTER 11
SIMPLE CURVES

INTRODUCTION

The purpose of this chapter is to help readers understand curves and develop the ability to compute solutions to problems relating to them, as well as to develop the skills necessary to field-locate these curves.

Straight-line surveying, usually associated with rural property surveys and older city lots, employs the field method of observing angles with a surveyor's instrument and measuring distances by tape or electronic means. Curved-line surveying is accomplished much the same way, substituting chords of the curve to produce the arc. Since chords represent the straight-line distances between any two points on the curve, by locating (mathematically or in the field) the end points of multiple chords, the arc of the curve is produced.

The early surveying profession was seldom involved with curved lines as we know them today. When it became necessary to survey the perimeter of rivers, streams, and lakes, meander lines were traversed in the field.[1] At each meander break point the distance from the meander line to the object was measured. Using this method to depict curved topographic features was common, but the size or area of a property bounded by such a feature was always questionable.

Modern solutions to curved lines have been used since the early 1900s and especially since the Interstate Highway System began in 1921. Precise property line locations, where the boundary followed the centerline of highways, as well as curved and spiraled offset right-of-way lines, have forced the introduction of complete and accurate "curve data."

This chapter will systematically introduce the "parts" of a curve and provide equations that will solve for each part. There are several mathematical methods used to compute different parts of a curve; however, this chapter proposes one method, and occasionally discusses two or more solutions.

Educational prerequisites can be satisfied with high school-level geometry, algebra, and trigonometry. College classes in these subjects will certainly offer an advantage, but the material is quite easily understood.

---

1 Meander lines are traverses run parallel to the object(s) being located and where distances are measured to the object from each break in the meander line. This permits curved objects to be plotted in a more accurate location.

This chapter concentrates on methods of solving problems when programmable computers are not available, but it also covers some personal computer methods and results. All solutions can be handled without the use of special programmable software. An inexpensive engineering hand-held calculator, designed for trigonometric relations, and a background in geometry will advance you to a level where you can understand horizontal curves in the office or in the field.

We will first review the geometry and trigonometry associated with curved lines and then show methods for solving the various parts of the curve.

TWO TYPES OF HORIZONTAL CURVES

There are two types of horizontal curves which require our attention: (1) the highway curve, and (2) the railroad curve.

We will see differences between these two curves on the next few pages. But first, you will need to review the curve "symbols" and vocabulary in Figure 11.1.

ARC DEFINITION

Highway curves, by definition, are those circular curves where distances are measured along the arc of the curve. A measure of whether the curve is sharp or slight is referred to as the "degree of curve." A 1° curve, by definition, shows a deflection angle at the intersection of the two tangents (and a radius angle at the center of the circle) of 1° for every 100 ft along the arc of the curve.

Example:

The following method outlines the procedure for determining an approximate degree of curve in the field: Set up a surveyor's instrument at the central point of a circle and turn an angle of 1°. This 1° angle will separate two points, measured around the arc, by 100 ft. Assuming you can physically occupy the central point of the circle just referred to, what would be the length of radius needed to satisfy these conditions? Suppose also, that you occupy the central point of another circle and, with the instrument, observe an angle of 2°. This 2° angle also separates two points, measured along the arc, by 100 ft, then this curve is a 2° curve. And what is this second curve's radius?

Answers to the above questions can be computed from the following equation:

$$R = \frac{360°}{D_c} \times \frac{100'}{2\pi} \tag{11.1}$$

Where:

R = radius
D_c = degree of curve
π = 3.141592654

Determine:

The radius of a 1° curve. By substitution:

$$R = \frac{360}{1} \times \frac{100}{6.2831853}$$

$$R = \frac{36000}{6.2831853}$$

$$R = 5729.58 \text{ ft}$$

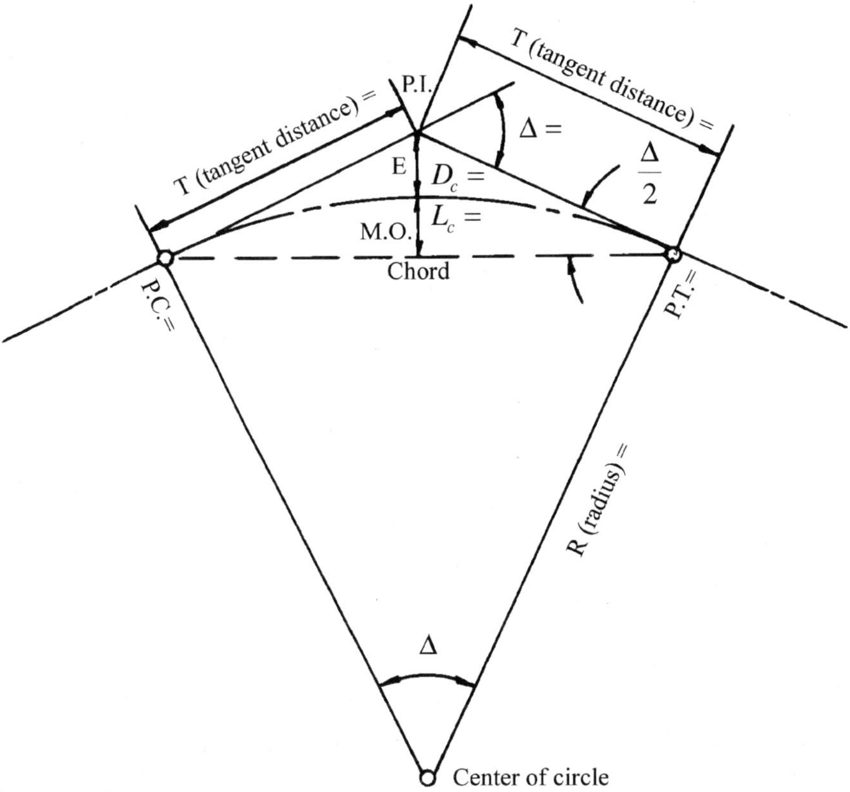

Δ = (delta) angle between the two local tangents, and central angle subtended by the curve length	L_c = length of curve
D_c = (degree of curve) angle subtended by an arc length of 100 ft	T = tangent distance from begin or end of curve to the P.I. (point of intersection)
P.C. = (point of curve) end of tangent and beginning of curve	R = radius
P.T. = (point of tangent) end of curve and beginning of tangent	E = (external) distance from P.I. to the middle of the curve
Ch = (chord) straight-line distance between begin and end of curve	M.O. = (middle ordinate) distance from the middle of the curve to the chord line
P.I. = (point of intersection) point where the two tangent lines intersect	$\frac{\Delta}{2}$ = (deflection angle) angle between each local tangent and the chord

FIGURE 11.1 Curve symbols

Determine:
The radius of a 2° curve. By substitution:

$$R = \frac{360}{2} \times \frac{100}{6.2831853} \tag{11.1}$$

$$R = \frac{36000}{12.5663706}$$

$$R = 2864.79 \text{ ft}$$

The above two computations show that our instrument in the original question had to be located at a distance of 5729.58 ft from the arc before a 1° radius angle could be observed which separates two points on the arc by 100 ft. And the instrument had to be located 2864.79 ft from the arc for a 2° radius angle to separate two points on the arc by 100 ft.

An even more simplified version of Eq. (11.1), with all the constants combined into one number, looks like this:

$$R = \frac{5729.58}{D_c} \tag{11.2}$$

CHORD DEFINITION

Railroad curves, by definition, are those circular curves where distances are measured along the chord of the curve. A railroad curve with a radius angle at the center of the circle of 1° will have a 100-ft long chord. And we call this a 1° railroad curve. However, a precise calculation of the degree of curve for a railroad curve is always slightly smaller than its radius angle. For a curve as slight as a 1° curve, however, the degree of curve would still calculate the same as the radius angle.

Most of the mathematics relating to *chord* definition can easily be computed by trigonometric means, taking advantage of the well-known "trig" relationships in 90° triangles. Two examples of trig solutions are shown in Figure 11.2, which depicts the difference between *arc* and *chord* definition.

ARC AND CHORD DEFINITIONS

Determine the length of radius for each of the sketches in Figure 11.2 by arc and chord definition.

Notice that the difference in the radii (resulting from using the two curve definitions) increases as the degree of curve increases. In the examples just given, the 1° curve shows a difference in radius length of only 0.07 ft, whereas the 7° curve shows radius length differences of 0.51 ft.

Do not be overly confused by the two curve definitions. Unless you are surveying a railroad centerline, or a curved right-of-way line along a railroad, you will probably never use a chord definition for computing curves. Before we leave the *chord* definition, however, we must return to the chord definition examples on the previous page to introduce one of the additional equations you will need in order to calculate the other parts of a circular curve. You probably are wondering how we arrived at the length of the curve of

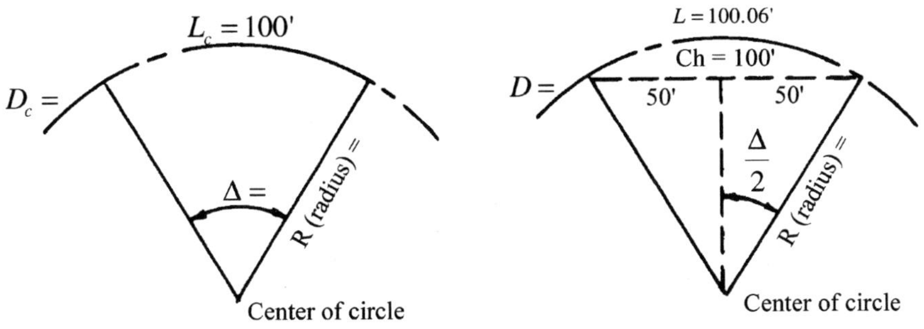

Arc definition	Chord definition
Assume: $\Delta = 1°$ $D_c = 1$	Assume: $\Delta = 1°$ $D = 1°$
Using Eq. (11.2): $R = \frac{5729.58}{1}$ $R = 5729.58$ ft	Using trigonometry: $R = \left(\sin\alpha = \frac{50'}{X}\right)$ where: $\alpha = \frac{\Delta}{2}$ $R = \left(\sin 0°\ 30' = \frac{50'}{X}\right)$ $R = 5729.65$ feet

Assume: $\Delta = 7°$; $D_c = 7°$	Assume $\Delta = 7°$
Using Eq. (11.2):	Using trigonometry: (Degree of curve calculated on the next page)
$R = \frac{5729.58}{7}$ $R = 818.51$ ft	D = 6° 59' 44" $R = \left(\sin\alpha = \frac{50'}{X}\right)$ $R = \left(\sin 3°\ 30' = \frac{50'}{X}\right)$ $R = 819.02$ ft

FIGURE 11.2 Arc and chord definition

100.06 ft in the 7° railroad curve example or how to calculate the degree of curve for this example. We already know how to calculate the degree of curve, from Eq. (11.1) or the simplified Eq. (11.2). The computation for the 7° railroad curve is as follows:

Known:

$$R = 819.02 \text{ ft}$$
$$\text{Ch} = 100.00 \text{ ft}$$

Where:

D = degree of curve

$$D = \frac{5729.58}{R} \tag{11.2}$$

D = 6.9956533° (shown as decimal parts of a degree)

D = 6° 59' 44" (converted to degrees, minutes, and seconds)

Then, with the degree of curve known, the length of the 7° railroad curve can be computed using the following new equation:

$$L = \frac{\Delta \times 100}{D} \tag{11.3}$$

$$L = \frac{7 \times 100}{6.9956533}$$

$$L = 100.06 \text{ ft}$$

From this text point onward, all the circular curves will be based on the *arc* definition used for highways. Most of the remaining equations you will need to solve for other parts of circular curves are listed on the next few pages, with examples shown using each equation.

SIMPLE CURVE EQUATIONS

Assume:

$$D_c = 6°$$
$$\Delta = 6°$$
$$L_c = 100 \text{ ft}$$
$$\pi = 3.141592654$$

1. $R = \dfrac{360°}{D_c} \times \dfrac{100}{2\pi}$ (11.1)

$= \dfrac{360°}{6} \times \dfrac{100}{6.2831853}$

$= 954.93$ ft

1a. $R = \dfrac{5729.58}{D_c}$ (11.2)

$= 954.93$ ft

2. $T = R \times \tan \dfrac{\Delta}{2}$ (11.4)

$= 954.93 \times \tan 3°$

$= 50.045$ ft

3a. $\Delta = \dfrac{L_c \times D_c}{100}$ (11.3)

$= \dfrac{100 \times 6}{100}$

$= 6°$

3b. $L_c = \dfrac{\Delta \times 100}{D_c}$ (11.3)

$= \dfrac{6 \times 100}{6}$

$= 100$ ft

3c. $D_c = \dfrac{\Delta \times 100}{L_c}$ (11.3)

$= \dfrac{6 \times 100}{100}$

$= 6°$

4. Ch $= 2R \times \sin\dfrac{\Delta}{2}$ (11.5)

$= 2(954.93)(\sin 3°)$

$= 1909.86 \times 0.052335965$

$= 99.954$ ft

5a. $E = T \tan \dfrac{\Delta}{4}$ (11.6)

$= 50.045' \times 0.026185922$

$= 1.31$ ft

5b. $E = \dfrac{R}{\cos\dfrac{\Delta}{2}} - R$ (11.7)

$= \dfrac{954.93 \text{ ft}}{\cos 3°} - 954.93$ ft

$= \dfrac{954.93 \text{ ft}}{0.998629535} - 954.93$ ft

$= 1.31$ ft

6. M.O. $= R \times \text{Ver}\dfrac{\Delta}{2}$ (11.8)

$= 954.93 \text{ ft} \times \left(1 - \cos\dfrac{\Delta}{2}\right)$ (see the alternate method below)

$= 954.93 \text{ ft} \times (1 - \cos 3°)$

$= 954.93 \text{ ft} \times (1 - 0.998629535)$

$= 954.93 \text{ ft} \times 0.001370465$

$= 1.31$ ft

(alternate method of solving for the versine when versine tables are not available)

$$\text{Ver}\,\frac{\Delta}{2} = (1 - \cos\frac{\Delta}{2})$$

You should notice that the external distance and the middle ordinate distance, in this example, is exactly the same. This situation will exist when the degrees of curve are rather flat. A comparison of this phenomenon is tabulated below:

Assume for test one:	**Assume for test two:**	**Assume for tat three:**
$D_c = 7°$	$D_c = 8°$	$D_c = 15°$
$\Delta = 7°$	$\Delta = 8°$	$\Delta = 15°$
$L_c = 100$ ft	$L_c = 100$ ft	$L_c = 100$ ft
R = 818.51 ft	$R = 716.20$ ft	$R = 381.97$ ft
$T = 50.06$ ft	$T = 50.08$ ft	$T = 50.29$ ft
Computed: $E = 1.53$ ft	Computed: $E = 1.75$ ft	Computed: $E = 3.29$ ft
Computed: M.O.= 153 ft	Computed: M.O. = 1.74 ft	Computed: M.O.= 3.27 ft

HIGHWAY STATIONING

See Chapter 4 for detail information concerning highway stationing.

STATION EQUATIONS

See Chapter 4 for detail information concerning station equations.

FIELD TEST FOR DEGREE OF CURVE

You have solved for the radius of a curved line, and learned that the radius and its subtended angle determines the degree of curve. Then, how can a degree of curve be determined by simple field methods?

Most surveyors will at one time or another find it necessary to approximate an existing curve for the purpose of establishing that curve or establishing a concentric curve (often referred to incorrectly as parallel curves).

Rarely (except on interstate projects or state highway supervised projects) will the surveyor find a single curved roadway that follows precisely the same degree of curve. The age of the roadway and whether the roadway was originally monumented for right-of-way control quite often help the surveyor predict whether the centerline will have curve uniformity. Older highways within the county or township always present unique situations. So, as a practical method, the surveyor can field test for the degree of curve by using the following technique:

1. The curve must exist on the ground.
2. Locate a minimum of three points, each separated by 100-ft intervals (arc length)
3. Occupy the middle of the three points and observe the deflection angle between these three points.

This observed angle becomes the degree of curve. Further tests along the curve are recommended to locate any changes in the degree of curve and thus point out variations in the radius.

The above "field test" is supported geometrically by knowing that the normal layout of any length of curve involves deflecting an angle with the transit or total station equal to one-half the known delta (Δ) angle of that curve and measuring the chord distance to locate the other end.

In this situation, the field test must use *arc* distances of 100 ft in order to comply with the premise of arc definition, which states that the angle must subtend a *curved* distance of 100 ft.

When occupying the middle of three points and observing the deflection angle between the two chords, you produce the total subtended angle that defines the degree of curve. Refer to Figure 11.3 and assume local tangents are drawn through each of the three points.[2] In this example, an 8° angle, off the local tangent, is observed in both directions (back-station and upstation) and, therefore, the individual deflection angles off each local tangent, must be added together, and this angular amount is equal to the degree of curve.

2. The local tangent at any point on a curve is a line perpendicular to any radius through the point.

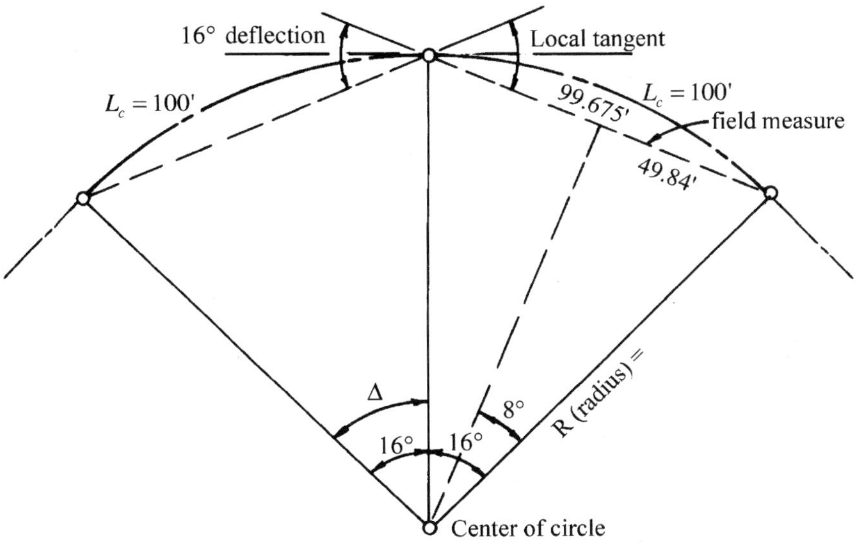

Compute the radius
(Knowing the $\Delta = 16°$)

$$R = \frac{5729.58}{D_c} \qquad (11.2)$$

$$R = \frac{5729.58}{16}$$

$$R = 358.098 \text{ ft}$$

Verify the computed radius
(Solve for the chord where $L_c = 100$ ft)

$$\text{Ch} = (2R) \times \text{Sin}\,\frac{\Delta}{2} \qquad (11.5)$$

$$\text{Ch} = 2(358.09)(\text{Sin } 8°)$$

$$\text{Ch} = 716.18 \times 0.139173101$$

$$\text{Ch} = 99.673 \text{ ft}$$

(Using trigonometry, solve for the radius)

$$\text{Sin } 8° = \frac{49.8375}{R}$$

$$R = 358.097 \text{ ft}$$

FIGURE 11.3 Field test for degree of curve

In our example, the total deflection angle (sighting along the chord to the opposite end of the first curve, and turning the angle to the chord at the end of the second curve) equals 16° and thus identifies the angle observed at the central point of a circle, which separates two points on a curve by an arc length of 100 ft. (See the explanation, as discussed earlier in the chapter, under "Arc Definition.")

REVIEW OF FIELD TESTING FOR DEGREE OF CURVE

Figure 11.3 illustrates the steps required to field test for a degree of curve. The individual steps are reviewed below:

1. Roughly locate the centerline of the curved roadway. If the curve is sharp, and the arc distance is difficult to obtain, simply reduce the increments that you use in your taping to lengths of 10 or 20 ft and continue around the center of the roadway setting points at 100 ft intervals.
2. Set the instrument at the middle of any three points and observe the angle of deflection between these three points. Figure 11.3 shows a deflection of 16°, therefore, the degree of curve is 16°. The radius can be computed using Eq. (11.1) or (11.2) as on the previous page. We also verified the results of that radius by using trigonometry, and that verification is reviewed below.
3. After the three points on the centerline are located, measure the *chord* length between points (straight-line distance between points). If the original measuring of 100 ft intervals around the arc was completed with any degree of precision, and if the curve has some degree of uniformity, you should measure chord distances between the three points that are somewhat equal. In Figure 11.3, the chord was measured at 99.68 ft. For verification purposes, drop a perpendicular from the center of this chord to the center of the circle. The result of this perpendicular creates a 90° triangle with the included angle (8°) known and the short leg of the triangle (49.84 ft) also known.

By computing the hypotenuse of the above triangle, which is equal to the radius of the curve, you can verify the length of the radius and thus verify that the field procedure for determining the degree of curve is valid. Remember, this is an approximate method for determining the degree of curve, and therefore, you should not expect the mathematics to produce identical results. There are two important reasons for this:

1. Your measurement around the arc of the curve is only approximate.
2. Your choice of the centerline of the existing roadway is also approximate.

The primary purpose is to quickly determine a degree of curve from which you can field-locate that curve, or if traffic will not permit this, you can locate a concentric (parallel) curve. This application can often be used when setting clearing limits or work-hubs off a busy side road that will later be improved but is not yet closed to traffic. The concentric curve allows the surveyor to remain at an approximate distance from the center of the roadway, and therefore, the distances to the clearing limits and/or work-hubs will remain within a reasonable tolerance. Regardless of the situation, this ability will prove valuable at some period in your career.

FIELD LAYOUT OF CURVES

The common practice of locating roadway curves will include one of the following:

1. Random curves along township or nonmonumented county roads
2. County, state, or federal highways, where engineering preceded the improvement

Random Curves

Random highway curves are those where the road probably developed without engineering design, or if designed in earlier years, the monumentation and reference ties of that road have long since been destroyed. To add to the problem, pavement reconditioning over the years could have completely altered the original curve alignment. However, if the roadway was designed, there will be engineering drawings that outline this design, and this information must be located. State highway road drawings are usually available in the appropriate district offices. County roadway plans are recorded in the county engineer's office and, as with the state plans, are always available for copying.

Township road plans, on the other hand, are occasionally recorded in the county engineer's office or available through the township trustees.

If the engineering drawings are available, the surveyor attempts to reconstruct the centerline, using all available monuments and references, being cognizant that all property acquisition was originally referenced to the centerline of survey. Therefore, the property boundaries are only as valid as the accuracy of the relocated centerline. If the roadway is being surveyed for a new design, several additional field procedures are required before the engineers can accurately compute and plot the old road in relation to the improvement. These field requirements are:

1. Estimate and field-locate the beginning and ending of each curve
2. Locate sufficient points along the curve so that the engineers can "office" compute a degree of curve that will most nearly fit this existing alignment

Highway curves that have been designed are those where the exact location of the "beginning" and "ending" of each curve has been computed, and all pertinent information relating to the curve is shown on a set of plans.

PRELIMINARY FIELD INVESTIGATION

Preliminary field investigation is the first step used to reestablish and field-locate the old alignment. This investigation includes locating the reference ties for the project, replacing the stations referred to in the references, reestablishing the "tangent lines," and finally setting the PIs. After the PIs are established, the surveyor locates the begin P.C. and end P.T. stations for each curve, followed by the intermediate points on the curve. Figure 11.4 shows the tangent lines on each end of a curve, the point of intersection, the degree of curve, the centerline stationing, and the location of one curve point to be used as an example of a single layout computation.

PREPARE SKETCHES FOR LAYOUT

Prepared drawings are always recommended before field layout, often eliminating simple errors that are committed when you try to envision the problem without the sketch. Geometric solutions, created by curves and spirals, are reduced to their simplest form when the geometric figure is drawn. Figure 11.4 is an example of a single curve-point location with a sketch.

After the centerline station is chosen, use Eq (11.3) $\left(\Delta = \dfrac{L_c \times D_c}{100}\right)$ to compute the delta angle between the P.C. or P.T. instrument setup and the point to be located. In this exam-

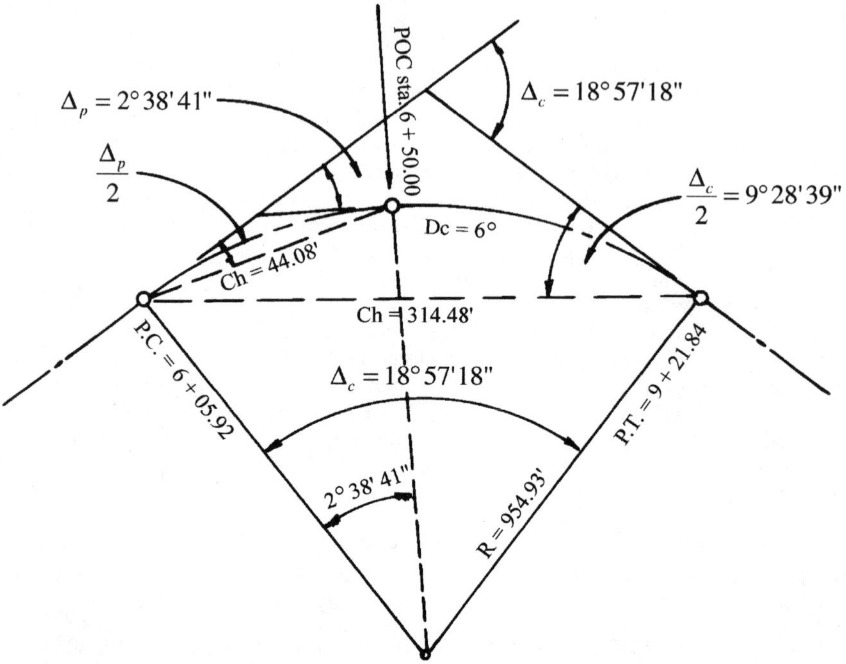

Locate station 6 + 50.00

$$\begin{aligned} POC &= 6 + 50.00 \\ PC &= 6 + 05.92 \\ L_c &= 44.08 \text{ ft} \\ \Delta_p &= \frac{44.08 \text{ ft} \times 6^\circ}{100} \\ \Delta_p &= 2^\circ\ 38'\ 41.2'' \\ \frac{\Delta_p}{2} &= 1^\circ\ 19'\ 20.5'' \\ Ch &= 2\ (954.93 \text{ ft})\ (\sin 1^\circ\ 19'\ 20.5'') \\ Ch &= 44.079 \text{ ft} \end{aligned}$$

Locate station 9 + 21.84

$$\begin{aligned} PT &= 9 + 21.84 \\ PC &= 6 + 05.92 \\ L_c &= 3 + 15.92 \\ \Delta_p &= \frac{315.92 \text{ ft} \times 6^\circ}{100} \\ \Delta_c &= 18^\circ\ 57'\ 18.7'' \\ \frac{\Delta_c}{2} &= 9^\circ\ 28'\ 39.3'' \\ Ch &= 2\ (954.93)\ (\sin 9^\circ\ 28'\ 39.3'') \\ Ch &= 314.48 \text{ ft} \end{aligned}$$

FIGURE 11.4 Locating a point on curve

ple the instrument setup is P.C. station 6 + 05.92. The computed delta angle of the point (Δ_p), locating station 6 + 50 is 2° 38' 41" as shown by the computations in Figure 11.4. By observing the geometry, you can determine that the delta angle of the point is not the layout angle of the point. Layout angles, by trigonometric definition, are equal to one-half of the delta angle. As support for this conclusion, observe the small triangle, where one leg of the triangle is made by the mainline tangent, the second leg of the triangle is the line representing the local tangent through station 6 + 50, and the third leg of the triangle is the chord (Ch) distance of 44.08 ft. If you subtract the delta angle of the point, which is 2° 38' 41", from a straight line (180°) at the P.I., the difference is 177° 21' 19". This angular difference becomes the large angle in the triangle.

From geometry studies, you know that all three-sided figures have a total of 180° within the triangle and, in this situation, the triangle is isosceles (a triangle with two equal sides) and also two equal angles. Each angle is therefore 1° 19' 20.5", and together all three angles equal 180° (177–21–19 + 1–19–20.5 + 1–19–20.5 = 180°)

Two equal angles remain within the triangle, and subtracting the large angle (177° 21' 19") from 180°, returns you to the delta angle of the point, which is 2° 38" 41". Each small angle within the triangle shares one-half of this total, or 1° 19' 21". Therefore, the layout angle to any point is equal to one-half of the computed delta.

As an example of field layout (Figure 11.4), use the following recommendations to locate the point on curve (POC) station 6 + 50.00. Setup the instrument (transit or total station) at P.C. station 6 + 05.92, back-sight along the mainline tangent with the horizontal plate set on 180°, return the instrument 180° horizontal back to zero on the plate, turn a deflection angle of one-half the delta (1° 19' 20.5"), measure the chord distance (44.08 ft), and set the POC. The instrument setup can also be handled by setting 0° on the horizontal plate, inverting the scope, back-sighting the mainline tangent, rotating the scope 180° vertically, and continuing the remainder of the process as outlined above.

MULTIPLE-POINT CURVE LAYOUT

Multiple-point curve layout is completed in a similar manner but requires individual deflection angle computations for each point to be located, chord computations for each point if the point is located from the beginning of the curve, or normal chord corrections if the point is set using a three-person crew (where two members are measuring as they move forward on the curve).

Curves can also be located by moving up on the centerline, occupying one of the previously established curve points, and continuing to observe deflection angles, setting additional stations (this will be discussed later in the chapter).

Quite often, multiple-point centerlines require replacement several times during the length of a construction project, and therefore, a horizontal "line-book" must be prepared that contains these curve computations. The line-book will eliminate duplicate sets of mathematics. Since curved roadways consume a considerable amount of field time. The surveyor must not add to this lengthy field procedure by recomputing the layout. One series of computations, well documented, is sufficient.

MULTIPLE-POINT CURVE COMPUTATIONS

The following steps will help organize the calculations for multiple curve locations.

1. Determine the first variable curve length from the P.C. station to the next quarter, half or full station (POC).
2. Calculate the delta of the point for the above POC.
3. Determine all successive equal curve lengths, such as 25 or 50 ft intervals.
4. Compute the representative delta of the point for the above chosen interval.
5. Determine the last variable curve length (distance from the last even station to the P.T.).

6. Prepare a summation of the computed delta angles to determine whether the sum is equal to the plan delta.
7. If, on the other hand, the angle summation represents deflection (layout) angles, they are by definition, equal to one-half of each computed delta, and therefore, the sum of the angles will be equal to one-half of the plan delta.

Using the suggested steps on the previous page, compute the curve deflection angles and prepare a field line-book showing the following: (1) stations, (2) deflection angles, and (3) chord distances for a multiple-point curve layout.

Refer to Figure 11.4 for the working drawing.

P.C. station = 6 + 05.92 $L_c = 315.92$ ft
P.T. station = 9 + 21.84 $D_c = 6°$, $R = 954.93$ ft

Compute 1st variable dist.	Compute 50-ft interval	Compute last variable dist.
POC = 6 + 50.00	POC = 7 + 00.00	P.T. = 9 + 21.84
$\text{P.C.} = \dfrac{6 + 05.92}{Lc = 44.08'}$	$\text{POC} = \dfrac{6 + 50.00}{L_c = 50.00'}$	$\text{POC} = \dfrac{9 + 00.00}{L_c = 21.84 \text{ ft}}$
$\Delta = \dfrac{44.08 \text{ ft} \times 6°}{100}$	$\Delta = \dfrac{50 \text{ ft} \times 6°}{100}$	$\Delta = \dfrac{21.84 \times 6°}{100}$
$\Delta = 2°\ 38'\ 41''$	$\Delta = 3°\ 00'\ 00''$	$\Delta = 1°\ 18'\ 37''$
$\dfrac{\Delta}{2} = 1°\ 19'\ 20.5''$	$\dfrac{\Delta}{2} = 1°\ 30'\ 00''$	$\dfrac{\Delta}{2} = 0°\ 39'\ 18.5''$

Field-book entries showing the early stations at the bottom of the table; they increase up the table.

Station	Deflection angle	Chord distance
P.T. = 9 + 21.84	9° 28' 39"	End of curve 21.84 ft
9 + 00.00	8° 49' 20" 49.99 ft	
8 + 50.00	7° 19' 20" 49.99 ft	
8 + 00.00	5° 49' 20" 49.99 ft	
7 + 50.00	4° 19' 20" 49.99 ft	
7 + 00.00	2° 49' 20" 49.99 ft	
6 + 50.00	1° 19' 20" 44.07 ft	
P.C. = 6 + 05.92	0° 00' 00"	Begin curve
Station	Deflection angle	Chord distance

CURVE DATA LINE-BOOK

The line-book, prepared above, uses three computations, the first variable distance, the 50 ft interval, and the last variable distance, to compute the various deflection angles. These computations are entered into your field-book (occasionally, office-prepared form sheet) to facilitate present and future field layout. The curve distance from P.C. station 6 + 05.92 to station 6 + 50.00 is 44.08 ft. The deflection angle to this point, computed using Eq. (11.3), is 1° 19' 20". The chord distance is 44.07 ft, computed using Eq. (11.5) and is inserted on the form between the two stations. The next layout station, and several beyond, are 50 ft intervals, where the deflection computation for each of these 50-ft stations is 1° 30' 00". Each single 50 ft deflection is added to the first, such as:

$$\begin{array}{r} 1^\circ\ 19'\ 20'' \\ +\ \underline{1^\circ\ 30'\ 00''} \\ =2^\circ\ 49'\ 20'' \end{array}$$

Stations 7 + 50, 8 + 00, 8 + 50, and 9 + 00 are all 50-ft intervals, and therefore, the deflection angle for a 50-ft distance must be added sequentially, such as:

$$\begin{array}{r} 2^\circ\ 49'\ 20'' \\ +\ \underline{1^\circ\ 30'\ 20''} \\ =4^\circ\ 19'\ 20'' \\ +\ \underline{1^\circ\ 30'\ 00''} \\ =5^\circ\ 49'\ 20'' \\ +\ \underline{1^\circ\ 30'\ 00''} \\ =7^\circ\ 19'\ 20'' \\ +\ \underline{1^\circ\ 30'\ 00''} \\ =8^\circ\ 49'\ 20'' \end{array}$$

The final variable deflection angle, from station 9 + 00.00 to the P.T. station 9 + 21.84 is 0° 39' 18.5". The addition of this angle will complete our entries in the line-book, such as:

$$\begin{array}{r} 8^\circ\ 49'\ 20'' \\ +\ \underline{0^\circ\ 39'\ 19''} \\ =9^\circ\ 28'\ 39'' \end{array}$$

To expedite the curve layout, the instrument person can remain at the beginning of the curve and turn off deflection angles, while the two other crew members are measuring chord distances, setting intermediate curve stations, and finally setting the end of the curve, P.T. station 9 + 21.84.

When the line of sight is broken between the beginning of the curve and any other point, the instrument can be moved to any usable position, noting the station, setting the instrument at 00°, back-sighting the beginning of the curve, inverting the scope, and turning-off an angle equal to the point where the instrument is situated. The scope is now oriented on the local tangent.

Continue locating each successive point by returning to the line-book and turning off the angle computed for each station beyond.

INSTRUMENT MOVE-UP WHEN LINE OF SIGHT IS BROKEN

As an example of instrument "move-up," refer to Figure 11.5, where we assume the line of sight is broken between P.C. station 6 + 05.92 and station 9 + 00.00. The centerline has been established up to and including station 8 + 50.00. We can set the instrument at any previously located station and continue to use the line-book as our source of deflection angles. The mathematical principle behind this field procedure is supported geometrically. As we continue with the explanation, refer to the line-book deflections and chord distances previously calculated.

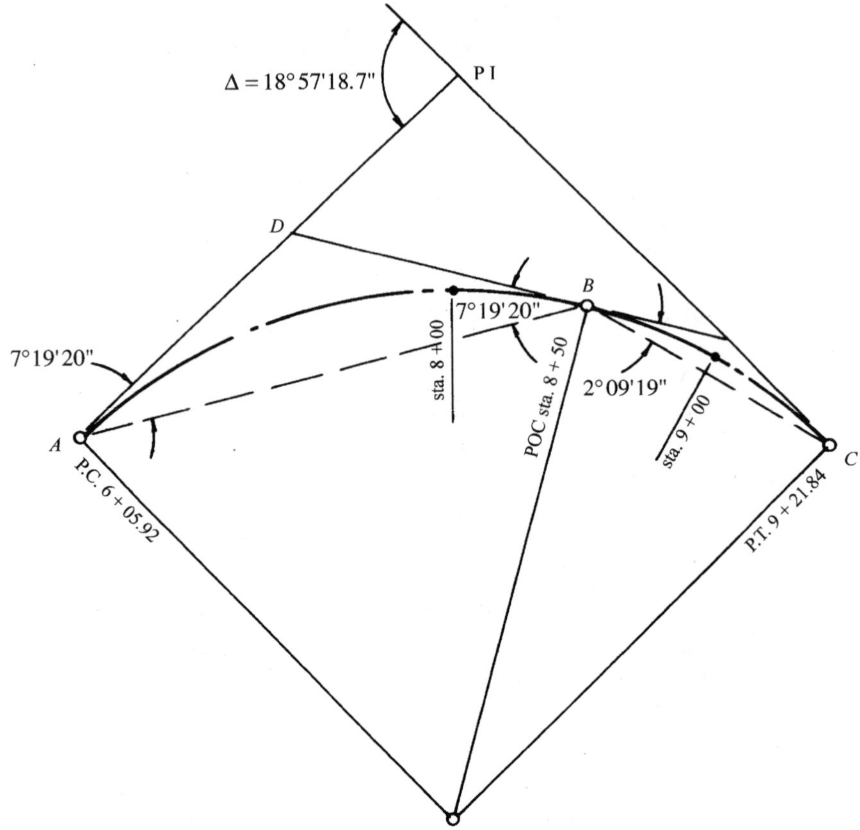

FIGURE 11.5 Instrument move-up on curves

In Figure 11.5, we have chosen to move the instrument to station 8 + 50.00. The line-book shows a deflection angle from point A to point B of 7° 19' 20". A review of the geometry in Figure 11.5 shows that the angle at point B in triangle ABD is also 7° 19' 20". Therefore, when the instrument is set at 00° 00' 00", with the scope inverted, back-sight the P.C. of the curve. By turning a horizontal plate angle of 7° 19' 20", the instrument line of sight is along the local tangent.[4] By inverting the instrument scope, the line of sight is directed forward along the local tangent and the line-book can be referred to for the next deflection angle. If station 9 + 00.00 is to be located, the instrument's horizontal plate is turned to a deflection angle of 8° 49' 20", and the chord distance of 49.99 ft is measured, locating the point. If the P.T. station 9 + 21.84 is to be located, the instrument's horizontal plate is turned to a deflection angle of 9° 28' 39", and the proper chord distance is measured, locating the point.

4 The local tangent at any point on the curve is a line perpendicular to any radius through the point.

INSTRUMENT MOVE-UP WHEN THE BACK-SIGHT IS NOT THE P.C.

There are many instances when the trip back to the beginning of a curve for sighting purposes is too time consuming and the centerline needs immediate location. In this situation, the surveyor can occupy any station on the curve and observe any station as a back-sight, providing the angle set on the instrument, before observing the back-sight, is equal to the deflection of the back-sight. Invert the instrument scope and come to the deflection angle where the instrument is situated. This line of sight is along the local tangent where the instrument is located, and you can continue using the line-book for additional stations. Assume, in Figure 11.5, that station 8 + 50 is the instrument setup, and that station 8 + 00.00 is to be used as the back-sight before proceeding with additional curve stations. Set the instrument's horizontal plate at 5° 49' 20", back-sight station 8 + 00.00, invert the instrument scope, and continue rotating the horizontal angle to 7° 19' 20". This places the instrument line of sight on the local tangent at station 8 + 50. Continue using the line-book computations to set additional stations.

CONCENTRIC CURVES

The concentric curve layout (Figure 11.6) has quite an advantage after the surveyor has calculated the *centerline* deflections. If the purpose of the layout is to set a concentric curve (referred to as parallel or offset) at a common distance from centerline, whether inside or outside of the centerline, the mainline deflection angles are used to produce these concentric curves.

When the concentric curve is field-located as a primary line, new chord distances must be calculated and measured to locate one point from another. It should be emphasized that

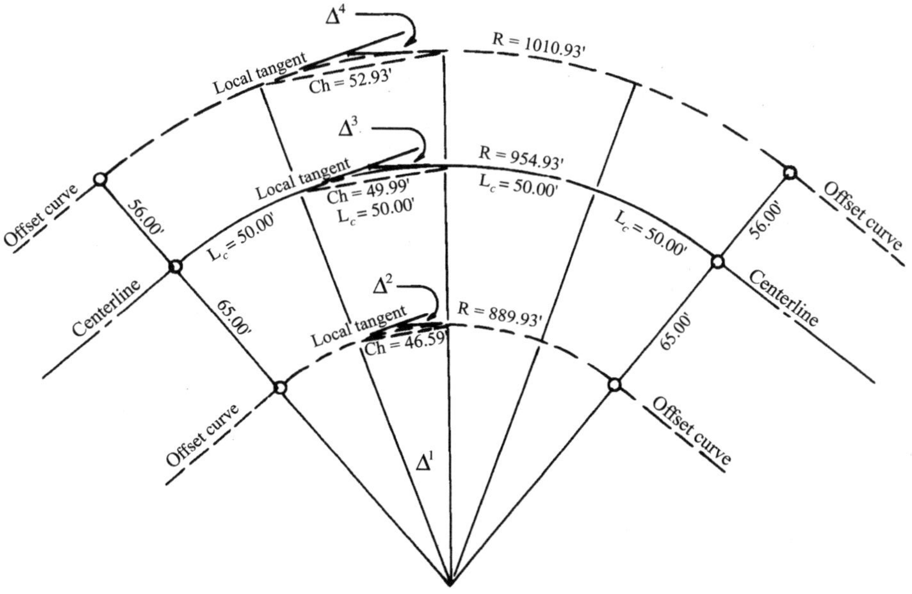

FIGURE 11.6 Concentric curves

concentric curves are actually reproducing 50-ft or 100-ft centerline stations at some specified distance off the mainline, and as the stations are set, chord distances between stations are subject to the radius of the offset. Concentric curves on the outside of the centerline will have longer chord distances, while curves on the inside of the centerline will have shorter chord distances.

Equation (11.5) is used to compute the new chord distance between offset points, where the equation is dependent on the new length of radius and where the delta angles were originally computed based on 50-ft or 100-ft centerline stations. The chord is computed using the following equation:

$$\text{Ch} = 2R \times \sin \frac{\Delta}{2} \tag{11.5}$$

You may be asked to locate a concentric curve where both the centerline and offset curve are referred to as being equal degrees of curve. Concentric and centerline curves can never have similar degrees of curve unless the two individual curves are separating or converging from each other. So, such a request becomes a geometric impossibility.

Example:

Use the computed curve data identified under Multiple-Point Curve Data in this chapter, and assume you are to locate a 56-ft outside concentric curve. The mainline curve radius is 954.93 ft (6° curve), and by adding the offset distance of 56 ft, the new radius is 1010.93 ft. Use Eq. (11.2) (simplified version), and compute the degree of curve for this radius:

$$D_c = \frac{5729.58}{R} \tag{11.2}$$

By substitution, the new radius of the concentric curve is 1010.93 ft, and computing the offset degree of curve, you have the following:

$$D_c = \frac{5729.58}{1010.93}$$

$$D_c = 5.667632774° \tag{11.2}$$

$$D_c = 5° \, 40' \, 03''$$

Therefore, the degree of curve for a 56-ft concentric offset, with a radius length of 1010.93 ft is 5° 40' 03". Obviously, the degrees of curve for the centerline and the offset are not identical.

FIELD-WORK CAN BE EXPEDITED BY USING CONCENTRIC CURVES

Concentric curves can shorten the layout procedure while construction staking, and especially during slope staking, taking cross sections, setting work-hubs, and occasionally for setting clearing limits. Remember, concentric offset points represent radial lines to the original 50-ft or 100-ft centerline stations and thus eliminate several field procedures. They help to eliminate the following:

- The replacement of the centerline, if destroyed
- Occupying each centerline station
- Becoming radial to each station
- Measuring from centerline to offset

THE MATHEMATICS OF CONCENTRIC CURVES

In addition to saving time by eliminating certain field procedures, the centerline survey occasionally cannot be occupied due to traffic or construction hazards, and when this situation occurs, the concentric curve becomes very useful.

You have already computed the centerline deflection angles in the section Multiple-Point Curve Layout, and as previously mentioned, the centerline deflection angles are used for the concentric curve layout, providing that you compute new chord lengths between offset points.

Figure 11.6 depicts a centerline with offset curves, one inside the original centerline and the other outside the original centerline. By reviewing the geometry of the triangles associated with each of the three different curves, you see that stations are projected on either side of the centerline, with the delta angle remaining equal for each of the three separate curves. Notice, delta 1 equals delta 2 equals delta 3 equals delta 4. The only variable in the layout is the chord distance.

Chord comparisons between centerline and offset curves are computed below, based on the working drawing in Figure 11.6.

Assume 50-ft station 65-ft offset (inside)	Assume 50-ft station centerline	Assume 50-ft station 56-ft offset (outside)
$D_c = 6° 26' 18''$	$D_c = 6° 00' 00''$	$D_c = 5° 40' 03''$
$R = 889.93$ ft	$R = 954.93$ ft	$R = 1010.93$ ft
$\text{Ch} = 2(R)\left(\sin\frac{\Delta}{2}\right)$	$\text{Ch} = 2(R)\left(\sin\frac{\Delta}{2}\right)$	$\text{Ch} = 2(R)\left(\sin\frac{\Delta}{2}\right)$
$\text{Ch} = 2(889.93)(\sin 1° 30')$	$\text{Ch} = 2(954.93)(\sin 1° 30')$	$\text{Ch} = 2(1010.93)(\sin 1° 30')$
$\text{Ch} = 1779.86 \times \sin 1° 30'$	$\text{Ch} = 1909.86 \times \sin 1° 30'$	$\text{Ch} = 2021.86 \times \sin 1° 30'$
$\text{Ch} = 46.59$ ft	$\text{Ch} = 49.99$ ft	$\text{Ch} = 52.93$ ft

ALTERNATE EQUATION FOR COMPUTING CONCENTRIC CHORDS

An alternate method of computing the chord distance between concentric curve stations can be setup as a proportional ratio, since the radius is the only variable in the equation.

Example:

Compute the chord distance between P.C. station 6 + 05.92 and station 6 + 50 on a 56-ft concentric offset with a radius of 1010.93 ft. The centerline chord distance between these two points was previously computed at 44.07 ft. The ratio is as follows:

1. Known chord is to centerline radius as
2. Unknown chord is to offset radius

Substituting:

$$\frac{44.07\text{ ft}}{954.93\text{ ft}} = \frac{x}{1010.93\text{ ft}}$$

$$x = 46.65\text{ ft}$$

Example:

Compute the chord distance between centerline station 6 + 50.00 and 7 + 00.00. The previously computed chord distance between these centerline points is 49.99 ft. Assume you are on a 56-ft outside concentric offset with a radius of 1010.93 ft. The ratio is as follows:

$$\frac{49.99 \text{ ft}}{954.93 \text{ ft}} = \frac{x}{1010.93 \text{ ft}}$$

$$x = 52.92 \text{ ft}$$

SPECIAL FIELD APPLICATION

Before leaving concentric curves, it should be noted that any curved line, when the end stations are known and the points are intervisible, can be field-located by computing the delta angle between the two end points, setting the instrument on the ground where it is estimated the concentric curve will arc, and using a method similar to "bucking-in" on a tangent line, locate the concentrlc curve.[5] The field procedure for this rarely used technique is:

1. Compute the total delta angle, remembering the deflection angle between any two curve points is one-half delta.
2. Set the instrument on the estimated arc of the curve, sight the beginning of the curve, and observe the horizontal angle to the end of the curve.
3. If the deflection angle does not match, move the instrument in the direction that seems appropriate.
4. When the instrument location is found on the arc of the curve, where the deflection angle does match one-half the delta angle, the instrument is precisely on the concentric curve.

INTERSECTION OF LINE AND CURVE

INTRODUCTION

The intersection of a line and a curve is one of the most basic problems that surveyors must address, both in the field and in the office. The early surveyors (from Egyptian time to the advent of the first railroad system) were seldom involved with precise curve layout. The old deed records of the 1800s reflect surveys along rivers, streams, and other curvilinear topographic features where curves must have existed, but the early record does not reflect a precise method of curve layout as surveyor's use today. The public accepted what was and is still known as "meander line" surveying. The meander line closely followed the curves of the topographic feature, and whenever the topography changed direction, the meander line also changed direction. The surveyor located the edge or centerline of the feature at each break, or when the changes were too numerous, distances to the curves were measured at

5 Bucking-in on a tangent line is used when the line of sight between end points is not visible. The instrument is set up at any location where the line of sight is possible, and one end point is sighted. The instrument scope is inverted, and the angle is observed to the other end of the curve, noting the offset difference. Several tests will probably be necessary before the instrument is set directly on an alignment representing a straight line.

intermediate points along the meander line and later plotted to produce a more accurate layout of the object.

During the past 40 or more years, some of these early surveys (mid-1800s) have been corrected, brought up to date, and the actual curved perimeters located using today's surveying science. The area (in acres or square feet) of most early surveys was an approximation, although in the 1900s, surveyors had at their command a form of mathematics, called "double meridian distance" that allowed acreage to be computed. Stories are told where surveyors calculated acreage by " throwing so many stone widths" or by "wrapping twine or leather (to help with the count) around a wagon wheel," and the number of revolutions the wheel covered determined the acreage. These are events of the past, and as the early railroad and highway systems began to emerge, the surveying profession had to update its technique.

EQUATIONS AND OTHER FORMS OF MATH

Several methods of solving for individual parts of the curve were suggested in the early part of this chapter, and by combining this specialty math with more traditional forms of mathematics, the intersection of lines and curves can be computed. Figure 11.7 is an example of this type of intersection and illustrates a property line intersecting a highway curve.

To complete the mathematics of this intersection, the surveyor must accurately locate the property line (either a field measurement or an assumed office distance), and supply sufficient angles that tie both lines together. In Figure 11.7, the mainline tangent was ex-

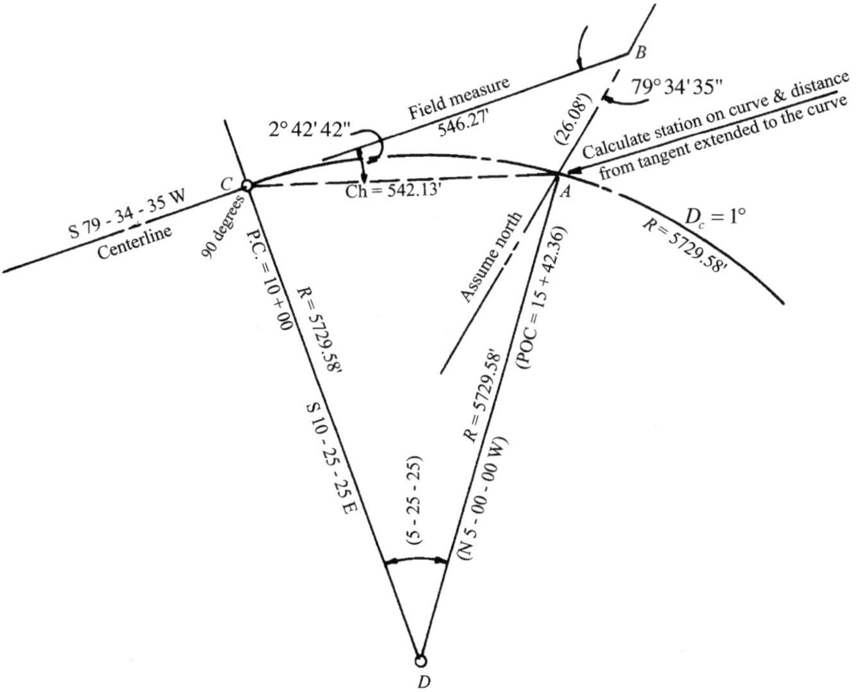

FIGURE 11.7 Line and curve intersection

tended beyond the P.C. of the curve, station 10 + 00, and made to intersect with the property line. An angle of 79° 34' 35" was observed at this intersection, and a distance of 546.27 ft was measured between the P.C. of the curve and the property line.

COMPUTING THE LINE AND CURVE INTERSECTION

Figure 11.7 shows the intersection of a highway curve and a tangent property line, where all known field and design data have been placed on the drawing in their proper locations. The bearings in this solution are assumed, but absolute bearings can be used and will be discussed later. Line *AB* has been assigned an arbitrary "due north" bearing, while the remaining bearings are calculated as described in Chapter 3, section "From Bearings to Angles."

When computing for two unknowns, gather all field and plan measurements (bearings and distances), placing this data on the illustration in their proper location. A final observation of the figure indicates that two measurements are missing:

1. The distance from points *A* to *B*, shown in parentheses
2. The bearing of the line from points *D* to *A*, also shown in parentheses

When the solution for both unknowns is complete, the centerline station at point *A* can be calculated, the chord length from P.C. station 10 + 00 to point *A* can be calculated, and finally, point *A* can be field located.

DEVELOPING A CLOSURE

We are going to develop a technique called "closure," where all known data is assembled in a prepared format (Figure 11.8) consisting of columns and rows, all survey information is entered, and by "process of elimination," the unknowns can be determined. "Closure" is a method of using data from triangles to solve these two unknowns. Not all manual closure solutions are possible; they are limited to the following:[6]

1. A bearing and distance on the same line
2. A bearing on one line and a distance on a second line
3. Two unknown distances, provided one of these unknown distances corresponds to a "true" geographical bearing (true bearings will be discussed later).

Most manual closures appear less complicated if you position the "letter" identification for each traverse corner so that the unknown *distance* or *bearing* is the last in the series. Notice in Figure 11.7, where the unknown bearing from points *D* to *A* is placed in this recommended position and is in parentheses.[7] The development of the manual closure con-

6 A manual closure is one in which the computations are completed using hand-held engineering calculators, rather than personal computer software.

7 Parentheses are placed along a traverse line showing that either a bearing or distance is missing.

Course	Bearing	Distance	Cosine	Sine	North	South	East	West

$\tan\alpha =$ -----------------------

$\tan\alpha =$

$\alpha =$

FIGURE 11.8 Manual "closure" form

sists of using a traverse, such as shown in Figure 11.9, and computing individual triangles, where each traverse line represents the hypotenuse of a triangle. Figure 11.9 has been rotated from the illustration in Figure 11.7 so that line AB is directed toward the top of the page representing due north. As the closure is being computed (Figure 11.8), invisible computations are being developed that represent the "legs" of each triangle. When the legs of all the triangles are assembled, such as in Figure 11.9, they form a rectangle. A rectangle, by definition, has two equal side distances, as well as equal top and bottom distances. In a closure, the sides of the rectangle represent *north* and *south* directions, while the top and bottom of the rectangle represent *east* and *west* directions.

When the data are assembled, in the larger closure form, as in Figure 11.8, in smaller versions of the closure form as used in this chapter; or assembled to form the rectangle, as in Figure 11.9, two unknown solutions can be computed.

The development of the manual closure consists of using a traverse, such as shown in Figure 11.9, and computing individual triangles, where each traverse line represents the

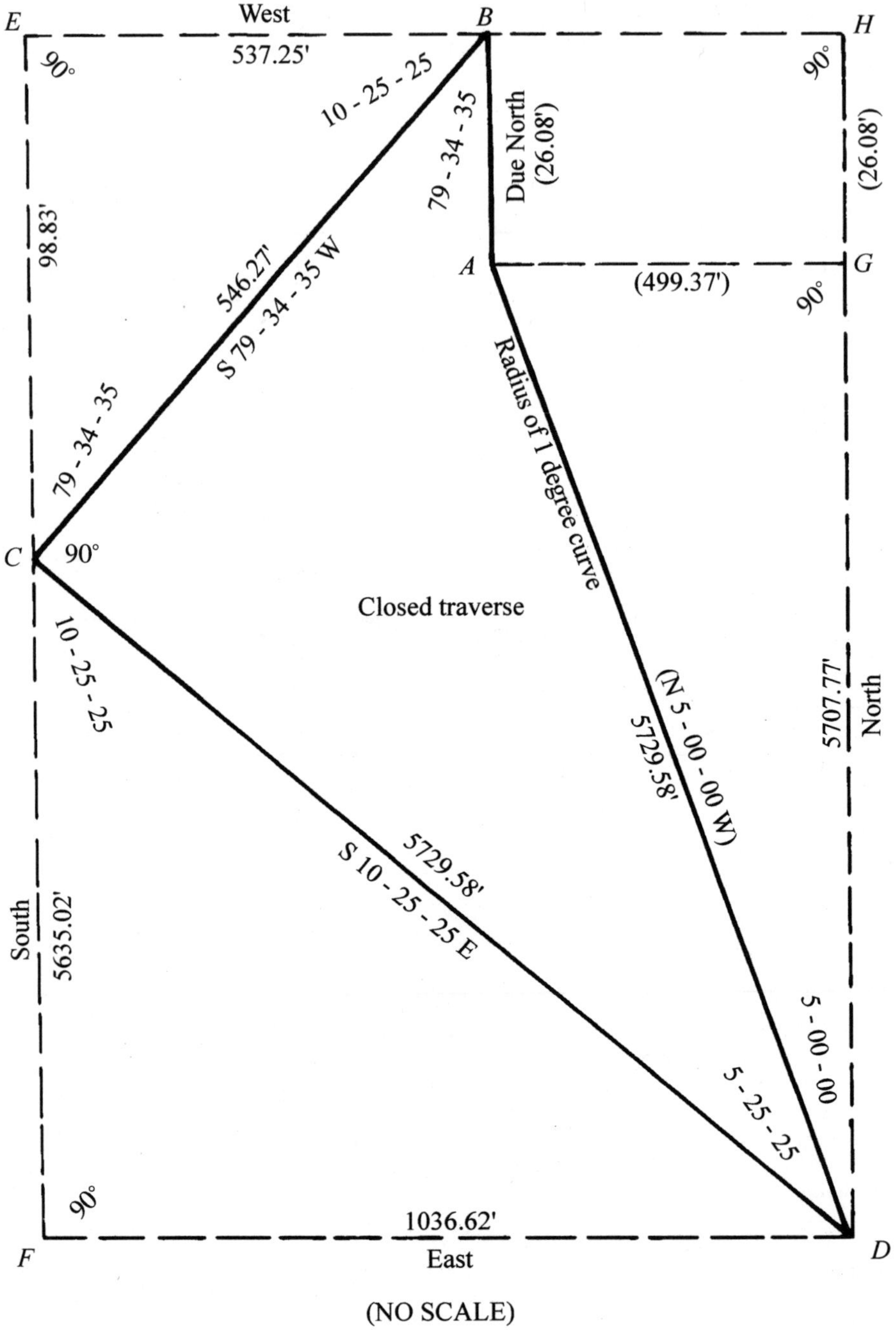

FIGURE 11.9 Traverse in rectangular form

hypotenuse of a triangle. Figure 11.9 has been rotated from the illustration in Figure 11.7 so that line *AB* is directed toward the top of the page representing due north. As the closure is being computed (Figure 11.8), invisible computations are being developed that represent the legs of each triangle. When the legs of all the triangles are assembled, such as in Figure 11.9, they form a rectangle. A rectangle, by definition, has two equal side distances, as well as equal top and bottom distances. In a closure, the sides of the rectangle represent *north* and *south* directions, while the top and bottom of the rectangle represent *east* and *west* directions. When the data are assembled, in the larger closure form, as in Figure 11.8, in smaller versions of the closure form as used in this chapter, or assembled to form the rectangle, as in Figure 11.9, two unknown solutions can be computed. In each of these instances, the process is leading to a solution for two unknowns by "process of elimination." The closure requires that we always compute for two unknowns, and this leads us to issue a warning: Never attempt to solve for less than two unknowns. If such an occasion arises, remove one known from the solution, and proceed to complete the mathematics as though there are two unknowns. When your final results are calculated, you must have one of your results equal to the value that was withheld. When only one unknown is calculated, you have made the assumption that all other field measurements and calculated data are error free, and that this single result balances the closure. Experience often tells us that, by examining the computation results more closely, you will find triangular inconsistencies, therefore, solving for one unknown can produce incorrect results.

COMPLETING THE CLOSURE FORM

Computing by *process of elimination* will be easier to handle if we describe a traverse, such as in Figure 11.7, and proceed to enter the calculated data, one unit at a time, into an abbreviated closure form. Continue referring to Figure 11.7, and notice that after placing all known field and plan data in their respective locations, there still remains two unknown elements in the figure, each identified in parentheses:

1. The distance from *A* to *B*
2. The bearing of the line from *D* to *A*

Both of these solutions can be computed by completing a series of trigonometric computations, whose function is to compute the legs of each triangle and place these distances in the closure form. By visual inspection of the form (Figure 11.9), you can see we made side distances (north and south) equal to each other; while also making the top and bottom distances equal to each other. By this process, two unknowns are computed.

To reduce this process to a more manageable concept, insert the bearing and distance information from Figure 11.7 into a representative closure form, where the form consists of rows and columns as you see in Figure 11.8.

COURSE

The course represents each traverse line and can be identified in any manner, such as Arabic numbers, lower or uppercase letters, Roman numerals, or other means of noting the beginning and ending of a line.

BEARING

The bearing represents the geographical direction of the traverse.

DISTANCE

The distance is the length of the traverse.

COSINE AND SINE

The cosine and sine of each course represent the trigonometric relations of angles within a triangle that permits us to solve for the legs of the triangle.

The cosine and sine columns are completed by obtaining the trigonometric function for each of the angles (synonymous with the bearing of the course). Your engineering calculator or an eight-place sine/cosine function table can be used to obtain these digits. Function tables (books) can be purchased at local and college book stores, and in past years from the government printing press. If you are using programmable computers, the function information is all internal, but the computing procedure and results are similar.

North, south, east, and west entries on the form represent the computed distances (legs) of the triangle for each traverse line, and their entry onto the form will be explained.

After each item (course, bearing, distance, cosine, and sine) is entered into its respective cell, the solution continues by multiplying the cosine of the angle (for each course) by the distance in the course and filling in the required cell. Each cell is occupied based on the direction of the bearing.

Example:

When a north-east bearing is traversed, only north and east cells are occupied. As another example: the course of line *BC*, Figure 11.7 (abbreviated below), with a bearing of S 79° 34' 35" W, is entered into the grid; the distance of line *BC* is also entered into the grid; cosine and sine functions can also be entered because the bearing is known and the function digits for that bearing can be found in the sine/cosine table; and finally, the triangle legs are computed and entered into their respective cells as shown below:

Course *BC* (S 79° 34' 35" W) has two directions, south and west, and therefore, the first cell to be filled is south (assuming you multiplied the cosine of the angle by the distance). West is the second cell to be filled (assuming you multiplied the sine of the angle by the distance).

The north and east cells remain blank and should be lined-through to indicate they have been considered and are found irrelevant to the solution. (For more information, see "Secondary instructions" later in this chapter.)

Course	Bearing	Distance	Cosine	Sine	North	South	East	West
AB	North	( )	1	0	( )	---------	---------	---------
BC	S 79 34 35 W	546.27'	0.1809244	0.983497	---------	98.833'	---------	537.255'

Course *CD* (S 10° 25' 25" E) has two directions, south and east, and therefore, the first cell to be filled is south, and the second cell to be filled is east, while the north and west cells should be lined-through.

Course	Bearing	Distance	Cosine	Sine	North	South	East	West
AB	North	( )	1	0	( )	---------	---------	---------
BC	S 79 34 35 W	546.27'	0.1809244	0.983497	---------	98.833'	---------	537.255'
CD	S 10 25 25 E	5729.58'	0.983497	0.180924	---------	5635.024	1036.621	---------

The shaded portion of the grid above shows the entries for line *CD*; this line also represents the beginning of a 1° curve, P.C. station 10 + 00.00, whose radius is 5729.58 ft. Multiply the cosine of the angle by the course distance, and the result is the north-south leg of the triangle but, in this situation, is entered into the south cell due to the direction of the bearing. Multiply the sine of the angle by the course distance, and the result is the east-west leg of the triangle but, in this situation, is entered into the east cell due to the direction of the bearing. The north and west cells remain blank and lined-through to indicate they have been considered but are irrelevant to the solution.

Course	Bearing	Distance	Cosine	Sine	North	South	East	West
AB	North	( )	1	0	( )	---------	---------	---------
BC	S 79 34 35 W	546.27'	0.1809244	0.983497	---------	98.833'	---------	537.255'
CD	S 10 25 25 E	5729.58'	0.983497	0.180924	---------	5635.024	1036.621	---------
DA	( )	5729.58'	( )	( )	( )	---------	---------	( )

Line *DA* is shaded in the above grid to show the next series of computations required before the final distance of line *AB* can be determined. Review the closure grid thoroughly, and you will find two missing cells in the north direction and one missing cell in the west direction.

Remember, east and west lines represent the top and bottom of the rectangle, which must be equal in distance to one another. The easterly distance, in this problem, is greater than the westerly distance, and a subtraction of the smaller distance from the larger will provide the amount needed to balance the east and west sides. The shaded grid below shows the entry in the west cell that balances the east and west directions of the rectangle:

Course	Bearing	Distance	Cosine	Sine	North	South	East	West
AB	North	( )	1	0	( )	---------	---------	---------
BC	S 79 34 35 W	546.27'	0.1809244	0.983497	---------	98.833'	---------	537.255'
CD	S 10 25 25 E	5729.58'	0.983497	0.180924	---------	5635.024'	1036.621	---------
DA	( )	5729.58'	( )	( )	( )	---------	---------	(499.36)

Although several cells remain blank for line *DA*, we need only remember that north, south, east, and west cells are completed by locating the cosine or sine function of the angle (bearing), multiplied by the distance of the line itself, and the result entered in the appropriate cell. In this situation, the bearing of line *DA* is missing, and therefore, it seems apparent that the cosine and sine functions cannot be located.

By reversing your reasoning process, you can complete one additional entry in the grid by dividing the distance of line *DA* (5729.58 ft) into the completed westerly cell (499.366 ft) and, in reverse order, compute the sine of an angle equal to the bearing of line *DA*. This computation is shown in the shaded grid below:

Course	Bearing	Distance	Cosine	Sine	North	South	East	West
AB	North	( )	1	0	( )	---------	---------	---------
BC	S 79 34 35 W	546.27'	0.1809244	0.983497	---------	98.833'	---------	537.255'
CD	S 10 25 25 E	5729.58'	0.983497	0.180924	---------	5635.024	1036.621	---------
DA	( )	5729.58'	( )	(0.08715)	( )	---------	---------	(499.36)

Now that the sine of the angle is known, you can enter your sine/cosine table or your engineering handheld calculator and determine the bearing of this function. The calculator or the table shows a bearing of 5° 00' 00". The cosine of this angle can be located from the same table or the engineering handheld calculator and inserted in its proper location (see the shaded grid below in the cosine column).

Course	Bearing	Distance	Cosine	Sine	North	South	East	West
AB	North	( )	1	0	( )	---------	---------	---------
BC	S 79 34 35 W	546.27'	0.1809244	0.983497	---------	98.833'	---------	537.255'
CD	S 10 25 25 E	5729.58'	0.983497	0.180924	---------	5635.024'	1036.621	---------
DA	( )	5729.58'	(0.99619)	(0.08715)	( )	---------	---------	(499.36)

You are now ready to multiply the cosine function of the angle (5°) by the distance of the line itself and enter this result in the appropriate grid cell. Will the result of this multiplication be entered in the north cell or in the south cell? Obviously, the grid is one step ahead by placing the parentheses in the north cell, but without this help, the direction can be determined in one of three ways:

1. The graphics of the illustration can be reviewed, and if the sketch has some order of scale, by rotating line *AB* in a northerly direction, line *DA* appears to be in the northwest quadrant.
2. The distances, previously entered in the grid, can be reviewed and their combined lengths will often suggest in which quadrant the unknown bearing is located.
3. The knowledge of the physical conditions on the ground will establish a range of distance for line *AB*, helping to decide where the cell entry might be placed.

The shaded grid below shows the cell entry produced by multiplying the cosine function of the angle by the course distance:

Course	Bearing	Distance	Cosine	Sine	North	South	East	West
AB	North	( )	1	0	( )	---------	---------	---------
BC	S 79 34 35 W	546.27'	0.1809244	0.983497	---------	98.833'	---------	537.255'
CD	S 10 25 25 E	5729.58'	0.983497	0.180924	---------	5635.024	1036.621	---------
DA	( )	5729.58'	(0.99619)	(0.08715)	(5707.77)	---------	---------	(499.36)

You probably have observed that some of the multiplication and division has not produced the exact digits you have computed. This is due to the size of the horizontal grid on these pages, where some of the significant figures are omitted due to the width of certain cells. For complete accuracy, the functions should be carried to eight places beyond the decimal, and the multiplication and division results should be carried to at least three places beyond the decimal, preferably four places.

The shaded grid below shows line *DA* completed to the point where the final distance of line *AB* can be completed by process of elimination.

Course	Bearing	Distance	Cosine	Sine	North	South	East	West
AB	North	( )	1	0	( )	---------	---------	---------
BC	S 79 34 35 W	546.27'	0.1809244	0.983497	---------	98.833'	---------	537.255'
CD	S 10 25 25 E	5729.58'	0.983497	0.180924	---------	5635.024	1036.621	---------
DA	(N 5 00 W)	5729.58'	(0.99619)	(0.08715)	(5707.77)	---------	---------	(499.36)

The closure grid, above, shows all cells completed except one, which represents the distance of line *AB*. To complete this missing element, combine the distances in the south column, subtracting the one entry in the north column. The final result is the missing distance in course *AB*. The closure grid, below, shows the final entry:

Course	Bearing	Distance	Cosine	Sine	North	South	East	West
AB	North	(26.08)	1	0	(26.08)	---------	---------	---------
BC	S 79 34 35 W	546.27'	0.1809244	0.983497	---------	98.833'	---------	537.255'
CD	S 10 25 25 E	5729.58'	0.983497	0.180924	---------	5635.024'	1036.621	---------
DA	(N 5 00 W)	5729.58'	(0.99619)	(0.08715)	(5707.77)	---------	---------	(499.36)

Both unknown elements are now completed by simply using this manual method of computation. We need now to turn our attention to "secondary solutions" that can also be used in the closure under other circumstances and often expedite the results.

SECONDARY SOLUTIONS

A secondary solution for computing the missing bearing in our sample problem can be found in the trigonometric relation of angles and distances and their relation to the hypotenuse of a 90° triangle. The triangle, shown below, uses the westerly (leg) distance of 499.366 ft, combined with the length of the hypotenuse (5729.58 ft), and solves for the angle between these two lines.

We know from trigonometry studies that the sine of an unknown angle is equal to the opposite leg of the triangle divided by the hypotenuse. The traditional equation for this trigonometric relation is as follows:

$$\sin \alpha = \frac{499.366 \text{ ft}}{5729.58 \text{ ft}} \tag{11.9}$$

$$\sin \alpha = 0.087155777°$$

$$\alpha = 5°\ 00'\ 00''$$

The angular solution in the above equation is identical to the bearing that was computed for line *DA* when using the closure method. In that computation, we divided the distance (5729.58 ft) of line *DA* into the westerly triangle leg (499.366 ft) and determined the sine function of that angle. By entering the function tables or using a hand-held calculator, we find that the sine function (0.0871557) represents an angle of 5° 00' 00". As a test, place these digits in your engineering hand-held calculator and ask for the angle whose sine function equals these digits. The result will be 4.99999754°, or 5° 00' 00". The triangle below, shows this computation.

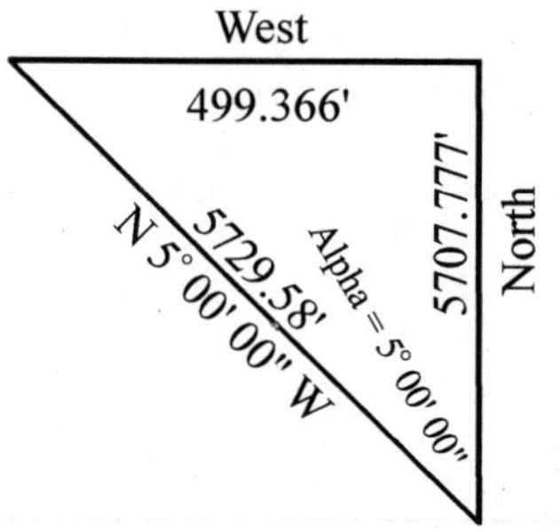

From this same sketch, we will determine the trigonometric equation that computes the second leg of the triangle, when the angle and the hypotenuse are known. The traditional equation is as follows:

$$\cos 5° = \frac{\text{adjacent}}{\text{hypotenuse}} \tag{11.10}$$

and rearranging the equation to solve for the adjacent distance gives:

$$\textit{adjacent} = \cos 5° \times \text{hypotenuse}$$
$$\textit{adjacent} = 0.996194698 \times 5729.58$$
$$\textit{adjacent} = 5707.777 \text{ ft}$$

An unknown bearing, as in the example below, can also be computed by trigonometric means when the closure cells have been completed, such as line *DA*.

In this situation, the north and west cells were unknown and solved by process of elimination. The bearing of the course represented by these two triangle legs can be computed by using the tangent function of a 90° triangle.

Where: The opposite distance is represented by either the west column (499.366 ft), and the adjacent distance is represented by the north or south column (5707.777 ft).

$$\tan \alpha = \frac{\text{opposite}}{\text{adjacent}} \tag{11.11}$$

$$\tan \alpha = \frac{499.366 \text{ ft}}{5707.777 \text{ ft}}$$

$$\tan \alpha = 0.087488702$$

$$\tan \alpha = 5° \, 00'00''$$

Course	Bearing	Distance	Cosine	Sine	North	South	East	West
AB	North	26.08'	1	0	26.08'	---------	---------	---------
BC	S 79 34 35 W	546.27'	0.1809244	0.983497	---------	98.833'	---------	537.255'
CD	S 10 25 25 E	5729.58'	0.983497	0.180924	---------	5635.024'	1036.621	---------
DA	(N 5 00 W)	( )	(0.99619)	(0.08715)	(5707.77)	---------	---------	(499.36)

After computing the bearing, as in the 5° result above, the bearing of the course is added to the closure form, leaving only the distance of the course unknown. Cosine and sine functions for a 5° bearing are obtained from the table, or hand-held calculator, as discussed before, and placed in their respective locations. At this level of completion, we need to reconstruct the sequence by which data is computed and entered into the cells.

Sequence

When the course, distance, and bearing is known, the computations are continued by multiplying the cosine of the angle by the distance, resulting in either a north or south cell entry, and multiplying the sine of the angle by the distance, resulting in either an east or west cell entry.

This process can be reversed by dividing the cosine function of the angle into the north-south cell and computing the distance of the course. Similarly, by dividing the sine function of the angle into the east-west cell, you also compute the distance of the same course. The distances from these two divisions should produce similar results, and in this example they do, if the functions are carried out to eight places beyond the decimal point. In general, distances should match within a tolerance of ±0.03 ft, but if errors are present, either through addition, subtraction, multiplication, or incorrect data, you must review the material and locate the problem (this will be discussed more on the following page).

CLOSURE TOLERANCES

A tolerance factor of ±0.03 ft was mentioned above, suggesting that occasionally the course distances obtained by dividing either the cosine or sine function into the appropriate leg of the triangle will not match. If the difference is greater than a few hundreds of a foot, you probably have an error and must review your input data. Whenever two computed distances fail to match, and you think the results are accurate, use the distance obtained by dividing through with the larger number. Dividing with digits that have a number of zeros immediately following the decimal point will produce small errors, and therefore the larger function numbers, those approaching 0.99999 will provide greater accuracy.

OFFICE COMPUTED LINE AND CURVE INTERSECTION

With the closure form completed, we find the distance of line *AB* is 26.08 ft, the bearing of line *DA* is N 5° 00' 00" W, but unresolved are the centerline station at the intersection of the property line and the chord distance from the P.C. of the curve to the intersection.

To complete the station and chord distance, we need to compute the delta angle between the bearing of line *CD* (S 10° 25' 25" E), Figure 11.7, and the bearing of line *DA* (N 5° 00' 00" W). This comparison results in a delta angle at point *D* of (5° 25' 25"). The length of the curve is then computed using Eq. (11.3).

$$L_c = \frac{\Delta \times 100}{D_c} \tag{11.3}$$

$$L_c = \frac{5.4236111 \times 100}{1}$$

$$L_c = 542.36\,ft$$

With the length of the curve computed, add this distance to P.C. station 10 + 00.00, and the curve intersecting station is 15 + 42.36. The length of the long chord is computed using Eq. (11.5)

$$\text{Ch} = 2(R) \times \left(\sin \frac{\Delta}{2}\right) \tag{11.5}$$

$$\text{Ch} = 2(5729.58) \times (\sin 2°\ 42'\ 42")$$

$$\text{Ch} = 11459.16 \text{ ft} \times 0.047309845$$

$$\text{Ch} = 542.13 \text{ ft}$$

With both computations completed, the point can be established by setting the instrument at the P.C. of the curve, foresighting the mainline tangent, turning one-half the delta angle (2° 42' 42") in the appropriate direction, measuring the chord (542.13 ft), and setting the point.

LINE AND CURVE INTERSECTION USING FIELD APPLICATIONS

The intersection of a line with a curve can be field-located. This will also establish the bearing of the property line or other topographic features. Surveyors, during the location of preliminary highway centerlines, often field-locate property lines, fences, tree rows, and other physical objects that intersect with the centerline of survey.

The field procedure for locating the property line or other topographic features is outlined below:

1. Establish the position of the property line, using minimum state and county standards, or if other physical objects, such as fences, trees rows, pipes, streams, etc., intersect the curved centerline, locate these objects with sufficient detail for line reconstruction (if required).
2. Estimate the arc of the curve in the vicinity of these intersecting objects, or if the curve is difficult to approximate, accurately field-locate several curve stations on each side of the intersection.
3. Return to the intersecting line and set a series of intermediate points (using nails and flagging), where one or more of these points are located on each side of the centerline curve.
4. Using the centerline stations (set previously), measure an approximate distance to a point short of the intersecting line and measure another distance to a point beyond the intersecting line.
5. Compute the delta angle and chord distance for each of these two new stations, setting the points.
6. Your field survey should now include nails and flagging along the intersecting tangent line, and nails and flagging at centerline stations, both ahead and back of the intersection.
7. Use two string lines, one held along the line representing the intersecting tangent and the other held along the line representing the arc of the curve.
8. Where the string lines intersect, place a hub and tack to represent the true intersection. Measure from the nearest centerline station, adding or subtracting as appropriate, and note the new intersecting station in your field book.
9. Instrument-occupy the new hub and tack and, if possible, observe the angle from a known backsight to the intersecting line.
10. Where an angle to the intersection cannot be observed, set a parallel offset line to the intersecting object and tie to the centerline using a short traverse, or if this is not sufficient, proceed to set new centerline curve points on each side of this offset line. Establish a similar string line intersection and set the new hub and tack at the offset intersection. Observe the angle to the offset line, which will establish the bearing of the intersecting tangent.

This field procedure is somewhat time consuming, but with experience, helpful shortcuts will be learned, or perhaps other field methods will be recommended by your employer.

Secondary Instructions

COMPUTATIONAL IMPASSE

We have mentioned previously in this chapter that certain manual closure computations cannot be made.

One of these computations occurs when there are two unknowns, each occupying separate courses, and where each unknown has a two-directional bearing. When courses are two-directional, such as northeast, southeast, southwest, or northwest, the process of balancing sides by process of elimination cannot be completed. Refer to the closure form below, and notice that the north column has two vacant cells, whereas the south column has one entry and no vacant cell. With two cells vacant in the north column and all entries complete in the south column, it becomes impossible to compute either entry in the north column. The east and west columns have a similar situation, in that neither column can have a summation, and therefore, no additional mathematics, such as balancing the sides, can take place.

Course	Bearing	Distance	Cosine	Sine	North	South	East	West
AB	N 14 05 34 W	( )	0.969902	0.243492	( )	---------	---------	( )
BC	S 04 20 23 W	8972.36	0.997132	0.075670	---------	8946.635	---------	678.938
CA	( )	5302.03	( )	( )	( )	---------	( )	---------

If you forget and try to complete a closure that has similar conditions, just rotate the course with the unknown distance to a "true" geographical direction, recompute the new bearings based on this adjustment, and the solution becomes possible. The closure grid below is rotated to show how the solution can be completed:

Course	Bearing	Distance	Cosine	Sine	North	South	East	West
AB	North	( )	1	0	( )	---------	---------	( )
BC	S 18 25 57 W	8972.36	0.9486968	0.3161872	---------	8512.049	---------	2836.945
CA	( )	5302.03	( )	( )	( )	---------	( )	---------

The next closure grid shows the completed mathematics, where the east column is balanced with the west column, the distance of the course divided into the east leg of the triangle, the sine function becomes known through this division, the angle equivalent to this function is obtained in the sine/cosine table, and the remainder of the closure completed with the following results.

Course	Bearing	Distance	Cosine	Sine	North	South	East	West
AB	North	(4032.85)	1	0	4032.851	---------	---------	( )
BC	S 18 25 57 W	8972.36	0.9486968	0.3161872	---------	8512.049	---------	2836.945
CA	(N 32 20 55 E)	5302.03	0.8448081	0.5350677	(4479.198)	---------	(2836.945)	---------

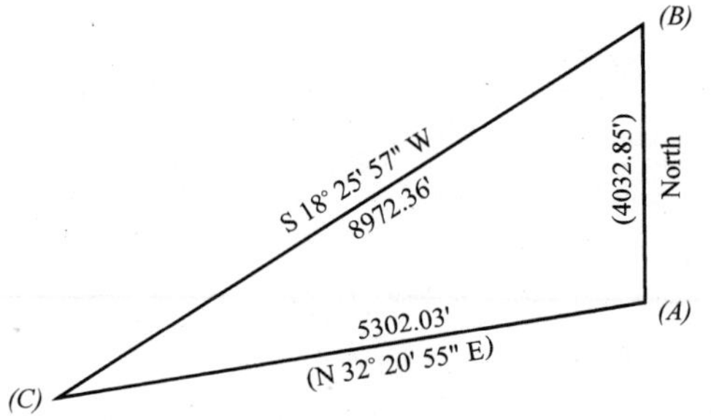

LOCATION OF SINE AND COSINE IN CLOSURES

The location of the sine/cosine column within the closure form requires some additional explanation. You are probably wondering why the cosine of the angle comes in advance of the sine when preparing the closure form entries. The answer lies in the trigonometric relation of angles and distances and their relation to the hypotenuse of a 90° triangle. Refer to line *BC* (Figure 11.9), which has a bearing of S 79° 34' 35" W. Concentrate on the triangle made from this hypotenuse, and construct a line due north beginning at point *C*. Also construct a line due west beginning at point *B*. These three legs form triangle *BEC*, where one angle, by definition is 90° and the other two angles are computed by comparing the bearings between two adjacent lines.

Example:

The hypotenuse bearing is S 79° 34' 35" W and the north line, *CE*, produces an angle at point *C* equal to the bearing of the hypotenuse, or 79° 34' 35". The west line produces an angle that is equal to the difference between 90° and the hypotenuse, which is 10° 25' 25". The angles are shown in their proper location on the illustration (Figure 11.9), and by using trigonometric equations, we can compute the distances for each leg in triangle *BCE*. From Eq. (11.9), we know that the sine of the angle, at point C for example, equals the opposite distance divided by the hypotenuse, where the opposite distance always refers to east-west directions. From Eq. (11.10), we know that the cosine of the angle, at point *C*, equals the adjacent distance divided by the hypotenuse, where the adjacent distance always refers to north-south directions.

The columns in the typical closure form, reading from left to right, commence with north-south entries, followed by east-west entries, and therefore, the computation form is set up to lessen the rate of error by placing the cosine function in a position on the form that when multiplied by the distance (hypotenuse) of the course, the product will automatically be placed in one of the first two columns. The sine function is placed further into the form, where its multiplication with the distance (hypotenuse) of the course will have no other choice than be entered into one of the last two columns. This probably seems insignificant, but whenever possible, sources of mathematical error should be. eliminated.

FUNCTIONS OF ZERO AND ONE

The entries of 0 and 1 in a typical closure form, where one or more bearings are situated in true geographical directions require an additional explanation.

Refer to the grid below, noticing the first course in the closure, line *AB*, has a bearing of due north, while the cosine function is entered as number 1, and the sine function is entered as 0.

Course	Bearing	Distance	Cosine	Sine	North	South	East	West
AB	North	(4032.85)	1	0	(4032.85)	---------	---------	---------
BC	S 18 25 57 W	8972.36	0.9486968	0.3161872	---------	8512.049	---------	2836.945
CA	(N 32 20 55 E)	5302.03	0.8448091	0.5350677	(4479.198)	---------	(2836.945)	---------

It is a good idea to memorize the following trigonometric functions.

1. Sine of 90°,
2. Cosine of 90°,
3. Sine of 0°,
4. Cosine of 0°,
5. Tangent of 0°, 45°, and 90°.

To refresh your memory, turn to a sine/cosine table or your hand-held engineering calculator to locate the function digits of these angles. Remember, all bearings are referenced to a north-south axis, and for this reason we have bearings that are either north, such as northeast or northwest, or bearings that are south, such as southeast or southwest. North-south bearings are, therefore, considered to be 0° off of the north-south axis, making cosine or sine functions a relation of 0°. Consequently, the cosine of 0° is 1, and the sine of 0° is 0.

Bearings that are true east-west directions are deflected 90° off of the north-south axis, making cosine or sine functions a relation of 90°. Therefore, the cosine function of 90° is 0, and the sine function of 90° is 1.

This probably appears confusing, but one explanation covers all four situations. Most closures have one true geographical bearing where the distance is missing, and choosing the correct function is not difficult, but if it is, the following should help:

North-south bearings:	The cosine function is 1 and the sine function is 0.
East-west bearings:	The cosine function is 0 and the sine function is 1.

Notice that the function digits (0.5350677) for the sine of N 32° 20' 55" E, in the closure above, does not exactly match the sine/cosine function table value for this same bearing.

As this closure was being completed, the easterly leg of the triangle for course *CA* had not yet been completed, and we know the east and west columns must equal one another. Therefore, the easterly cell in course *CA*, was made to equal the westerly cell in course *BC* The distance of the course (5302.03 ft) was divided into the easterly leg of the triangle, computing the bearing of the course. When the cosine of course *CA* is placed into its proper cell, however, the correct table function is used, and these small variances in function digits do not change the final results. Most, if not all, closures will show this type of discrepancy when columns are made to balance and a bearing is computed from this result.

COMPUTING INACCESSIBLE CURVES

Surveyors are often required to field-locate one curve, relating the mathematics to a second, more monumented centerline. Locating inaccessible curves in this manner might seem remote, but the computations can be rewarding and time saving when the need arises. A remedial project, such as relocating destroyed right-of-way fence lines on an interstate highway ramp is an example of this problem. Highway centerlines are most often easier to reconstruct due to the existence of original monumentation and especially when these controls are in the median. Entrance and exit ramp baselines are generally more difficult to reconstruct since the engineering plan was designed without monuments or the monuments have long been destroyed.

Often, a remedial fence located adjacent to a ramp will be designed with fence corners that have been mathematically related to the ramp baseline.

In some instances, where the differences in ramp elevation and mainline elevation are considerable, the main highway centerline could be closer to the fence, and there are other instances where the surveyor prefers to complete the layout from the main centerline. When this situation occurs, the surveyor can compute duplicate stations and offsets to each fence break by relating one curve with another.

If the surveyor has a portable programmable computer that retains coordinates, these computations can be completed in a minimum length of time. However, surveyors must consider their own calculating experiences before deciding whether to field or offfice compute.

Figure 11.10 is an example of this situation, where the fence break point is set at a distance of 80.00 ft left of the ramp, station 11 + 95.86. The break point is obviously closer to the ramp baseline, but let's assume a resurvey of this ramp baseline would require an unusual amount of field work. However, if the ramp baseline could be related to the main

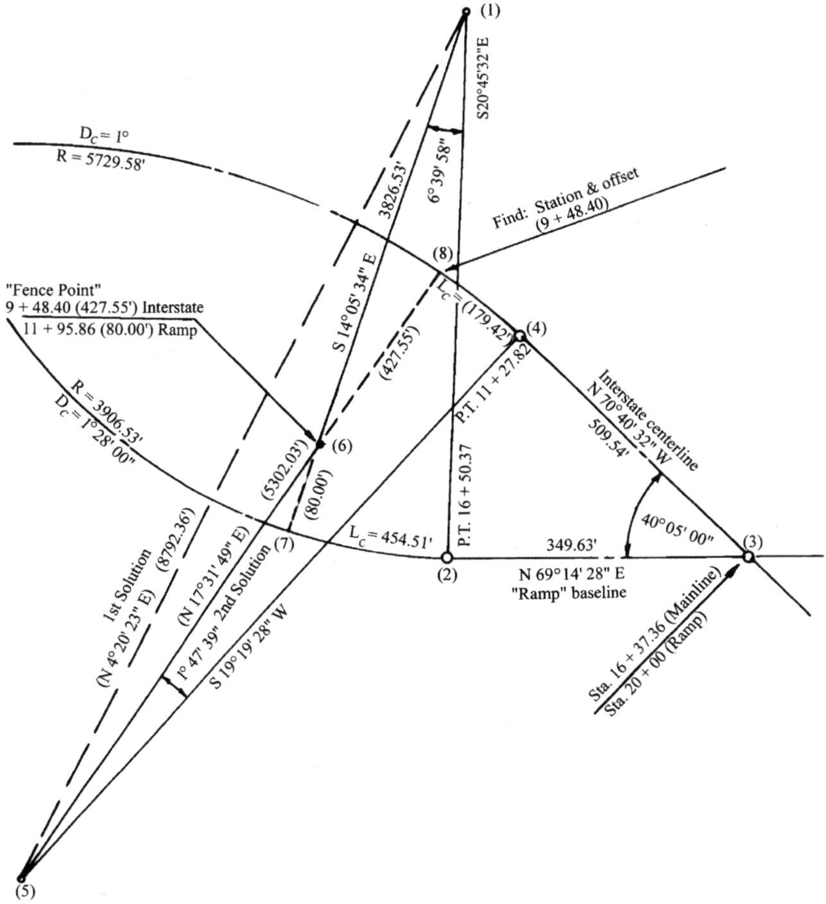

FIGURE 11.10 Locating one curve from another

centerline, the field work would be considerably less. The list below directs you through the various steps required when relating one curve to another:

- Refer to the highway centerline survey plat or the geometric control drawings, where the mainline and all intersecting roads and ramps are horizontally related.
- Sketch the problem, as in Figure 11.10, placing all known mathematical information in its proper location.
- The solution (solving for two unknowns) should be evident after thoroughly reviewing the sketch.
- Frequently, however, when searching out a solution that involves two curves and the manual closure form is to be used, you must first compute the bearing and distance between the radii points of both curves and then finish the computation.

The closure form below will be used to compute the bearing and distance between curve radii points 5 and 1 (Figure 11.10). All plan information is placed in its respective cell, and we propose to use plan bearings rather than rotating the figure since both unknowns are situated on the same course.

Course	Bearing	Distance	Cosine	Sine	North	South	East	West
1–2	S20-45-32E	3906.53'	0.93508023	0.35443611	---------	3652.918	1384.615	---------
2–3	N69-14-28E	349.63'	0.35443611	0.93508023	123.921	---------	326.932	---------
3–4	N70-40-32W	509.54'	0.33091702	0.94365986	168.615	---------	---------	480.832
4–5	S19-19-28W	5729.58'	0.94365986	0.33091702	---------	5406.774	---------	1896.015
5–1	(N4-20-23E)	(8792.36)	0.99713291	0.07567004	(8767.15)	---------	(665.300)	---------

$$\tan \alpha = \frac{665.3005}{8767.1566}$$

$$\tan \alpha = 0.0758855 \tag{11.11}$$

$$\alpha = 4° \, 20' \, 23''$$

After balancing the north and south columns and the east and west columns and entering the proper distances into their respective cells, our only option for continuing this solution was to use the tangent of the angle to compute the bearing. No other information was available. After the bearing (N 4° 20' 23" E) is computed, we can obtain the cosine and sine functions, entering these digits into the form, and finally, the distance of line 5 to 1 can be computed. You will remember that dividing the cosine function of an angle into the north leg of the triangle produces the distance of the line, and similarly, dividing the sine function of the bearing into the east leg of the triangle produces the same distance of the line. Both distances should maintain a tolerance factor of +0.03 ft, but in this computation, the cosine function digits commence with 0.997 and the sine function digits commence with 0.075, therefore, rigid tolerances cannot be held. The error produced by these two divisions exceeds the recommended tolerance (8792.36 versus 8792.13 ft) but, in this situation, we can be confident that the cosine function digits approaching the value of 1 will produce the more accurate result.

Note: This computation was verified by using personal computer programmable software and inversing between two known coordinates.[8]

With the completion of the bearing and distance of course 5 to 1 (Figure 11.10), we must turn our attention to finding a method whereby the corresponding station and offset of the fence break point can be computed. We read earlier in this chapter that when degrees of curve and stations are known, we can solve the subtended angle between any two curve points. In Figure 11.10, the P.T. of the curve, station 16 + 50.37 and the fence break, station 11 + 95.86, are both known. When stations are known, the L_c can be determined by subtraction, the delta angle can be computed, and from this subtended angle the bearing of line 1 to 6 can be determined. The computation for the subtended angle is shown below:

$$\begin{aligned} \text{P.T. station} &= 16 + 50.37 \\ \text{Fence break point} &= \underline{11 + 95.86} \\ L_c &= 4 + 54.51 \text{ ft} \end{aligned}$$

$$\Delta = \frac{L_C \times D_C}{100} \tag{11.3}$$

$$\Delta = \frac{454.51 \times 1.466666}{100}$$

$$\Delta = 6° \ 39' \ 58'' \text{ (Subtended angle)}$$

The subtended angle at point 1 is used to calculate the bearing of line 1 to 6 by preparing a sketch that includes the approximate bearing of line 2 to 1 (N 20° 45' 32" W) and the approximate location of line 1 to 6, as shown below.

When the lines in the sketch are extended 180°, where both lines are in quadrant 2, a subtraction of the subtended angle (6° 39' 58") from the bearing of line 1 to 2 (S 20° 45' 32" E) produces a bearing on line 1 to 6 of S 14° 05' 34" E. The final solution is that of computing the two unknowns (a bearing and a distance) in triangle 6,1,5, and from this solution, the corresponding station of the fence break point can be finalized.

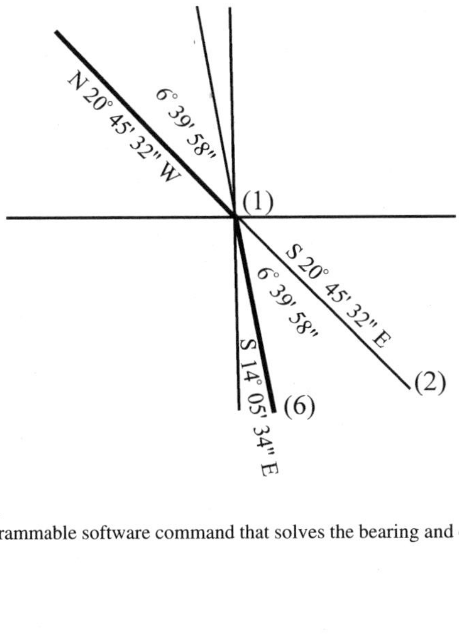

8 "Inverse" is a programmable software command that solves the bearing and distance between any two known coordinates.

Having completed the bearing of line 6 to 1 in the sketch above, you can compute the distance from point 6 to 1 by subtracting the fence offset (80.00 ft) from the radius of the 1° 28' 00" curve, such as (3906.53 – 80.00 ft), leaving a distance of 3826.53 ft. The bearing of line 5 to 1 was calculated earlier and found to be N 4° 20' 23" E with a distance of 8792.36 ft. The closure form below traverses this same course, but in the opposite direction, and therefore, the bearing of the line has been reversed. Continue to enter the known data into the proper cells. Course 5 to 6 represents the two missing elements (bearing and distance) in the closure and are computed below:

Course	Bearing	Distance	Cosine	Sine	North	South	East	West
6 - 1	N14 05 34W	3826.53'	0.96990272	0.24349276	3711.362	---------	---------	931.7323
1 - 5	S04 20 23W	8792.36'	0.99713291	0.07567004	---------	8767.151	--------	665.3182
5 - 6	(N17 31 49E)	(5302.03')	0.95355791	0.30120975	(5055.79)	---------	1597.05	---------

$$\tan \alpha = \frac{1597.0505}{5055.7987}$$

$$\tan \alpha = 0.315885468$$

$$\tan \alpha = 17^\circ\ 31'\ 49''$$

Remember to place the unknowns in parentheses, not only to make the calculation appear less complicated, but because the documentation can become very important in future years to reconstruct the solution sequence if called upon for court testimony.

With the completion of the above closure, the bearing and distance of the line that intersects with point 6 is known. By Subtracting this distance (5302.03 ft) from the radius of the 1° curve (5729.58 ft), you will have determined the offset distance is 427.55 ft from the 1° curve to the fence.

One final computation remains before the corresponding fence station is completed. Refer to Figure 11.10 and prepare a sketch showing the two bearings, one from 5 to 4 and the other from 5 to 8. Both bearings are in a north east direction, or in quadrant 1, and a subtraction of one bearing from the other provides the subtended angle at point 5. Thus, N 19° 19' 28" E (line 5 to 4) is the most easterly bearing in quadrant 1, and a subtraction of the bearing N 17° 31' 49" E (line 5 to 8) gives a subtended angle of 1° 47' 39", from which the length of curve can be computed. Equation (11.3) is used to solve the unknown curve length and is shown below:

$$L_c = \frac{\Delta \times 100}{D_c} \tag{11.3}$$

$$L_c = \frac{1^\circ\ 47'\ 39'' \times 100}{1}$$

$$L_c = \frac{1.7941666 \times 100}{1}$$

$$L_c = 179.42 \text{ ft}$$

With the length of curve completed, and knowing that point 8 is back-station of the P.T., station 11 + 27.82, subtract the 179.42 ft, and the result will be the station (9 + 48.40) of the fence break. The distance from the 1° curve to the fence break is completed by subtracting the length of line 5 to 6 from the radius (5729.58 – 5302.03 ft) and this equals the offset distance of 427.55 ft.

This distance is obviously greater than the original 80.00 ft offset from the ramp, but with the use of electronic measuring equipment, the fence break can be located rather quickly, depending on the terrain. The problem, however, was an exercise is curve computation.

Each surveyor requires mathematical practice in the field, whether using programmable computers or solving with manual closures, but relating one curve line with another opens new possibilities in your everyday surveying practice. The use of personal computers and programmable software will always make field computations easier, but knowledge and understanding of the background in these computations, will make your future computer applications more meaningful.

INTERSECTION OF TWO CURVES

This computation is the last in the series on simple curves, and involves the *intersection* of two curves. As mentioned earlier, the manual closure solution for two intersecting curves is generally advanced more quickly by computing the bearing and distance between both curve radii points. Figure 11.11 outlines the double-curve intersection, where line 5 to 1 completes the tie between these points. The cross-hatched triangle in the figure outlines the "parts" of the double curve that require solutions before the centerline stationing can be completed.

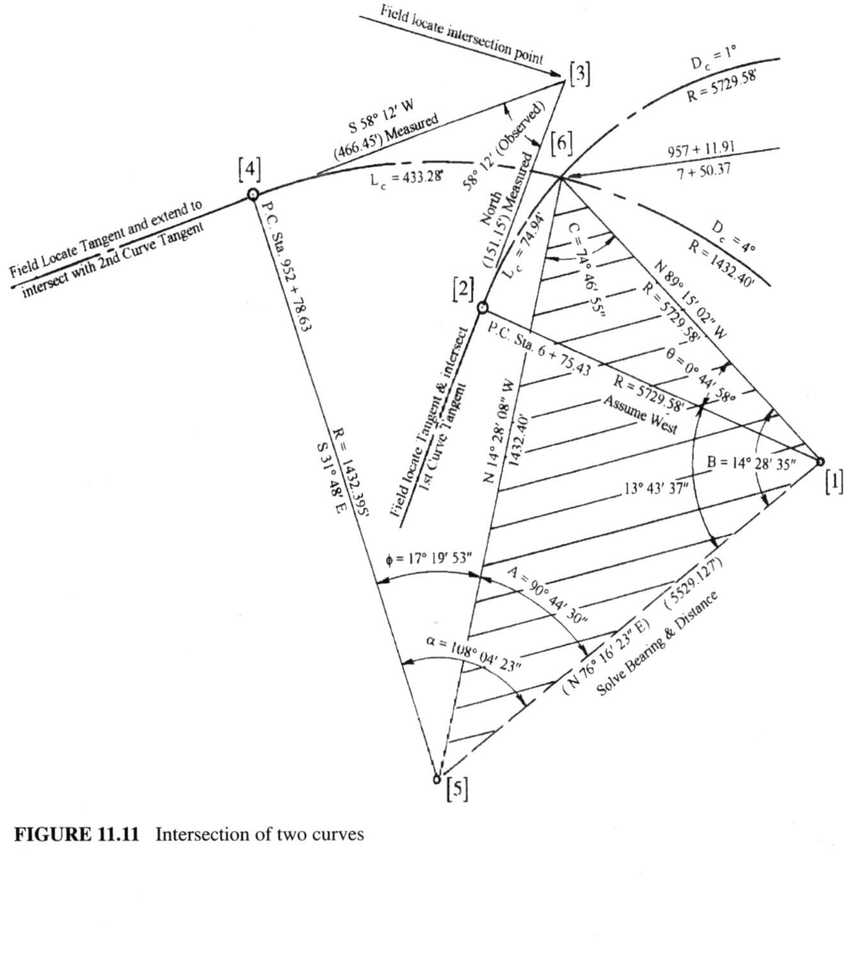

FIGURE 11.11 Intersection of two curves

With the bearing and distance of line 5 to 1 computed and with the remaining two legs of the triangle, line 1 to 6 and line 5 to 6, both representing radius lines for each of the curves, you then have triangle 5,1,6 consisting of one bearing and all three sides known.

Most programmable computer software programs can solve for the angles within this triangle using a command called "bearing-bearing" intersect; however, in keeping with the theme of this book, which is to solve surveying problems without programmable computers, we will introduce a manual method of computing the three interior angles in the cross-hatched triangle.

When these angles are computed, you can develop bearings on each of the radius lines as was outlined earlier in the chapter under, "Computing Inaccessible Curves," where subtended angles were computed by comparing the difference between two known bearings, and completing the solution by using Eq. (11.3) to compute the length of curve and establish the station.

Before continuing with each individual computation, a "solution sequence" is listed below, and this outline will be followed as the mathematics continues.

1. Compute the bearing and distance between points 5 and 1.
2. Compute all three interior angles for triangle 5,6,1.
3. Compute the angle between points 4,5,1 (angle α).
4. Compute angle phi (ϕ) by subtracting angle A from α.
5. Compute the length of the 4° curve using phi (ϕ) angle.
6. Establish the station on the 4° curve.
7. Compute the subtended angle at point 1 in triangle 6,1,2.
8. Establish the station on the 1° curve.

SEQUENCE ITEM NO. 1

The closure below completes item 1 in the solution sequence by solving the bearing and distance of line 5 to 1.

Course	Bearing	Distance	Cosine	Sine	North	South	East	West
1–2	West	5729.58'	0	1	---------	---------	---------	5729.58'
2–3	North	151.15'	1	0	151.15'	---------	---------	---------
3–4	S 58 12 00 W	466.45'	0.526955	0.849892	---------	245.798'	---------	396.432'
4–5	S 31 48 00 E	1432.39'	0.849892	0.526955	---------	1217.377'	754.8062'	---------
5–1	(N 76 16 23 E)	(5529.13')	0.237295	0.971437	(1312.026')	---------	5371.206'	---------

$$\tan \alpha = \frac{5371.2062}{1312.0263}$$

$$\tan \alpha = 4.0938251$$

$$\alpha = 76° \ 16' \ 23''$$

Remember, due to the page size in this book, the number of digits within each cell has been reduced. For your personal computation, you will want to carry all function numbers to eight places beyond the decimal point and carry all products of function and distances to at least three places beyond the decimal point.

With the bearing of line 5 to 1 completed, the sine and cosine functions for this angle (bearing) are located in the function tables or by using a hand-held calculator, the function digits are placed in the closure form, and the distance of course 5 to 1 is computed by dividing the cosine function into the north triangle leg and the sine function into the east triangle leg. Compare both distances to determine whether they match within a tolerance of ±0.03 ft. When the distances do not match and exceed the recommended tolerance, you can generally use the result obtained by dividing with the larger number; however, a thorough check of all input data, including multiplication, division, addition, and subtraction, should be made before proceeding.

SEQUENCE ITEM NO. 2

Compute all three interior angles for triangle 5,6,1 (Figure 11.11) using the "law of cosines." When all three sides of an obtuse triangle are known, the interior angles are completed using the law of cosines for each of the three angles, or the interior angles can be computed by using the law of cosines for the first angle and then the remaining two angles completed by using the "law of sines." This later method has a "time" advantage, but it also has a disadvantage, which we will discuss toward the end of the chapter.

As mentioned above, when the three sides in an obtuse triangle are known, the angles represented between any two adjacent sides can be computed using the law of cosines.[9] This equation involves a considerable amount of hand calculations, and therefore, a second equation, the law of sines, is used as a supplement to solve additional angles within the triangle. The equation is as follows:

$$a = \sqrt{b^2 + c^2 - 2bc(\cos A)} \tag{11.12}$$

Where:

a, b, and c represent sides of the triangle
cos *A* represents the angle opposite side *a*
cos *B* represents the angle opposite side *b*
cos *C* represents the angle opposite side *c*

Equation (11.12) is the base equation for the law of cosines and is rearranged in Eq. (11.13) so that cos *A* is removed from the square root by squaring both sides of the equation, with the following result:

$$a^2 = b^2 + c^2 - 2bc(\cos A) \tag{11.13}$$

By another rearrangement, using Eq. (11.13), we can isolate the cosine of angle *A* in the following manner:

$$\text{Cos } A = \frac{b^2 + c^2 - a^2}{2bc} \tag{11.14}$$

Figure 11.12 represents a triangle with three known sides:

Where:

side a = 5729.58 ft (the radius of a 1° curve)
side b = 1432.39 ft (the radius of a 4° curve)
side c = 5529.13 ft [the solution of line 5 to 1 (Figure 11.11)]

9 An obtuse angle is any angle greater than 90°. An oblique angle is any angle other than a right angle.

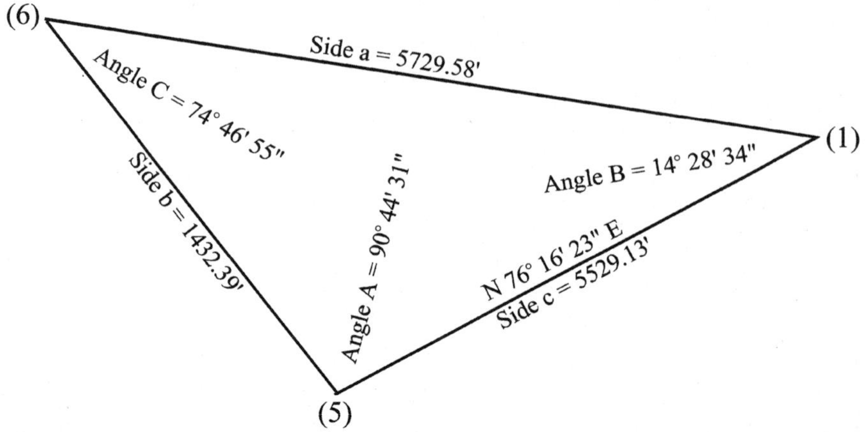

FIGURE 11.12 Computing interior angles in an obtuse triangle, using the Law of Cosines

The Law of Cosines

The solution for angle A (Figure 11.12), using the reduced equation for the law of cosines, is:

$$\text{Cos } A = \frac{b^2 + c^2 - a^2}{2bc} \tag{11.14}$$

Substituting:

$$\text{Cos } A = \frac{(1432.39)^2 + (5529.13)^2 - (5729.58)^2}{2(1432.39 \times 5529.13)}$$

$$= \frac{2{,}051{,}741.11 + 30{,}571{,}278.56 - 32{,}828{,}086.98}{2(7{,}919{,}870.521)}$$

$$= \frac{-205{,}067.31}{15{,}839{,}741.04}$$

$$= -0.012946380$$

$$A = 90°\ 44'\ 31''$$

Caution: The positive and negative signs must be strictly observed, and in this example, the cosine function of angle A is a negative number. This negative number must be used to determine the cosine of the angle that satisfies this function. Natural, eight decimal-place sine/cosine tables provide negative and positive functions for all angles. Hand held engineering calculators, however, do not permit the entry of negative function numbers, and therefore, your result does not recognize triangles with one obtuse angle (greater than 90°). You will, however, receive a positive function for the angle, such as the angle (89° 15' 29") if you place the negative function numbers in your calculator. Upon using this result in your summation of all three angles in the triangle, you will find an error, in that the angles do not total 180°. The error is the difference between an angle with a cosine negative function and one with a cosine positive function, as shown below:

Cosine negative function of –0.01294638 = 90° 44' 31"
Cosine positive function of + 0.01294638 = 89° 15' 29"

Sperling (1957) observed:

If a > b, ∠ A < B, which makes ∠ B less than 90°, and allows for only one solution.
If a = b, ∠ A = ∠ B, both angles are less than 90°, and only an isosceles triangle can be formed.
If a < b and ∠ A is acute, two triangles are possible.
If a = b sin A, the figure is a right triangle and only one solution is possible.
If a < b sin A, no triangle is possible.

Before doing a problem of this type you can generally determine the number of possible solutions by making an approximate small-scale drawing of the given parts.

The angle *B* solution (Figure 11.12), using the reduced equation for the law of cosines, is:

$$\text{Cos } B = \frac{a^2 + c^2 - b^2}{2ac} \tag{11.14}$$

Substituting

$$\text{Cos } B = \frac{(5729.58)^2 + (5529.13)^2 - (1432.39)^2}{2(5729.58 \times 5529.13)}$$

$$= \frac{32{,}828{,}086.98 + 30{,}571{,}278.56 - 2{,}051{,}741.11}{2(31{,}679{,}592.67)}$$

$$= \frac{61{,}347{,}623 \text{ rounded}}{63{,}359{,}184 \text{ rounded}}$$

$$= 0.9682514$$

$$B = 14° \; 28' \; 34''$$

The angle *C* solution (Figure 11.12), using the reduced equation for the law of cosines, is:

$$\text{Cos } C = \frac{b^2 + a^2 - c^2}{2ba} \tag{11.14}$$

Substituting:

$$\text{Cos } C = \frac{(1432.39)^2 + (5729.58)^2 - (5529.13)^2}{2(1432.39 \times 5729.58)}$$

$$= \frac{2{,}051{,}741.11 + 32{,}828{,}086.98 - 30{,}571{,}278.56}{2(8{,}206{,}993.09)}$$

$$= \frac{4{,}308{,}549.53}{16{,}413{,}986.18}$$

$$= 0.262492577$$

$$C = 74° \; 46' \; 55''$$

The computation of angle *C* completes the interior angles in obtuse triangle 5, 6, 1, using the law of cosines. Although the angles have been computed, you must yet verify the accuracy of the results. This verification is made by adding the three angles together, knowing that all three-sided figures have a total of 180°.

Summation of angles in Figure 11.12

$$\begin{array}{r} A = 90° \ 44' \ 31'' \\ B = 14° \ 28' \ 34'' \\ \underline{C = 74° \ 46' \ 55''} \\ 178° \ 118' \ 120'' \ or \ 180° \end{array}$$

Angular Totals Within Multiple-Sided Figures

The interior angles within multiple-sided figures can be computed using the following, Where n = number of sides:

Total degrees within a figure = *n* diminished by 2, multiplied by 180°

Example:
Using a seven-sided figure, n = 7, diminished by 2 = 5, times 180° = 900°.

The Law of Sines

The law of sines can be used as a supplement for computing interior angles within an obtuse triangle but only after the first angle is computed by the law of cosines. The equation for the law of sines uses the following ratio: One side is to the sine of the opposite angle as a second side is to the sine of the opposite unknown angle. Therefore, the first angle in the ratio must be known.

Example:
Refer to Figure 11.12, and compute the three angles in obtuse triangle 5, 6, 1 assuming that angle *C* is known. The law of sines states that the sides of a triangle are proportional to the sines of their opposite angles. Therefore:

$$\frac{a}{\sin A} = \frac{b}{\sin B} = \frac{c}{\sin C}$$

Where:
side a = 5729.58 ft
side b = 1432.39 ft
side c = 5529.13 ft
angle C = 74° 46' 55"

Computing angle *B*:

$$\frac{\text{side } c}{\sin \textit{angle } C} = \frac{\text{side } b}{\sin \textit{angle } B}$$

$$\frac{5529.13}{0.964933823} = \frac{1432.39}{\text{Sin } B}$$

$$\text{Sin } B = \frac{1382.161559}{5529.13}$$

$$\text{Sin } B = 14.47621783°$$

$$B = 14° \ 28' \ 34''$$

Computing angle A:

$$\frac{\text{side } c}{\text{Sin } angle\ C} = \frac{\text{side } a}{\text{Sin } angle\ A}$$

$$\frac{5529.13}{0.964933823} = \frac{5729.58}{\text{Sin } A}$$

$$\text{Sin } A = \frac{0.964933823 \times 5729.58}{5529.13}$$

$$= \frac{5{,}528.665534}{5529.13}$$

$$= 0.999915996$$

$$= 89.25733892°$$

$A = 89° 15' 26"$ (refer to the "caution" earlier in the chapter)

Note: Angles A, B, and C does not add up to 180°. This looks like an error if you do not realize that angle A is the supplement of the true angle. This was discussed earlier in the chapter.

SEQUENCE ITEM NO. 3

Compute the angle between points 4, 5, and 1 (angle α). Angle α is computed by comparing the bearing of line 4 to 5 (S 31° 48' 00" E) with the bearing of line 5 to 1 (N 76° 16' 23" E). The most secure method is to sketch the intersection of two perpendicular lines, representing the four quadrants, and to draw the approximate location for each of these two bearings. One bearing (N 76-16-23 E) is in the northeast quadrant, and the second bearing (N 31-48-00 W) is in the northwest quadrant. In this situation, both angles are added together.

	76-16-23
	+ 31-48-00
Angle alpha α equals =	107-64-23 (reduces to 108-04-23)

SEQUENCE ITEM NO. 4

Compute angle ϕ by subtracting angle A from α:

	108-04-23 (α)
	–90-44-30 (A)
Angle ϕ equals =	17-19-53 (ϕ)

SEQUENCE ITEM NO. 5

Compute the length of the 4° curve using ϕ angle. The length of a curve is computed using Eq. (11.3), as shown below:

$$L_c = \frac{(\Delta) \times (100)}{D_c}$$

$$L_c = \frac{(17^\circ\ 19'\ 53'') \times (100)}{4^\circ}$$

$$L_c = \frac{1733.138889}{4}$$

$$L_c = 433.28 \text{ ft}$$

SEQUENCE ITEM NO. 6

Establish the station on the 4° curve.

P.C. station	952 + 78.63
Length of curve =	4 + 33.28
POC =	957 + 11.91

SEQUENCE ITEM NO. 7

Compute the subtended angle in triangle 6, 1, 2. The subtended angle is computed by comparing the bearing of line 2 to 1 (assume west) with the bearing of line 1 to 6 (N 89-15 02 W). Prepare the quadrant sketch and draw the approximate locations for each of the bearings. One bearing is due west and the other is also in the northwest quadrant. Subtract the N 89-15-02 W bearing from 90° and the subtended angle is 00° 44' 58".

SEQUENCE ITEM NO. 8

Establish the station on the 1° curve. The subtended angle (item 7) produces a curve length of 74.94 ft. Add this distance to the P.C. station 6 + 75.43, and the intersection of the two curves is POC station 7 + 50.37.

CHAPTER 12
SIMPLE SPIRALS

INTRODUCTION

This chapter is dedicated to the training of entry-level engineers, surveyors, contractors, and technicians who want to improve their computing and field layout skills. Spirals, their calculations and field location, is a most fascinating study. During the latter part of the 1950s, the highway engineer began to *expand* the use of spirals in the design of roads, and when the state highway departments commenced recording the "centerline survey plat" as a legal document, greater responsibility was placed on the surveyor. The major responsibility was, of course, to accurately locate the centerline in a position that represented the control of all property acquisitions required for the construction of the highway. Many registered professional surveyors were not prepared for the introduction of spirals and the mathematics involved with this transition. The computations and field layout methods as presented in this chapter have become standard practice during the last 40 years.[1]

EARLY RECOGNITION

The spiral was first recognized during early design and construction of our railroad system, where it provided a smooth transition between straight trackage and curved alignments. The spiral, during this early period, was designed to minimize the centrifugal force brought about by deflecting from a tangent to a curve. This method is now called "superelevation" and was added to the trackage to facilitate the better movement in and out of the curve.[2]

---

1 I wish to acknowledge Joseph Barnett, author of Transition Curves for Highways, published by the United States Government Printing Press, Public Roads Administration, in 1940, while he was Senior Highway Design Engineer, Division of Design, Public Roads Administration. This chapter could not have been written without Mr. Barnett's text. It is in the public domain.

---

2 Superelevation refers to the mathematics of raising the outside edge of highway pavement and railroad track to lessen centrifugal force when transitioning to a curve.

Early highway engineers began using the transition curve (spirals) when the speed of vehicles and trains increased, and in the late 1930s, Barnett wrote his highway transition text, incorporating roadway design speed with the length of the spiral, and he computed superelevation rates to be used with degrees of curve ranging from 1° 28' 00" to 25° 00' 00".

DESIGNING SPIRALS

When engineers began designing spirals within the highway centerline, the surveying profession quickly became proficient in the mathematics and layout of this unusual curve. The 1960s and 1970s saw more interstate highways being designed and constructed than all the previous decades since 1921, when Congress enacted the first Federal Aid Highway Act. This congressional enactment not only furthered the design and construction of our highway system but provided the finances necessary to carry on this task.

THE SPIRAL

Spirals are the means of transitioning highways and railroads from straight-line alignments to curved alignments. Transition curves have become the method by which vehicles and trains can safely and smoothly change directions. Barnett, (1940) states "the transition curve is one in which the degree of curve D or D_c varies directly as the length of transition," being 0° of curve at the beginning of the transition and equal to the radius of the adjacent curve at the end of the transition. This explanation requires some additional input.

To simplify the sentence, concentrate on the zero deflection angle of the spiral, which is the commencing, or flat, part of the spiral. The spiral begins and ends at the moment of tangency. If you occupy this tangent point, not entering into the spiral, you will have no deflection angle and will be observing a straight line. The length of radius at this point is equal to infinity. When you enter the spiral, minute deflection angles appear and continually increase throughout the length of spiral. At the common point between spiral and curve, their "degree of curve," or "radius," is equal. Therefore, the spiral *radius* commences with a length equal to infinity and gradually diminishes in length until the spiral *radius* equals the curve *radius*.

DEGREE OF CURVE

The degree of curve can be explained if we use a field setting such as described below:

- Set a survey instrument at the central point of a curve
- Turn an angle of 1°
- If the observed angle separates two points (on the arc) by 100 ft, you have a 1° curve.

(The only radius satisfying this condition is 5729.58 ft, and we will verify this in a moment.)

A second example is:

- Set a survey instrument at the central point of a curve
- Turn an angle of 2°
- If the observed angle separates two points (on the arc) by 100 ft, you have a 2° curve.

(The only radius satisfying this condition is 2864.79 ft, and we will verify this in a moment.)

In each of these examples, an angle separates two points (on the arc) by 100 ft and thus determines the degree of curve. Therefore, the observed angle controls the degree of curve and the length of radius is computed using Eq. (12.2).

SPIRAL EQUATIONS

The first computation in this chapter is to solve for the radius of any point on the spiral. This can be computed after calculating the D or D_c, defined as degree of curve in spirals as well as in simple curves.

$$D = \frac{L}{L_s} \times D_c \tag{12.1}$$

Where:

D = degree of curve at each individual point
L = the length you have chosen along the spiral
L_s = the length of the whole spiral
D_c = the degree of curve adjacent to the spiral

Example:

Assume the total spiral is 200 ft in length, the degree of curve (D or D_c) is 4°, and the length of spiral required is 100 ft. The reduced equation is as follows:

$$D = \frac{100}{200} \times 4^\circ$$

$$D = 2^\circ$$

Therefore, the degree of curve at a point 100 ft into the spiral is 2°. Use the following equation to convert degrees of curve to a radius length:

$$R = \frac{5729.58}{D_c} \tag{12.2}$$

$$R = \frac{5729.58}{2}$$

$$R = 2864.79 \text{ ft}$$

(Therefore, any spiral radius can be computed in a similar manner.)

Perhaps this explanation will provide a preliminary understanding of the spiral and an insight into the statement, "the transition curve is one in which the (D or D_c), degree of curve, varies directly as the length of the spiral" (Barnett, 1940).

SPIRAL SYMBOLS

Several of the following pages outline the various equations used to compute the spiral and its parts. The parts are identified by universal symbols generally accepted throughout the years by engineers and surveyors. The symbols used in this chapter are exclusive to the

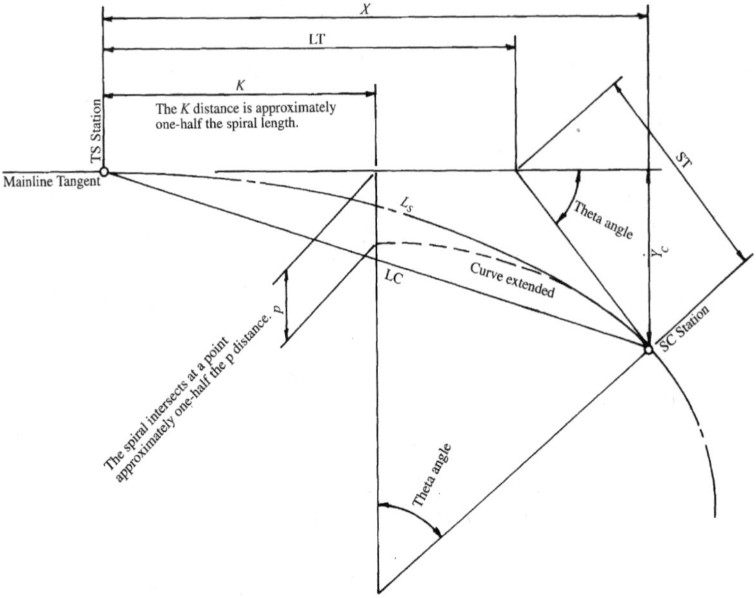

FIGURE 12.1 Spiral symbols and location of parts

spiral located in its normal position between a tangent and a curve. You will, however, be introduced to different symbols in Chapter 13. These differences are for the purpose of spiral identity, letting the surveyor know immediately the concept under which to compute.

The list below should be used as you review Figure 12.1. This figure shows the location of each spiral "part." Figure 12.2 shows the spiral located in its overall position, transitioning from a tangent to a curve and returning from the curve toward the tangent. The locations for the T_s and E_s distances are shown only in Figure 12.2. Figure 12.3 is an enlarged view of a curve being substituted for a spiral.

D = the degree of curve at any point on the spiral; Eq. (12.1)
R = radius at any point on the spiral; Eq. (12.2)
L_s = length of the total spiral; Eq. (12.3)
L = length of the partial spiral required; assumed and/or set
θ_s = theta angle between the tangents of the total spiral; Eq. (12.4)
θ = theta angle between the tangents of the partial spiral; Eq. (12.5)
LT = long tangent of the total spiral or partial spiral; function value
ST = short tangent of the total spiral or partial spiral; function value
LC = long chord between any two points on the spiral; function value
K = distance to the midpoint of the spiral; function value
p = distance from the spiral tangent to the mainline curve extended; function value
X_c = distance from the beginning of the spiral to a perpendicular through the end of the spiral; function value
Y_c = perpendicular distance from the spiral tangent extended to the end of the spiral; function value
T_s = distance from the beginning of the spiral to the mainline PI; Eq. (12.6)
E_s = distance from the mainline PI to the middle of the mainline curve; Eq. (12.7)
Δ = intersecting angle (delta) between mainline tangents

Δ_c = intersecting angle (delta) between the tangents of the curve only
TS = tangent to the spiral
SC = spiral to the curve
CS = curve to the spiral
ST = spiral to the tangent

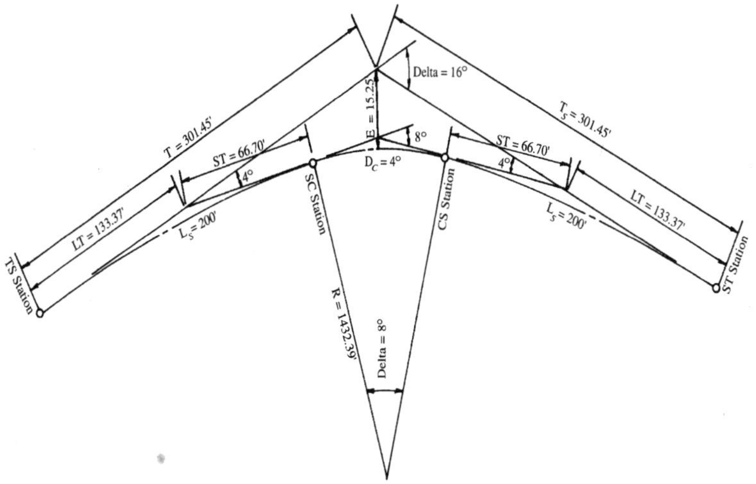

FIGURE 12.2 Curve with transition, both sides

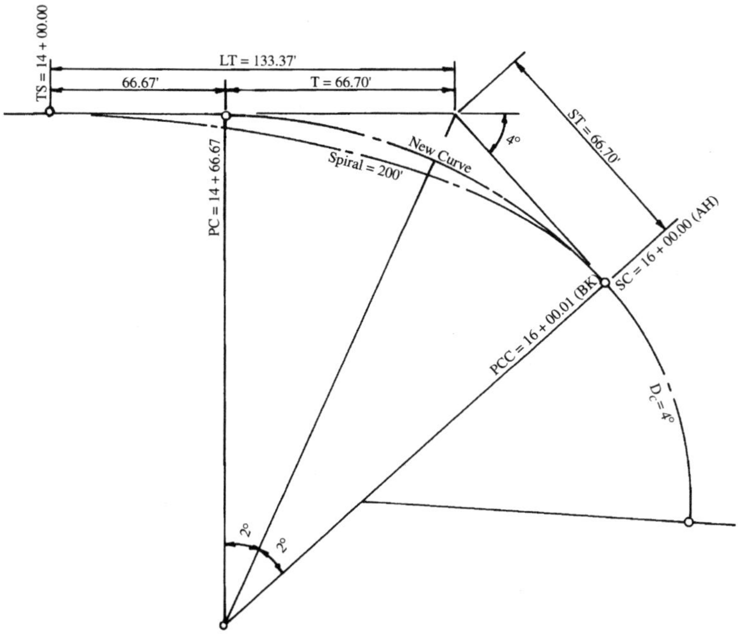

FIGURE 12.3 Substituting a curve for a spiral

The next several pages in this chapter address the individual computations of the various parts of the spiral. Use Figures 12.1, 12.2, and 12.3 as the equations are being introduced.

SPIRALS AS AN EXACT SCIENCE

Since the early 1940s, several authors have completed studies and written material covering this subject, and from this material we can deduce that spirals cannot be considered an exact mathematical science. This is supported by the fact that various authors have slightly different interpretations and function values of the spiral, spiral computations cannot be completed in one solution, but first require the length of the spiral to be chosen before the remaining parts of the spiral can be solved, all indicating that spiral transitions are not an exact science. You will find as the study continues that, when a length of spiral is chosen, all the remaining parts can be solved, and this nearly always leads to conflicting mathematical data between the spiral and surrounding criteria. This will be discussed more fully later in the Chapter.

The standard method of spiral layout is to locate all tangent lines, locate the intersecting PI of these tangents, measure the distance from the PI to the beginning of the spiral, locate the long and short spiral tangents, and finally, fill in the arc of the spiral. The errors that are accumulated during the layout of spiral stations must always be contained within the spiral. Never layout the arc of the spiral in anticipation of locating the next forward tangent. This statement is also true for the layout of simple curves.

SOLVING COMPONENTS

Assume the following mainline curve data for the following solutions:

$$\begin{aligned} L_s &= 200 \text{ ft} \\ D_c &= 4^\circ \\ R &= 1432.39 \text{ ft} \\ \Delta &= 16^\circ \end{aligned}$$

1. Find D at any point on the spiral. Assume for this computation that the length (L) of spiral required is 100 ft:

$$D = \frac{L}{L_s} \times D_c \tag{12.1}$$

$$D = \frac{100}{200} \times 4$$

$$D = 2^\circ$$

Assume for the following computation that the length (L) of spiral required is 150 ft:

$$D = \frac{150}{200} \times 4$$

$$D = 3^\circ$$

The radius for each of the previously computed degrees of curve is solved using an equation that was outlined in Chap. 11. The first computation is:

$$R = \frac{5729.58}{2} \tag{12.2}$$

$$R = 2864.79 \text{ ft (where } D = 2°)$$

The second computation is:

$$R = \frac{5729.58}{3}$$

$$R = 1909.86 \text{ ft (where } D = 3°)$$

Therefore, the radius at any point on the spiral can be determined by using Eq. (12.2).

2. Find the length (L_s) of the total spiral. The length of spiral was discussed in Chapter 6, where the design engineers use vehicle velocity (in miles per hour), the radius of the adjacent curve, and a factor based on centrifugal force to compute the minimum spiral length for their project. The surveyor, on the other hand, is usually provided the spiral length but often needs to compute the length when the θ angle is known. The computation is as follows:

Assume:

$$\theta_s = 4°.$$

Therefore:

$$L_s = \frac{\theta_s}{D_c} \times 200 \tag{12.3}$$

$$L_s = \frac{4}{4} \times 200$$

$$L_s = 200 \text{ ft}$$

3. Find the θ_s angle for the total spiral:

$$\theta_s = \frac{L_s}{200} \times D_c \tag{12.4}$$

$$\theta_s = \frac{L_s}{200} \times 4$$

$$\theta_s = \frac{200}{200} \times 4$$

$$\theta_s = 4°$$

4. Find the θ angle for the partial spiral. Assume that the length of the partial spiral is 100 ft.

$$\theta = \frac{L^2}{L_s^2} \times \theta_s \tag{12.5}$$

$$\theta = \frac{(100)^2}{(200)^2} \times 4$$

$$\theta = 1°$$

The following seven solutions are computed based on the spiral function tables (Table 12.1) at the end of this chapter, where the function digits represent a spiral length of 1 ft. Therefore, to complete the solution, the function digits must be multiplied by the length of the spiral.

TABLE 12.1 Spiral Functions (0° to 50° Theta Angle)

0°-5°	TABLE II- FUNCTIONS OF TRANSITION FOR L_s=1							
θ	p	k	x	y	L.T.	S.T.	L.C.	θ
0.0	.00000	.50000	1.00000	.00000	.66667	.33333	1.00000	0.0
.1	015	000	1.00000	058	67	33	1.00000	.1
.2	029	000	1.00000	116	67	34	1.00000	.2
.3	044	000	.99999	175	67	34	1.00000	.3
.4	058	000	999	233	67	34	1.00000	.4
.5	073	000	999	291	67	34	1.00000	.5
.6	088	000	999	349	67	34	1.00000	.6
.7	102	000	998	407	68	34	.99999	.7
.8	117	000	998	465	68	34	.99999	.8
0.9	.00131	.50000	.99997	.00524	.66668	.33334	.99999	0.9
1.0	.00146	.49999	.99997	.00582	.66668	.33334	.99999	1.0
.1	161	999	996	640	68	35	98	.1
.2	175	999	995	698	68	35	98	.2
.3	190	999	995	756	69	35	98	.3
.4	204	999	994	814	69	35	97	.4
.5	219	999	993	873	69	36	97	.5
.6	233	999	992	931	69	36	97	.6
.7	248	998	991	.00989	70	36	96	.7
.8	262	998	990	.01047	70	37	96	.8
1.9	.00277	.49998	.99989	.01105	.66671	.33337	.99995	1.9
2.0	.00291	.49998	.99988	.01163	.66671	.33337	.99995	2.0
.1	305	998	987	222	71	38	94	.1
.2	320	997	985	280	72	38	93	.2
.3	334	997	984	338	72	39	93	.3
.4	349	997	982	396	73	39	92	.4
.5	363	997	981	454	73	39	92	.5
.6	377	996	979	512	74	40	91	.6
.7	392	996	978	571	75	40	90	.7
.8	406	996	976	629	75	41	90	.8
2.9	.00421	.49996	.99975	.01687	.66676	.33341	.99989	2.9
3.0	.00435	.49995	.99973	.01745	.66676	.33342	.99988	3.0
.1	450	995	971	803	77	43	87	.1
.2	464	994	969	861	78	43	86	.2
.3	479	994	967	919	78	44	85	.3
.4	493	994	965	.01978	79	45	84	.4
.5	508	994	963	.02036	80	45	83	.5
.6	523	993	961	094	81	46	82	.6
.7	537	993	958	152	81	47	81	.7
.8	552	993	956	210	82	47	80	.8
3.9	.00566	.49992	.99953	.02268	.66683	.33348	.99979	3.9
4.0	.00581	.49992	.99951	.02326	.66684	.33349	.99978	4.0
.1	596	991	948	384	85	50	77	.1
.2	610	991	946	443	86	50	76	.2
.3	625	991	943	501	86	51	75	.3
.4	639	990	941	559	87	52	74	.4
.5	654	990	938	617	88	53	73	.5
.6	669	989	935	675	89	54	71	.6
.7	683	989	932	733	90	55	70	.7
.8	698	988	930	791	91	56	69	.8
4.9	.00712	.49988	.99927	.02849	.66692	.33357	.99967	4.9
5.0	.00727	.49987	.99924	.02907	.66693	.33358	.99966	5.0

FUNCTIONS OF TRANSITION FOR $L_S = 1$ TABLE II 5°-10°

θ	p	k	x	y	L.T.	S.T.	L.C.	θ
5.0	.00727	.49987	.99924	.02907	.66693	.33358	.99966	5.0
.1	742	987	921	.02965	694	359	965	.1
.2	756	986	918	.03023	696	360	963	.2
.3	771	986	914	082	697	361	962	.3
.4	785	985	911	140	698	362	961	.4
.5	800	985	908	198	699	363	959	.5
.6	814	984	904	256	700	364	958	.6
.7	829	984	901	314	701	365	956	.7
.8	843	983	897	372	703	366	954	.8
5.9	.00858	.49983	.99894	.03430	.66704	.33367	.99953	5.9
6.0	.00872	.49982	.99890	.03488	.66705	.33368	.99951	6.0
.1	887	981	886	546	706	369	950	.1
.2	901	981	882	604	708	371	948	.2
.3	916	980	879	662	709	372	946	.3
.4	930	979	875	720	710	373	944	.4
.5	945	979	871	778	712	374	943	.5
.6	960	978	867	836	713	376	941	.6
.7	974	977	863	894	715	377	939	.7
.8	989	976	859	.03952	716	378	937	.8
6.9	.01003	.49976	.99855	.04010	.66717	.33380	.99936	6.9
7.0	.01018	.49975	.99851	.04068	.66719	.33381	.99934	7.0
.1	033	974	846	126	720	382	932	.1
.2	047	973	842	184	722	384	930	.2
.3	062	973	838	242	724	385	928	.3
.4	076	972	833	300	725	386	926	.4
.5	091	971	829	358	727	388	924	.5
.6	105	970	824	416	728	389	922	.6
.7	120	969	819	474	730	391	920	.7
.8	134	969	815	532	732	392	918	.8
7.9	.01149	.49968	.99810	.04590	.66733	.33394	.99916	7.9
8.0	.01163	.49967	.99805	.04648	.66735	.33395	.99913	8.0
.1	178	966	800	706	737	397	911	.1
.2	192	965	795	764	738	399	909	.2
.3	207	965	790	822	740	400	907	.3
.4	221	964	785	879	742	402	904	.4
.5	236	963	780	937	744	403	902	.5
.6	250	962	775	.04995	745	405	900	.6
.7	265	961	770	.05053	747	407	897	.7
.8	279	961	764	.05111	749	409	895	.8
8.9	.01294	.49960	.99759	.05169	.66751	.33410	.99893	8.9
9.0	.01308	.49959	.99754	.05227	.66753	.33412	.99890	9.0
.1	323	958	748	285	755	414	888	.1
.2	337	957	742	342	757	416	885	.2
.3	352	956	737	400	759	417	883	.3
.4	366	955	731	458	761	419	880	.4
.5	381	954	725	516	763	421	878	.5
.6	395	953	719	574	765	423	875	.6
.7	410	952	713	632	767	425	873	.7
.8	424	951	708	690	769	427	870	.8
9.9	.01439	.49950	.99702	.05747	.66771	.33428	.99867	9.9
10.0	.01453	.49949	.99696	.05805	.66773	.33430	.99865	10.0

10°–15° TABLE II – FUNCTIONS OF TRANSITION FOR $L_s = 1$

θ	p	k	x	y	L.T.	S.T.	L.C.	θ
10.0	.01453	.49949	.99696	.05805	.66773	.33430	.99865	10.0
.1	468	948	690	863	776	432	862	.1
.2	482	947	684	921	778	434	859	.2
.3	497	946	677	.05978	780	436	856	.3
.4	511	945	671	.06036	782	438	854	.4
.5	526	944	665	094	784	440	851	.5
.6	540	943	658	152	787	442	848	.6
.7	555	942	652	210	789	444	845	.7
.8	569	941	645	267	791	447	842	.8
10.9	.01584	.49940	.99639	.06325	.66794	.33449	.99839	10.9
11.0	.01598	.49939	.99632	.06383	.66796	.33451	.99836	11.0
.1	613	938	625	440	798	453	833	.1
.2	627	937	618	498	801	455	830	.2
.3	642	935	612	556	803	457	827	.3
.4	656	934	605	614	806	460	824	.4
.5	671	933	598	671	808	462	821	.5
.6	685	932	591	729	811	464	818	.6
.7	700	931	584	787	813	466	815	.7
.8	714	929	576	844	816	469	812	.8
11.9	.01729	.49928	.99569	.06902	.66818	.33471	.99808	11.9
12.0	.01743	.49927	.99562	.06959	.66821	.33473	.99805	12.0
.1	757	926	555	.07017	823	476	802	.1
.2	772	924	547	075	826	478	799	.2
.3	786	923	540	132	828	480	795	.3
.4	801	922	532	190	831	483	792	.4
.5	815	921	525	248	834	485	789	.5
.6	829	919	517	305	836	488	785	.6
.7	844	918	509	363	839	490	782	.7
.8	858	917	502	420	842	493	778	.8
12.9	.01873	.49915	.99494	.07478	.66845	.33495	.99775	12.9
13.0	.01887	.49914	.99486	.07535	.66847	.33498	.99771	13.0
.1	902	913	478	593	850	500	768	.1
.2	916	911	470	650	853	503	764	.2
.3	931	910	462	708	856	505	761	.3
.4	945	909	454	765	859	508	757	.4
.5	960	908	446	823	862	511	753	.5
.6	974	906	438	880	865	513	750	.6
.7	.01989	905	430	938	868	516	746	.7
.8	.02003	904	421	.07995	871	519	742	.8
13.9	.02018	.49902	.99413	.08053	.66874	.33521	.99739	13.9
14.0	.02032	.49901	.99405	.08110	.66877	.33524	.99735	14.0
.1	046	900	396	168	880	527	731	.1
.2	061	898	387	225	883	530	727	.2
.3	075	897	379	282	886	532	723	.3
.4	090	895	370	340	889	535	720	.4
.5	104	894	362	397	892	538	716	.5
.6	118	892	353	455	895	541	712	.6
.7	133	891	344	512	898	544	708	.7
.8	147	889	335	569	901	547	704	.8
14.9	.02162	.49888	.99326	.08627	.66904	.33550	.99700	14.9
15.0	.02176	.49886	.99317	.08684	.66908	.33553	.99696	15.0

FUNCTIONS OF TRANSITION FOR $L_s=1$ TABLE II								15°-20°
θ	p	k	x	y	L.T.	S.T.	L.C.	θ
15.0	.02176	.49886	.99317	.08684	.66908	.33553	.99696	15.0
.1	190	884	308	741	911	556	692	.1
.2	205	883	299	799	914	559	688	.2
.3	219	881	289	856	918	561	683	.3
.4	234	880	280	913	921	564	679	.4
.5	248	878	271	.08970	924	567	675	.5
.6	262	876	261	.09028	928	571	671	.6
.7	277	875	252	085	931	574	667	.7
.8	291	873	242	142	934	577	662	.8
15.9	.02306	.49872	.99233	.09200	.66938	.33580	.99658	15.9
16.0	.02320	.49870	.99223	.09257	.66941	.33583	.99654	16.0
.1	335	868	213	314	945	586	649	.1
.2	349	867	203	371	948	589	645	.2
.3	364	865	194	428	952	593	641	.3
.4	378	864	184	485	955	596	636	.4
.5	393	862	174	543	959	599	632	.5
.6	407	860	164	600	962	602	627	.6
.7	422	859	154	657	966	606	623	.7
.8	436	857	143	714	970	609	618	.8
16.9	.02451	.49856	.99133	.09771	.66973	.33612	.99614	16.9
17.0	.02465	.49854	.99123	.09828	.66977	.33615	.99609	17.0
.1	479	852	113	885	981	619	605	.1
.2	494	850	102	942	984	622	600	.2
.3	508	849	092	.09999	988	626	595	.3
.4	522	847	081	.10056	992	629	591	.4
.5	537	845	071	113	995	632	586	.5
.6	551	843	060	170	.66999	636	581	.6
.7	565	841	050	227	.67003	639	576	.7
.8	579	840	039	284	.67007	643	572	.8
17.9	.02594	.49838	.99029	.10341	.67011	.33646	.99567	17.9
18.0	.02608	.49836	.99018	.10398	.67015	.33650	.99562	18.0
.1	622	834	.99007	455	019	654	557	.1
.2	637	832	.98996	512	023	657	552	.2
.3	651	830	985	569	027	661	547	.3
.4	666	828	974	626	031	664	542	.4
.5	680	827	962	683	035	668	537	.5
.6	694	825	951	740	039	672	532	.6
.7	709	823	940	797	043	675	527	.7
.8	723	821	929	854	047	679	522	.8
18.9	.02738	.49819	.98917	.10910	.67051	.33683	.99517	18.9
19.0	.02752	.49817	.98906	.10967	.67055	.33687	.99512	19.0
.1	766	815	894	.11024	059	690	507	.1
.2	781	813	883	081	063	694	502	.2
.3	795	811	871	138	067	698	497	.3
.4	810	809	860	194	072	702	491	.4
.5	824	808	848	251	076	706	486	.5
.6	838	806	836	308	080	709	481	.6
.7	853	804	824	364	084	713	476	.7
.8	867	802	812	421	089	717	470	.8
19.9	882	800	.98800	.11478	.67093	.33721	.99465	19.9
20.0	.02896	.49798	.98788	.11535	.67097	.33725	.99460	20.0

20°-25°	FUNCTIONS OF TRANSITION FOR $L_s = 1$						TABLE II	
θ	p	k	x	y	L.T.	S.T.	L.C.	θ
20.0	.02896	.49798	.98788	.11535	.67097	.33725	.99460	20.0
.1	910	796	776	591	102	729	454	.1
.2	925	794	764	648	106	733	449	.2
.3	939	791	752	705	111	737	443	.3
.4	954	789	740	761	115	741	438	.4
.5	968	787	728	818	119	745	432	.5
.6	982	785	715	875	124	749	427	.6
.7	.02997	783	703	931	128	753	421	.7
.8	.03011	781	690	.11988	133	758	415	.8
20.9	.03026	.49779	.98678	.12044	.67137	.33762	.99410	20.9
21.0	.03040	.49777	.98665	.12101	.67142	.33766	.99404	21.0
.1	054	775	652	157	147	770	399	.1
.2	068	773	639	214	151	774	393	.2
.3	083	770	627	270	156	779	387	.3
.4	097	768	614	327	161	783	381	.4
.5	111	766	601	383	165	787	376	.5
.6	125	764	588	439	170	791	370	.6
.7	140	762	575	496	175	796	364	.7
.8	154	759	562	552	180	800	358	.8
21.9	.03169	.49757	.98549	.12609	.67184	.33804	.99352	21.9
22.0	.03183	.49755	.98536	.12665	.67189	.33809	.99346	22.0
.1	197	753	523	721	194	813	340	.1
.2	211	751	509	777	199	818	334	.2
.3	226	749	496	834	204	822	328	.3
.4	240	747	482	890	208	826	322	.4
.5	254	745	469	.12946	213	831	316	.5
.6	268	743	455	.13002	218	835	310	.6
.7	283	740	442	059	223	840	304	.7
.8	297	738	428	115	228	844	298	.8
22.9	.03312	.49735	.98415	.13172	.67233	.33849	.99292	22.9
23.0	.03326	.49733	.98401	.13228	.67238	.33854	.99286	23.0
.1	340	731	387	284	243	858	279	.1
.2	354	728	373	340	248	863	273	.2
.3	369	726	359	396	254	868	267	.3
.4	383	723	345	452	259	872	261	.4
.5	397	721	331	508	264	877	254	.5
.6	411	719	316	564	269	882	248	.6
.7	426	716	302	621	274	886	242	.7
.8	440	714	288	677	280	891	235	.8
23.9	.03455	.49711	.98274	.13733	.67285	.33896	.99229	23.9
24.0	.03469	.49709	.98260	.13789	.67290	.33901	.99222	24.0
.1	483	707	245	845	295	906	216	.1
.2	497	704	231	901	301	910	209	.2
.3	512	702	216	.13957	306	915	203	.3
.4	526	699	202	.14012	311	920	196	.4
.5	540	697	187	068	317	925	190	.5
.6	554	694	172	124	322	930	183	.6
.7	568	692	157	180	328	935	177	.7
.8	583	689	143	236	333	940	170	.8
24.9	.03597	.49687	.98128	.14292	.67339	.33945	.99163	24.9
25.0	.03611	.49684	.98113	.14348	.67344	.33950	.99157	25.0

FUNCTIONS OF TRANSITION FOR $L_s = 1$					TABLE II			25°-30°
θ	p	k	x	y	L.T.	S.T.	L.C.	θ
25.0	.03611	.49684	.98113	.14348	.67344	.33950	.99157	25.0
.1	625	681	098	404	350	955	150	.1
.2	640	679	083	459	355	960	143	.2
.3	654	676	068	515	361	965	136	.3
.4	669	674	053	571	366	970	129	.4
.5	683	671	038	627	372	975	123	.5
.6	697	668	022	682	378	981	116	.6
.7	711	666	.98007	738	383	986	109	.7
.8	725	663	.97991	794	389	991	102	.8
25.9	.03739	.49661	.97976	.14849	.67395	.33996	.99095	25.9
26.0	.03753	.49658	.97960	.14905	.67400	.34001	.99088	26.0
.1	767	656	945	.14961	406	007	081	.1
.2	782	653	929	.15016	412	012	074	.2
.3	796	651	913	072	418	017	067	.3
.4	811	648	898	128	424	023	060	.4
.5	825	646	882	183	430	028	053	.5
.6	839	643	866	239	435	033	046	.6
.7	853	640	850	294	441	039	038	.7
.8	868	638	834	350	447	044	031	.8
26.9	.03882	.49635	.97818	.15405	.67453	.34049	.99024	26.9
27.0	.03896	.49632	.97802	.15461	.67459	.34055	.99017	27.0
.1	910	629	786	516	465	060	.99009	.1
.2	924	626	770	571	471	066	.99002	.2
.3	939	624	753	627	477	071	.98995	.3
.4	953	621	737	682	483	077	987	.4
.5	967	618	721	738	490	083	980	.5
.6	981	615	704	793	496	088	973	.6
.7	.03995	613	688	848	502	094	965	.7
.8	.04009	610	671	903	508	099	958	.8
27.9	.04023	.49608	.97655	.15959	.67514	.34105	.98950	27.9
28.0	.04037	.49605	.97638	.16014	.67520	.34111	.98943	28.0
.1	051	602	621	069	527	116	935	.1
.2	065	599	604	124	533	122	928	.2
.3	080	596	588	180	539	128	920	.3
.4	094	593	571	235	546	134	913	.4
.5	108	590	554	290	552	139	905	.5
.6	122	587	537	345	558	145	897	.6
.7	136	584	520	400	565	151	890	.7
.8	151	582	503	455	571	157	882	.8
28.9	.04165	.49579	.97486	.16510	.67578	.34163	.98874	28.9
29.0	.04179	.49576	.97469	.16565	.67584	.34169	.98866	29.0
.1	193	573	452	620	591	165	859	.1
.2	207	570	434	675	597	171	851	.2
.3	222	567	417	730	604	177	843	.3
.4	236	564	399	785	610	193	835	.4
.5	250	561	382	840	617	199	827	.5
.6	264	558	364	895	623	205	819	.6
.7	278	555	346	.16950	630	211	811	.7
.8	293	552	329	.17005	637	217	803	.8
29.9	.04307	.49549	.97311	.17060	.67643	.34223	.98795	29.9
30.0	.04321	.49546	.97293	.17114	.67650	.34229	.98787	30.0

30°-35° FUNCTIONS OF TRANSITION FOR $L_s = 1$ TABLE II

θ	p	k	x	y	L.T.	S.T.	L.C.	θ
30.0	.04321	.49546	.97293	.17114	.67650	.34229	.98787	30.0
.1	335	543	275	169	657	235	779	.1
.2	349	540	257	224	664	241	771	.2
.3	363	537	239	279	670	248	763	.3
.4	377	534	221	333	677	254	755	.4
.5	391	531	203	388	684	260	746	.5
.6	405	528	185	443	691	266	738	.6
.7	419	525	167	498	698	273	730	.7
.8	434	522	148	552	705	279	722	.8
30.9	.04448	.49519	.97130	.17607	.67712	.34285	.98713	30.9
31.0	.04462	.49516	.97112	.17661	.67719	.34292	.98705	31.0
.1	476	513	094	716	726	298	697	.1
.2	490	510	075	770	733	304	688	.2
.3	504	506	057	825	740	311	680	.3
.4	518	503	038	879	747	317	672	.4
.5	532	500	020	934	754	324	663	.5
.6	546	497	.97001	.17988	761	330	655	.6
.7	560	494	.96982	.18043	768	337	646	.7
.8	574	490	.96963	.18097	775	343	638	.8
31.9	.04588	.49487	.96944	.18152	.67783	.34350	.98629	31.9
32.0	.04602	.49484	.96926	.18206	.67790	.34356	.98621	32.0
.1	616	481	907	260	797	363	612	.1
.2	630	478	887	315	804	370	603	.2
.3	645	475	868	369	812	376	595	.3
.4	659	472	849	424	819	383	586	.4
.5	673	469	830	478	826	390	577	.5
.6	687	466	811	532	834	397	569	.6
.7	701	462	791	586	841	403	560	.7
.8	715	459	772	640	849	410	551	.8
32.9	.04729	.49455	.96752	.18694	.67856	.34417	.98542	32.9
33.0	.04743	.49452	.96733	.18748	.67863	.34424	.98534	33.0
.1	757	449	713	803	871	431	525	.1
.2	771	445	694	857	878	438	516	.2
.3	785	442	674	911	886	444	507	.3
.4	799	438	655	.18965	894	451	498	.4
.5	813	435	635	.19019	901	458	489	.5
.6	827	432	615	073	909	465	480	.6
.7	841	429	595	127	916	472	471	.7
.8	855	425	576	181	924	479	462	.8
33.9	.04869	.49422	.96556	.19234	.67932	.34486	.98453	33.9
34.0	.04883	.49419	.96536	.19288	.67939	.34493	.98444	34.0
.1	897	415	516	342	947	500	435	.1
.2	911	412	496	396	955	508	425	.2
.3	925	408	475	450	963	515	416	.3
.4	939	405	455	504	971	522	407	.4
.5	953	401	435	557	979	529	398	.5
.6	967	398	414	611	987	536	389	.6
.7	981	395	394	665	.67994	544	379	.7
.8	.04995	391	373	718	.68002	551	370	.8
34.9	.05009	.49388	.96353	.19772	.68010	.34558	.98361	34.9
35.0	.05023	.49385	.96332	.19826	.68018	.34565	.98351	35.0

FUNCTIONS OF TRANSITION FOR $L_s = 1$ TABLE II								35°-40°
θ	p	k	x	y	L.T.	S.T.	L.C.	θ
35.0	.05023	.49385	.96332	.19826	.68018	.34565	.98351	35.0
.1	037	381	311	879	026	573	342	.1
.2	051	378	291	933	034	580	333	.2
.3	065	374	270	.19987	042	587	323	.3
.4	079	371	250	.20040	051	595	314	.4
.5	093	367	229	.20094	059	602	304	.5
.6	107	363	208	147	067	610	295	.6
.7	121	360	187	201	075	617	285	.7
.8	135	356	166	254	083	625	276	.8
35.9	.05149	.49353	.96145	.20307	.68092	.34632	.98266	35.9
36.0	.05163	.49349	.96124	.20361	.68100	.34640	.98257	36.0
.1	177	345	103	414	108	647	247	.1
.2	191	342	081	467	116	655	237	.2
.3	204	338	060	521	125	663	227	.3
.4	218	335	038	574	133	670	218	.4
.5	232	331	.96017	627	141	678	208	.5
.6	246	327	.95996	680	150	686	198	.6
.7	260	324	.95974	734	158	693	188	.7
.8	273	320	.95953	787	167	701	179	.8
36.9	.05287	.49317	.95931	.20840	.68175	.34709	.98169	36.9
37.0	.05301	.49313	.95910	.20893	.68184	.34717	.98159	37.0
.1	315	309	888	.20946	192	725	149	.1
.2	329	306	866	.20999	201	732	139	.2
.3	343	302	844	.21052	210	740	129	.3
.4	357	299	822	105	218	748	119	.4
.5	371	295	800	158	227	756	109	.5
.6	385	291	778	211	236	764	099	.6
.7	399	287	756	264	244	772	089	.7
.8	413	284	734	317	253	780	079	.8
37.9	.05427	.49280	.95712	.21370	.68262	.34788	.98069	37.9
38.0	.05441	.49276	.95690	.21423	.68271	.34796	.98059	38.0
.1	455	272	668	475	279	804	049	.1
.2	469	268	645	528	288	812	038	.2
.3	482	264	623	581	297	820	028	.3
.4	496	260	601	634	306	829	018	.4
.5	510	256	578	686	315	837	.98008	.5
.6	524	252	556	739	324	845	.97997	.6
.7	538	249	533	792	333	853	987	.7
.8	551	245	511	844	342	861	977	.8
38.9	.05565	.49242	.95488	.21897	.68351	.34870	.97967	38.9
39.0	.05579	.49238	.95466	.21949	.68360	.34878	.97956	39.0
.1	593	234	443	.22002	369	886	946	.1
.2	607	230	420	054	379	895	935	.2
.3	620	226	397	107	388	903	925	.3
.4	634	222	374	159	397	911	914	.4
.5	648	218	351	212	406	920	904	.5
.6	662	214	328	264	415	928	893	.6
.7	676	210	305	316	424	937	683	.7
.8	690	207	281	369	434	945	872	.8
39.9	.05704	.49203	.95258	.22421	.68443	.34954	.97861	39.9
40.0	.05718	.49199	.95235	.22473	.68452	.34962	.97851	40.0

40°-45°	Functions of Transition for $L_s = 1$ Table II							
θ	p	k	x	y	L.T.	S.T.	L.C.	θ
40.0	.05718	.49199	.95235	.22473	.68452	.34962	.97851	40.0
.1	732	195	212	526	462	971	840	.1
.2	745	191	188	578	471	980	829	.2
.3	759	187	165	630	481	988	819	.3
.4	772	183	141	682	490	.34997	808	.4
.5	786	179	118	734	500	.35006	797	.5
.6	800	175	094	786	509	014	786	.6
.7	814	171	071	838	519	023	775	.7
.8	827	167	047	890	528	032	765	.8
40.9	.05841	.49163	.95023	.22942	.68538	.35041	.97754	40.9
41.0	.05855	.49159	.95000	.22994	.68547	.35049	.97743	41.0
.1	869	155	.94976	.23046	557	058	732	.1
.2	883	151	952	098	567	067	721	.2
.3	896	146	928	150	577	076	710	.3
.4	910	142	904	202	586	085	699	.4
.5	924	138	880	254	596	094	688	.5
.6	938	134	856	306	606	103	677	.6
.7	952	130	832	358	616	112	666	.7
.8	965	126	807	409	626	121	655	.8
41.9	.05979	.49122	.94783	.23461	.68635	.35130	.97643	41.9
42.0	.05993	.49118	.94759	.23513	.68645	.35139	.97632	42.0
.1	.06007	114	734	564	655	148	621	.1
.2	020	110	710	616	665	158	610	.2
.3	034	105	685	667	675	167	599	.3
.4	047	101	661	719	685	176	587	.4
.5	061	097	636	771	695	185	576	.5
.6	075	093	612	822	706	194	565	.6
.7	089	088	587	874	716	204	553	.7
.8	102	084	562	925	726	213	542	.8
42.9	.06116	.49079	.94538	.23976	.68736	.35222	.97531	42.9
43.0	.06130	.49075	.94513	.24028	.68746	.35232	.97519	43.0
.1	144	071	488	079	756	241	508	.1
.2	157	067	463	130	767	250	496	.2
.3	171	062	438	182	777	260	485	.3
.4	184	058	413	233	787	269	473	.4
.5	198	054	388	284	798	279	462	.5
.6	212	050	363	335	808	288	450	.6
.7	226	045	337	387	818	298	438	.7
.8	239	041	312	438	829	307	427	.8
43.9	.06253	.49036	.94287	.24489	.68839	.35317	.97415	43.9
44.0	.06267	.49032	.94262	.24540	.68850	.35327	.97404	44.0
.1	281	028	236	591	860	336	392	.1
.2	294	024	211	642	871	346	380	.2
.3	308	019	185	693	882	356	368	.3
.4	321	015	160	744	892	365	357	.4
.5	335	011	134	795	903	375	345	.5
.6	349	007	108	846	914	385	333	.6
.7	362	.49003	082	896	924	395	321	.7
.8	376	.48998	057	947	935	405	309	.8
44.9	.06389	.48994	.94031	.24998	.68946	.35415	.97297	44.9
45.0	.06403	.48990	.94005	.25049	.68957	.35424	.97285	45.0

FUNCTIONS OF TRANSITION FOR $L_s = 1$ TABLE II 45°-50°								
θ	p	k	x	y	L.T.	S.T.	L.C.	θ
45.0	.06403	.48990	.94005	.25049	.68957	.35424	.97285	45.0
.1	416	985	.93979	100	967	434	273	.1
.2	430	981	953	150	978	444	261	.2
.3	443	976	927	201	.68989	454	249	.3
.4	457	972	901	251	.69000	464	237	.4
.5	470	967	875	302	011	474	225	.5
.6	484	963	849	353	022	485	213	.6
.7	497	958	823	403	033	495	201	.7
.8	511	954	796	454	044	505	189	.8
45.9	.06524	.48949	.93770	.25504	.69055	.35515	.97177	45.9
46.0	.06538	.48945	.93744	.25555	.69066	.35525	.97165	46.0
.1	552	940	717	605	078	535	152	.1
.2	565	936	691	655	089	546	140	.2
.3	579	931	664	705	100	556	128	.3
.4	592	927	638	756	111	566	116	.4
.5	606	922	611	806	122	576	103	.5
.6	620	918	584	856	134	587	091	.6
.7	633	913	558	907	145	597	079	.7
.8	647	909	531	.25957	156	608	066	.8
46.9	.06660	.48904	.93504	.26007	.69168	.35618	.97054	46.9
47.0	.06674	.48900	.93477	.26057	.69179	.35629	.97041	47.0
.1	688	895	450	107	191	639	029	.1
.2	701	890	423	157	202	650	016	.2
.3	715	885	396	207	214	660	.97004	.3
.4	728	881	369	257	225	671	.96991	.4
.5	742	876	342	307	237	681	979	.5
.6	755	871	315	357	248	692	966	.6
.7	768	866	288	407	260	703	954	.7
.8	782	862	260	457	272	714	941	.8
47.9	.06795	.48857	.93233	.26506	.69283	.35724	.96928	47.9
48.0	.06809	.48852	.93206	.26556	.69295	.35735	.96916	48.0
.1	822	847	178	606	307	746	903	.1
.2	836	843	151	656	318	757	890	.2
.3	849	838	123	705	330	768	877	.3
.4	863	834	096	755	342	778	865	.4
.5	876	829	069	805	354	789	852	.5
.6	890	824	041	854	366	800	839	.6
.7	903	819	.93013	904	378	811	826	.7
.8	917	815	.92985	.26953	390	822	813	.8
48.9	.06930	.48810	.92958	.27003	.69402	.35833	.96800	48.9
49.0	.06944	.48805	.92930	.27052	.69414	.35844	.96787	49.0
.1	957	800	902	101	426	855	774	.1
.2	970	796	874	151	438	867	761	.2
.3	983	791	846	200	450	878	748	.3
.4	.06997	786	818	250	463	889	735	.4
.5	.07010	782	790	299	475	900	722	.5
.6	023	777	762	348	487	911	709	.6
.7	037	772	733	397	499	923	696	.7
.8	050	767	705	446	512	934	683	.8
49.9	.07064	.48762	.92677	.27495	.69524	.35945	.96670	49.9
50.0	.07078	.48757	.92649	.27544	.69536	.35957	.96656	50.0

5. Find the LT (long tangent) when the $\theta_s = 4°$. Refer to the first page of Table 12.1, which covers θ angles from 0.10° to, but not including 5°. Scan down the column on either side of the page to the 4° function. Follow this row across the page until you are directly below the LT column, reading a function of 0.66684 per each foot of spiral. The length of the LT is computed as follows:

$$\text{LT} = 0.66684 \times 200 = 133.37 \text{ ft}$$

6. Find the ST (short tangent) when the $\theta_s = 4°$. Refer to Table 12.1, moving across the page until directly below the ST column, reading a function of 0.33349 per each foot of spiral. The length of the ST is computed as follows:

$$\text{ST} = 0.33349 \times 200 = 66.70 \text{ ft}$$

7. Find the LC (long chord) when the $\theta_s = 4°$. Refer to Table 12.1, moving across the page until directly below the LC column, reading a function of 0.99978 per each foot of spiral. The length of the LC is computed as follows:

$$\text{LC} = 0.99978 \times 200 = 199.96 \text{ ft}$$

8. Find the K distance when the $\theta_s = 4°$. Refer to Table 12.1, moving across the page until directly below the K column, reading a function of 0.49992 per each foot of spiral. The length of K is computed as follows:

$$K = 0.49992 \times 200 = 99.98 \text{ ft}$$

9. Find the p distance when the $\theta_s = 4°$. Refer to Table 12.1, moving across the page until directly below the p column, reading a function of 0.00581 per each foot of spiral. The length of p is computed as follows:

$$p = 0.00581 \times 200 = 1.16 \text{ ft}$$

10. Find the X distance when the $\theta_s = 4°$. Refer to Table 12.1, moving across the page until directly below the X column, reading a function of 0.99951 per each foot of spiral. The length of X is computed as follows:

$$X = 0.99951 \times 200 = 199.90 \text{ ft}$$

11. Find the Y distance when the $\theta_s = 4°$. Refer to Table 12.1, moving across the page until directly below the Y column, reading a function of 0.02326 per each foot of spiral. The length of Y is computed as follows:

$$Y = 0.02326 \times 200 = 4.65 \text{ ft}$$

12. Find the T_s distance of the spiral-curve relation (see Fig. 12.2) using the previously assumed curve data and computed spiral parts:

$$T_s = (R_c + p)\left(\tan \frac{\Delta}{2}\right) + K \qquad (12.6)$$

$$T_s = (1432.39 + 1.16)(\tan 8°) + K$$

$$T_s = (1433.55)(0.14054083) + 99.98$$
$$T_s = 201.47 + 99.98$$
$$T_s = 301.45 \text{ ft}$$

13. Find the E_s distance of the spiral-curve relation (see Figure 12.2) using the previously assumed curve data and computed spiral parts:

$$E_s = (R_c + p)(\text{exsec } \frac{\Delta}{2}) + p \tag{12.7}$$

$$E_s = (1432.39 + 1.16)(0.0098276) + p$$
$$E_s = 14.09 + 1.16$$
$$E_s = 15.25 \text{ ft}$$

Note: When exsecant tables are not available, the exsecant value can be computed using the following equation:

$$\text{Exsec } \frac{\Delta}{2} = \frac{1}{\frac{\cos \Delta}{2}} - 1 \tag{12.8}$$

$$\text{Exsec } \frac{\Delta}{2} = \frac{1}{\cos 8°} - 1$$

$$\text{Exsec } \frac{\Delta}{2} = \frac{1}{0.99026807} - 1$$

$$\text{Exsec } \frac{\Delta}{2} = 0.00982$$

The following examples are spiral solutions where the computation consists of non-traditional values.

1. Find the degree of curve for any point on the spiral:

$$D = \frac{L}{L_s} \times D_c \tag{12.1}$$

Where:

D = the degree of curve (at the point required)
L = 160 ft
L_s = 350 ft
D_c = 7° 30'

$$D = \frac{160}{350} \times 7° \, 30'$$

$$D = 0.4571418 \times 7.5$$
$$D = 3.4285714°$$
$$D = 3° \, 25' \, 42''$$

Substituting D from Eq. (12.1) to obtain the radius:

$$R = \frac{5729.58}{3.4285714} \tag{12.2}$$

Therefore:

$$R = 1671.13 \text{ ft}$$

(If you need to determine the radius at any point on the spiral, use this method)

2. Find the length of spiral when the θ_s angle and the D_c are known:

$$L_s = \frac{\theta_s}{D_c} \times 200 \qquad (12.3)$$

Where:

L_s = the length of the total spiral
θ_s = 9° 52'
D_c = 6° 30' 30"
200 = constant

Substituting:

$$L_s = \frac{9° \ 52'}{6° \ 30' \ 30''} \times 200$$

$$L_s = \frac{9.8666666}{6.5083333} \times 200$$

$$L_s = 303.20 \text{ ft}$$

3. Find θ_s when the length of spiral and degree of curve are known.

$$\theta_s \frac{L_s}{200} \times D_c \qquad (12.4)$$

Where:

L_s = 475 ft
D_c = 1° 28' 30"

Substituting:

$$\theta_s = \frac{475}{200} \times 1° \ 28' \ 30''$$

$$\theta_s = 2.375 \times 1.475$$

$$\theta_s = 3.503125°$$

$$\theta_s = 3° \ 30' \ 11''$$

IMPORTANT SPIRAL CALCULATION

The following spiral calculation Eq. (12.5) is perhaps the most important in all spiral mathematics. By using this equation, and substituting a known or assumed spiral length, you solve for the θ angle between the spiral's two tangents, and by entering the function table (Table 12.1) with this angle, all other parts of the spiral can be computed. Abbreviations for the various parts of the spiral are located horizontally across each function page, with values assigned. It is extremely important to remember that the functions for each spiral part represent a value of 1 ft. Therefore, the function must be multiplied by the spiral distance to obtain the correct result.

4. Find the θ angle when the required point on the spiral is less than the total spiral:

$$\theta = \frac{L^2}{L_s^2} \times \theta_s \tag{12.5}$$

Where:

L = the required spiral length = 325 ft
L_s = the total designed spiral = 600 ft
θ_s = 4° 25' 30"

Substituting:

$$\theta = \frac{(325)^2}{(600)^2} \times 4° \, 25' \, 30''$$

$$\theta = \frac{105625.0}{360000.0} \times 4.425$$

$$\theta = 0.29340277 \times 4.425$$

$$\theta = 1.298307292°$$

$$\theta = 1° \, 17' \, 54''$$

Therefore, all of the remaining spiral parts can be calculated by using the function tables when the θ angle computation is completed. Let's continue our study by solving for the values of each spiral part whose θ angle is 1° 17' 54".

5. Find the LT for the above: Note: You will occasionally find θ and θ_s used interchangeably; however, the θ angle is interpreted as the deflection angle of a spiral whose length is less than the total spiral, and the θ_s is interpreted as the deflection angle of the original (total) spiral.

Refer to the abbreviated function table below or Table 12.1 locating a θ angle of 1° 17' 54" (and reduced to tenths of a degree), the angle is 1.3°. When you have found the angle, move horizontally across the page until you are directly under the LT column, selecting the function digits of 0.66669 per each foot of spiral. Multiply this function by the length of spiral, for example, 0.66669 × 325 = 216.67 ft.

θ	*p*	*k*	*x*	*y*	LT	ST	LC	θ
1.0	.00146	.49999	.99997	.00582	.66668	.33334	.99999	1.0
.1	. 161	. 999	. 996	. 640	. 68	. 35	. 98	.1
.2	. 175	. 999	. 995	. 698	. 68	. 35	. 98	.2
.3	. 190	. 999	. 995	. 756	. 69	. 35	. 98	.3
.4	. 204	. 999	. 994	. 814	. 69	. 35	. 97	.4

6. Find the ST for the above. Follow the same procedure as in the previous computation, locating the θ angle of 1.3°, moving horizontally across the page until directly below the ST column, and reading the function digits of 0.33335 per each foot of spiral. Multiply this function by the length of spiral, for example, 0.33335 × 325 = 108.34 ft. The ST is therefore 108.34 ft.

7. Find the LC for the above. Follow the same procedure as in the previous computation, locating the θ angle of 1.3°, moving horizontally across the page until directly under the LC column, and reading the function digits of 0.99998 per each foot of spiral. Multiply this function by the length of spiral, for example, 0.99998 × 325 = 324.99 ft. The LC is therefore 324.99 ft.

8. Find the *K* distance for the above. Follow the same procedure as in the previous computation, locating the θ angle of 1.3°, moving horizontally across the page until directly under the *K* column, and reading the function digits of 0.49999 per each foot of spiral. Multiply this function by the length of spiral, for example, 0.49999 × 325 = 162.50 ft. The *K* distance is therefore 162.50 ft.

   Note: Refer to Figure 12.1, where the *K* distance is shown as extending to a point, that where a perpendicular is dropped to the radius of the adjacent curve extended that subtends an angle equal to the θ angle of the spiral. The *K* distance nearly bisects the spiral, and the spiral arc is located approximately one-half the distance between the tangent and the curve produced.

9. Find the *p* distance for the above. Follow the same procedure as in the previous computation, locating the θ of 1.3°, moving horizontally across the page until directly under the *p* column, and reading the function digits of 0.00190 per each foot of spiral. Multiply this function by the length of spiral, for example, 0.00190 × 325 = 0.62 ft. The *p* distance is therefore 0.62 ft.

10. Find the *x* distance for the above. Follow the same procedure as in the previous computation, locating the θ angle of 1.3°, moving horizontally across the page until directly under the *x* column, and reading the function digit of 0.99995 per each foot of spiral. Multiply this function by the length of the spiral, for example, 0.99995 × 325 = 324.98 ft. The *x* distance is therefore 324.98 ft.

11. Find the y distance for the above. Follow the same procedure as in the previous computation, locating the θ angle of 1.3°, moving horizontally across the page until directly under the *y* column, and reading the function digit of 0.00756 per each foot of spiral. Multiply this function by the length of the spiral, for example, 0.00756 × 325 = 2.46 ft. The *y* distance is therefore 2.46 ft.

12. Find the T_s distance (refer to Figure 12.2) using the following curve data:

   L_s = 200 ft
   D_c = 4°
   R = 1432.39 ft
   Δ = 16°
   p = 1.16 ft
   k = 99.98 ft

   $$T_s = (R_c + p)\left(\tan\frac{1}{2}\ \Delta\right) + k \tag{12.6}$$

   Substituting:

   $$T_s = (1432.39 + 1.16)(\tan 8°) + 99.98$$
   $$T_s = (1433.55)(0.14054083) + 99.98$$
   $$T_s = 301.45 \text{ ft}$$

   Therefore, the distance from the beginning or end of the spiral to the mainline PI is 301.45 ft.

13. Find the E_s distance as shown in Figure 12.2. (The E_s is the distance from the mainline PI to the middle of the curve.) Assume the same curve data as noted above.

   $$E_s = (R_c + p)\left(\text{exsec}\frac{1}{2}\ \Delta\right) + p \tag{12.7}$$

Substituting:

$$E_s = (1432.39 + 1.16)\ (\text{exsecant } 8°) + p$$
$$E_s = (1433.55)(0.0098276) + 1.16$$
$$E_s = 15.25 \text{ ft}$$

Therefore, the distance from the mainline PI to the middle of the curve is 15.25 ft. Note: When exsecant tables are not available, the exsecant value can be computed as shown in Eq. (12.8), earlier in this chapter.

The remainder of this chapter will be devoted to solving spiral problems, and covering the technique of field layout, for the following subjects:

1. Substituting a curve for a spiral
2. Calculating any point on the spiral
3. Method of field layout
4. Intersections of tangents and spirals
5. Spiral layout, backward (using closures)
6. Spiral layout, backward (using trigonometry)

SUBSTITUTING A CURVE FOR A SPIRAL

INTRODUCTION

The substituting of a curve for a spiral in the preparation of highway plans commenced in the early 1960s, when spiral computations were difficult for engineering technicians and surveyors to perform, as was the field layout. Substituting curves began during the preparation of right-of-way plans, where the highway engineers designed the centerline to include the spiral-curve-spiral configuration, such as shown in Figure 12.2. Occasionally, right-of-way acquisition lines were, and still are, designed concentric with the mainline curves. The *curve* computations were not difficult to perform; however, when *concentric spirals* were introduced, the computation problems and field layout became unnecessarily complicated.

When attempts were made (during these early years) to show the concentric spiral on the right-of-way drawings, it nearly always became a (scaled) presentation, and therefore, was not computable for layout, nor was it usable to compute acreage.

The solution to this situation was envisioned early-on, when Joseph Barnett, wrote his textbook, *Transition Curves for Highways*: this text included a method of substituting curves for spirals. This substituting method will be outlined below, to provide better understanding of this method and a recognition that many older highway plans contain scaled concentric spirals.

Calculations

Figure 12.3 shows a 200 ft spiral lying between a 4° curve and the mainline tangent. The spiral begins at TS station 14 + 00.00 and ends at SC station 16 + 00.00. The substitution exercise is to place a curve of unknown radius between the SC station 16 + 00 and the mainline tangent extended. The substituted curve on Figure 12.3 is identified as "new curve."

Before continuing, however, you need to solve the LT, ST, and the θ_s angle for the original spiral. This is computed as follows:

Where:

D_c = 4°
L_c = 200 ft

Solution:

$$\theta_s = \frac{L_s}{200} \times D_c \tag{12.4}$$

Substituting:

$$\theta_s = \frac{200}{200} \times 4°$$

$$\theta_s = 4°$$

Computing the LT and ST, we have:

LT = 0.66684 × 200 = 133.37 ft
ST = 0.33349 × 200 = 66.70 ft

Note: The functions used in the above calculations came from Table 12.1, near the beginning of the chapter. These results are obtained by computing the θ_s above, entering the function table, scanning down the page until you reach a θ of 4°, moving horizontally across the page until directly under the proper designation, and multiplying the function digits by the length of spiral.

Return to Figure 12.3 and notice that the short tangent (66.70 ft) is forced to become the *T* distance (tangent) of the substituted curve. With these two curve parts known, the computation continues by solving the radius of a curve whose delta angle is 4°, and whose *T* distance is 66.70 ft. The equation for this solution is as follows:

$$R = \frac{\text{ST}}{\tan\left(\frac{\theta_s}{2}\right)} \tag{12.9}$$

Where:

R = radius
ST = short tangent
$\frac{\theta_s}{2}$ = 2°

Substituting:

$$R = \frac{66.70 \text{ ft}}{\tan 2°}$$

$$R = \frac{66.70}{0.034920769}$$

$$R = 1910.04 \text{ ft}$$

Therefore, the radius of the substituted curve is 1910.04 ft.

Careful observation of the computation for the radius of a substituted curve (Figure 12.3) shows that the ST is considered one leg of a 90° triangle, a line drawn from the PI to the central point of the radius is considered the hypotenuse of the triangle, and the hypotenuse line bisects the θ angle, providing an angle of 2° in the triangle. Thus, the figure supports the definition of a right triangle. The traditional equation for a 90° triangle, where the adjacent leg is unknown, is computed as follows:

$$\text{tangent of the } \angle = \frac{\text{opposite}}{\text{adjacent}}$$

Substituting:

$$\tan 2^\circ = \frac{66.70\text{ ft}}{\text{adjacent}}$$

$$0.034920769 = \frac{66.70\text{ ft}}{\text{adjacent}}$$

$$R\text{ (or adjacent)} = \frac{66.70\text{ ft}}{0.034920769}$$

$$R = 1910.04\text{ ft}$$

The degree of curve is computed using Eq. (12.2) and is as follows:

$$D_c = \frac{5729.58}{R} \tag{12.2}$$

$$D_c = \frac{5729.58}{1910.04}$$

$$D_c = 2.999717283^\circ \text{ (called } 3^\circ \text{ for field layout)}$$

With the D_c computed, we can solve the L_c (length of curve) by using the following equation:

$$L_c = \frac{\Delta_c}{D_c} \times 100 \tag{11.3b}$$

Substituting:

$$L_c = \frac{4}{2.999717283} \times 100$$

$$L_c = 133.34\text{ ft}$$

With the curve computations complete, we can now proceed to make a comparison between the length of the new curve and the length of the original spiral. Notice that the mainline tangent has been extended up-station to a new PC station. This movement along the extended tangent is the result of subtracting the T distance (66.70 ft) of the curve from the LT distance (133.37 ft) of the spiral. The result of this subtraction is 66.67 ft, and therefore, the beginning of the new curve slides forward along the tangent by this difference, making station 14 + 66.67 the PC of the new curve.

One final comparison must be made, that of relating the total spiral length (200 ft) with the combined length of the new curve, and the extended tangent. Therefore, the new curve length is 133.34 ft plus the extended tangent length of 66.67 ft, making a total of 200.01 ft. This difference (0.01 ft), although minute in this situation, is recognized as a station equation, and the difference is noted at the point of compound curve (PCC).

The PC of the new curve, station 14 + 66.67, remains in its computed position, and the new length of curve (133.34 ft) is added to produce PCC station 16 + 00.01. The difference in the two stations (16 + 00.01 versus 16 + 00.00) is contained at this single PCC point, where one station is identified as BK (back), and the other station is identified as AH (ahead). This means that when surveying down-station from the PCC, station 16 + 00.01 is used as control; while surveying up-station from the PCC, station 16 + 00.00 is used as control.

CONCLUSION

Substituting curves for spirals along the right-of-way acquisition lines has made the computing and surveying process easier; however, we must be mindful that many older highway plans are recorded with graphically depicted concentric spirals representing acquisition lines.

CALCULATING ANY POINT ON THE SPIRAL

INTRODUCTION

Calculating one or many points on the spiral has great importance in that an independent station on the spiral can be located using the tangent to spiral (TS) as the point of instrument setup. In earlier years, the 10-chord spiral was used as a replacement for locating individual stations, and where the 10-chord spiral did not align with the stations, the chord points were adjusted to match a station.[3] This was not an acceptable practice, but due to the construction contractor's inexperienced personnel, it was accepted. Figure 12.4 shows the graphics and method of layout for locating single points along the spiral. This computation is also used to locate multiple stations on the spiral. Toward the end of the chapter, we will study a method that uses the spiral to curve (SC) as the control for the instrument setup, thereby setting spiral points in a backward direction.

CALCULATING THE POINT

Equation (12.5) was previously identified as the most important spiral computation because it computes the θ angle when a length of spiral is known. With the θ angle known, all remaining parts of the spiral can be calculated by entering the function tables (Table 12.1), locating the specific function for a certain spiral part, and multiplying the function value by the length. As an example, assume the following curve data:

D_c = 4° (adjacent curve)
L_s = (total spiral)
θ_s = 4° (the deflection angle between tangents of the total spiral)
L = 149.67 ft (assumed length of the partial spiral)

Find θ, the deflection angle between tangents of the partial spiral:

$$\theta = \frac{L^2}{L_s^2} \times \theta_s \qquad (12.5)$$

3 The 10-chord spiral consists of precalculated deflection angles for each of 10 points, the tenth point being the SC of the spiral. Deflection angles to points other than a chord point are interpolated and adjusted with no apparent error. The precomputed lengths of spiral have a range of 150 to 600 ft.

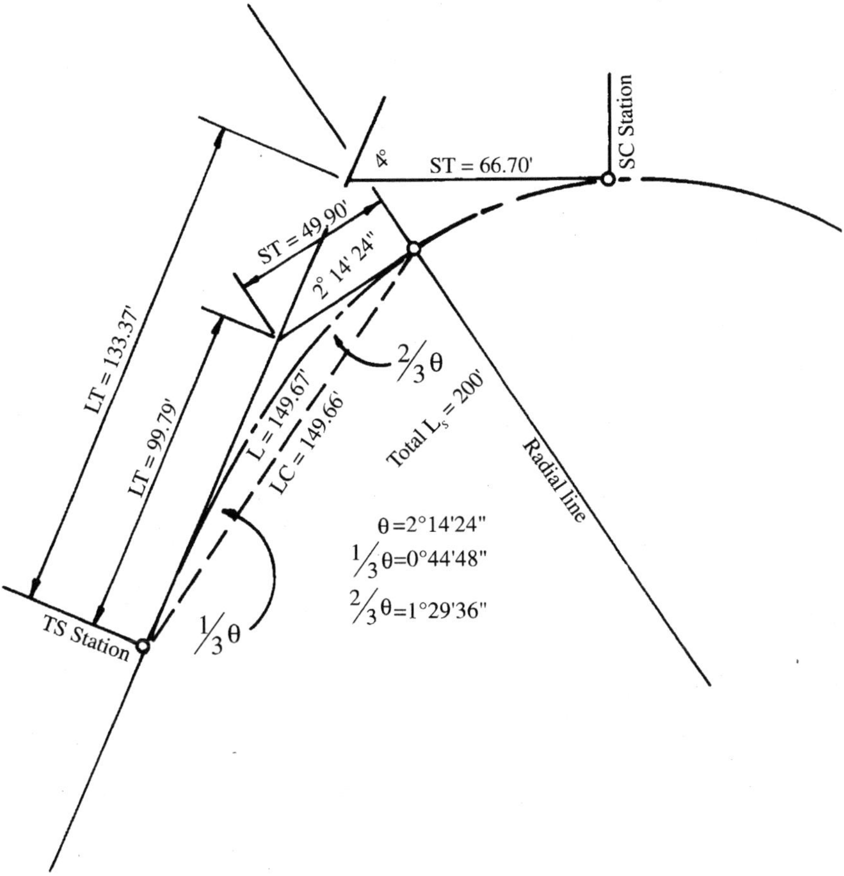

FIGURE 12.4 Calculating any point on spiral

Substituting:

$$\theta = \frac{(149.67)^2}{(200)^2} \times 4$$

$$\theta = \frac{22401.1089}{40000} \times 4$$

$$\theta = 0.5600277 \times 4°$$

$$\theta = 2.2401108°$$

$$\theta = 2°\ 14'\ 24''$$

Therefore, the θ angle for this example is 2°14' 24" and, in tenths of a degree (2.2°), is used to locate the functions, compute the long tangent, short tangent, and long chord as follows:

$$LT = 0.66672 \times 149.67 = 99.79 \text{ ft}$$
$$ST = 0.33338 \times 149.67 = 49.90 \text{ ft}$$
$$LC = 0.99993 \times 149.67 = 149.66 \text{ ft}$$

METHOD OF FIELD LAYOUT

There are two primary methods of field-locating points along a spiral. A third method, that of using the *X* and *Y* distances to form a coordinate, is occasionally used but only in circumstances where the chord line-of-sight to the point is obstructed by construction or other debris. The length of the *Y* distance (often less than 5 ft) contributes to the inability of adjusting the instrument's focal length, and the point cannot be set with certainty using one instrument setup.

The primary methods of locating the spiral follows:

Method 1

1. Measure from the TS point along the tangent extended a distance equal to the LT (or 99.79 ft) in the example.
2. Mark the point with a hub or nail.
3. Set the instrument at this PI point.
4. Backsight along the mainline tangent.
5. Invert the instrument scope or rotate the horizontal plate 180° and turn a deflection angle equal to the θ angle.
6. Measure the ST (or 49.90 ft) and set the point.

Angular measurements are often observed by setting 180° on the instrument plate before making the backsight, completing the backsight, turning the instrument back to zero, and then deflecting the θ angle. This method of turning horizontal angles is superior to inverting (flopping) the scope, since greater instrument accuracy can be expected in the horizontal plate versus the vertical rotation of the scope. With the instrument at zero in the forward direction, angles can be turned either left of right with most instruments.

Method 2

1. Set the instrument at the beginning of the spiral.
2. Backsight the main tangent.
3. Rotate the scope 180° back to zero.
4. Observe a deflection angle of one-third θ theta.
5. Measure the long chord distance and set the end of the spiral.

Both methods of locating the end of the spiral are equally effective and offer the same degree of accuracy. One-third θ is used because when three sides of a triangle (the long tangent, short tangent, and long chord) and the central angle are known, you have sufficient data to compute an obtuse triangle. The angular computation, in all instances, will nearly (within seconds of 1°) represent one-third θ for the small angle and two-thirds θ for the larger angle. The obtuse angle in the triangle is the compliment angle after subtracting the θ angle from 180°. Review Figure 12.4, and notice the locations and computations of these two angles.

FIELD APPLICATION

Being able to locate one or multiple points (stations) on a spiral has considerable significance when these stations are used to obtain cross sections, to locate topography, in commencing the slope-staking process, or to locate storm and sanitary pipes, bridge abutments or piers, and many other surveying fundamentals. The majority of these surveys require that radial lines (90° to the local tangent of a curve) be established before continuing with the work.

The field method used to locate a radial line is:

1. Set the instrument at the required spiral station (called *point on spiral* or POS).
2. Backsight the beginning (TS or ST) of the spiral (TS, if surveying up-station, or ST, if surveying back-station).
3. Observe an angle of two-thirds θ plus 90° if working toward the outside of the spiral, or 90° minus two-thirds θ if working toward the inside of the spiral.

OUTSIDE RADIAL LINE TO THE SPIRAL

Refer to Figure 12.4 and assume a centerline instrument set-up at the POS, backsight the TS and turn an angle equal to two-thirds θ (1° 29' 36"). This will place the instrument line of sight on the local tangent of the curve. Continue turning the horizontal plate an additional 90° (1° 29' 36" + 90° = 91° 29' 36"), which will place the instrument line of sight on the radial line.

INSIDE RADIAL LINE TO THE SPIRAL

When the line of sight is toward the inside of the spiral, subtract two-thirds θ from 90° (90° – 1° 29' 36" = 88° 30' 24"), and the instrument line of sight is on the radial line.

HAND-HELD PRISMS

Radial lines can also be established with hand-held prisms; however, the prism produces a more reliable line when the centerline is on tangent and, is less reliable when the centerline is on curve. The prism is used for many survey applications.

INTERSECTIONS OF TANGENTS AND SPIRALS

The intersection of a tangent property line and a spiral is a classic example of surveys related to interstate highway projects. Until the interstate system flourished in the early 1960s, surveyors had little opportunity (except on railroad projects) to compute or field-locate this type of intersection. Figure 12.5 shows the graphics associated with this computation, where we will concentrate on two ways of handling the problem:

1. Combined office and field applications
2. Field location only

When the office and field disciplines are combined, the surveyor proceeds to field-locate the property, connecting the property line to the highway centerline using proper field technique, and lets the office staff complete the intersection mathematics. Figure 12.5 portrays this intersection, where the mainline long tangent is used to make this field-tie. In this example, both lines are in tangent, and therefore, the lines are easily made to intersect with each other. The tangent and property intersection is referenced with a hub and tack or a nail with flagging. In this example, the property line intersects at a distance of 95.00 ft from the *tangent to spiral* and the observed angle at the intersection is 60°.

OFFICE COMPUTATION METHOD

Prepare a graphic sketch of the property intersection with the highway centerline, using the field survey data, and choose an exaggerated scale. The more exaggerated the scale, the easier it will be to choose the centerline intersection and then mathematically verify this assumption. Twenty-scale (1 in = 20 ft) drawings are recommended if the spiral length is not too long. The longer the spiral, the more difficult it becomes to accurately position the arc of the spiral between the tangent and the next adjacent curve when normal drafting equipment (French curves) is used. However, if several individual stations are computed, using your previous calculating skills, each of these stations can be graphically located on the sketch using a more precise method than drawing with a French curve, and the spiral

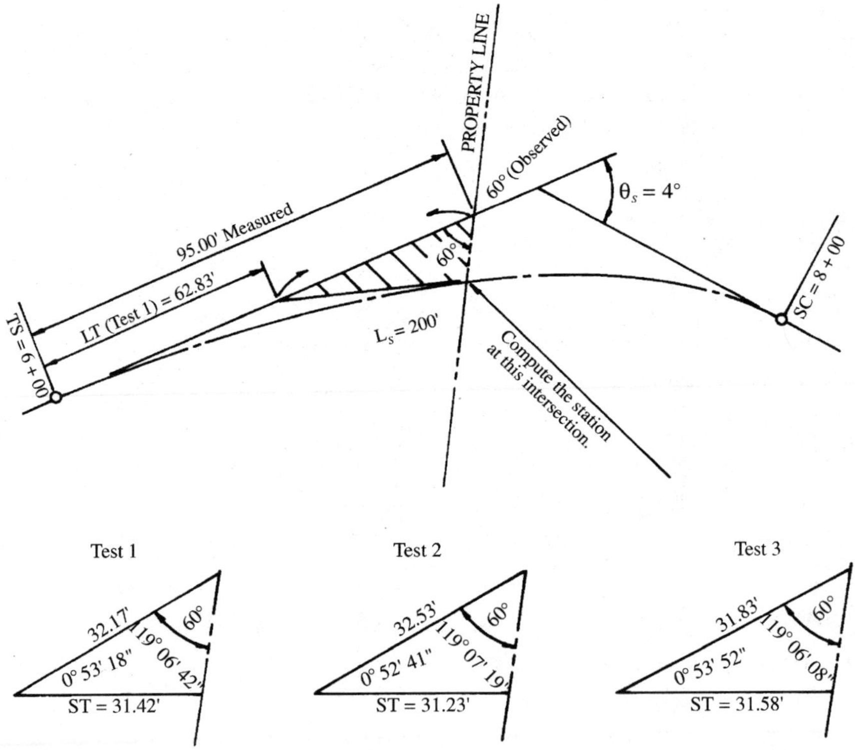

FIGURE 12.5 Property line and spiral intersection

arc can be made to flow through these points. This provides a more accurate location of the spiral, from which you can better estimate the property line intersection.

One excellent office spiral-locating procedure is to compute the X and Y distances, using this coordinate method to better establish the points on the arc. Multiple spiral coordinate points should be computed and plotted before proceeding to draw the arc.

THE UNIQUENESS OF SPIRALS

Spirals are unique in that a single geometric solution cannot be made. You have just learned, that when a spiral length is chosen, all remaining parts of the spiral can be computed. With all the parts computed, a mathematical situation often arises where some of the calculated spiral parts fail to be in accordance with other criteria within the same solution. In other words, the "parts don't fit."

Before continuing, let's temporarily return to the exaggerated graphics, where the field survey ties and centerline curve data were used to draw the property and spiral intersection.

This drawing will permit you to choose an approximate station for the intersection, which must be mathematically verified before continuing. With both stations known (1) the station at the beginning (flat part) of the spiral and (2) the estimated station at the property intersection, by subtracting one station from the other, you can determine the length of the partial spiral. Use Eq. (12.5) to solve for the θ angle and, knowing this angle (to the nearest tenth of a degree), enter the function tables (Table 12.1) and solve for the remaining parts of the spiral.

Although you now can compute all spiral parts, the station chosen may not be correct. You must review the figure and choose a geometric solution, whereby your assumed intersecting station can be verified with surrounding mathematical data. Each intersection will have its own unique method of verification, or some "proof" that the spiral station chosen is correct. If the choice of your intersecting station does produce an error and, it probably will after your first try, you will want to know the magnitude of this error so that the next assumed station can be set by prorating the original error and arriving at a closer estimate for the second "test." In Figure 12.5, you will notice a triangle, whose distances have been chosen as the method of verifying this example. The triangle contains one known angle (the 60° angle), one leg of the triangle that is coincident with the main tangent, one leg of the triangle that is coincident with the property line, and one leg of the triangle that is coincident with the local tangent through the station chosen for the test.

Find:

The spiral data for test 1:

Where:

Centerline curve data

$L_s = 200$ ft

$D_c = 4°$

$\theta_s = 4°$

Assume:

The station intersection is 6 + 94.25.

Therefore, the length of partial spiral is 94.25 ft, and the θ angle is computed as follows:

$$\theta = \frac{(L)^2}{(L_s)^2} \times \theta_s \qquad (12.5)$$

$$\theta = \frac{(94.25)^2}{(200)^2} \times 4$$

$$\theta = 0.88830625$$

$$\theta = 0° \; 53' \; 18''$$

Using the function tables (Table 12.1) solve for the long tangent and short tangent of the spiral whose θ angle (0.9°) you just computed.

$$\text{LT} = 0.66668 \times 94.25 = 62.83 \text{ ft}$$
$$\text{ST} = 0.33334 \times 94.25 = 31.42 \text{ ft}$$

In Figure 12.5, no additional spiral computations need to be made before verifying the results of test 1. Figure 12.5 shows a measured distance along the mainline tangent of 95.00 ft from TS station 6 + 00 to the property line. The long tangent, computed above, is a part of this distance, and by subtracting one station from the other, you have the distance (32.17 ft) for one leg of the cross-hatched triangle. Also, by computing the short tangent, you have the distance (31.42 ft) along the local tangent for the second leg of the triangle, and by computing the θ angle (0° 53' 18") for the partial spiral, you have the second angle within the triangle.

By definition, we know that all triangles have an inside angular total of 180°, and therefore, by adding the two known angles (60° 00' 00" + 0° 53' 18") and subtracting from 180°, the third angle in the triangle becomes 119° 06' 42".

The verification process begins by using the *law of sines t*o determine whether the distances in the triangle, based on test 1, are compatible with other known mathematical data.

Where:

x	=	the distance of the triangle leg coincident with the mainline tangent
119° 06' 42"	=	the angle remaining in the triangle after adding the observed angle with the θ angle (0° 53' 18") and subtracting from 180°
60° 00' 00"	=	the observed angle between the mainline tangent and the property line
32.17 ft	=	the difference between the long tangent and the measured distance

Law of sines:

$$\frac{32.17 \text{ ft}}{\sin 119° \; 06' \; 42''} = \frac{x}{\sin 60°}$$

$$\frac{32.17 \text{ ft}}{0.873673176} = \frac{x}{0.866025404}$$

$$x = 31.888 \text{ ft}$$

Therefore, the solution for x (31.89 ft) in test 1 does not match with the short tangent distance (31.42 ft) computed when station 6 + 94.25 was assumed to be the intersection. Test 2 will, therefore, be required and is shown below:

Assume:

The station intersection is 6 + 93.70.

Therefore, the length of spiral is 93.70 ft, and the θ angle is computed as follows:

$$\theta = \frac{(L)^2}{(L_s)^2} \times \theta_s \tag{12.5}$$

$$\theta = \frac{(93.70)^2}{(200)^2} \times 4$$

$$\theta = 0.877969°$$

$$\theta = 0° \ 52' \ 41''$$

Using Table 12.1, solve for the long tangent and short tangent of the partial spiral whose angle is also 0.9° (when rounded):

$$\text{LT} = 0.66668 \times 93.70 = 62.47 \text{ ft}$$

$$\text{ST} = 0.33334 \times 93.70 = 31.23 \text{ ft}$$

The same comparisons, as suggested in test 1, must be followed to determine whether the assumed station in test 2 is correct. The comparisons are:

1. Subtract the long tangent (62.47 ft) from the measured distance (95.00 ft) to obtain the distance (32.53 ft) for one leg of the triangle.
2. Add the observed angle (60°) to the θ angle (0° 52' 41") and subtract from 180° to obtain the obtuse angle (119° 07' 19") in the triangle.
3. Setup the law of sines to determine whether the distances in test 2 are compatible with other known mathematical data. The law of sines is setup as follows:

$$\frac{32.53 \text{ ft}}{\sin 119° \ 07' \ 19''} = \frac{x}{\sin 60°}$$

$$\frac{32.53 \text{ ft}}{0.873585891} = \frac{x}{0.866025404}$$

$$x = 32.25 \text{ ft}$$

Therefore, the solution for x (32.25 ft) in test 2 does not match with the short tangent distance (31.23 ft) computed when station 6 + 93.70 was assumed to be the intersection. You should note that changing the spiral to a lesser length made the situation worse. Test 3 is as follows:

Assume:
The station intersection is 6 + 94.75.

Therefore, the length of spiral is 94.75 ft, and the θ angle is computed as follows:

$$\theta = \frac{(L)^2}{(L_s)^2} \times \theta_s \tag{12.5}$$

$$\theta = \frac{(94.75)^2}{(200)^2} \times 4$$

$$\theta = 0.89775625°$$

$$\theta = 0° \ 53' \ 52''$$

Using Table 12.1, solve for the long tangent and short tangent of the partial spiral whose θ angle is 0.9° (rounded).

$$LT = 0.66668 \times 94.75 = 63.17 \text{ ft}$$

$$ST = 0.33334 \times 94.75 = 31.58 \text{ ft}$$

Complete the comparison suggested in tests 1 and 2 to determine whether the assumed station in test 3 is correct. The comparison is:

1. Subtract the long tangent (63.17 ft) from the measured distance (95.00 ft) to obtain the distance (31.83 ft) for one leg of the triangle.
2. Add the observed angle (60°) to the θ angle (0° 53' 52") and subtract from 180° to obtain the obtuse angle (119° 06' 08") in the triangle.
3. Setup the law of sines to determine whether the distances in test 3 are compatible with other known mathematical data.

$$\frac{31.83 \text{ ft}}{\sin 119° \; 06' \; 08''} = \frac{x}{\sin 60°}$$

$$\frac{31.83 \text{ ft}}{0.87375336} = \frac{x}{0.866025404}$$

$$x = 31.55 \text{ ft}$$

Note: This distance (31.55 ft) matches the short tangent distance (31.58 ft) within 0.03 ft, and therefore, test 3 is compatible with all surrounding mathematical data and is acceptable for normal field accuracy.

Summary

Tests 1, 2, and 3 were setup to determine which of the three spiral station intersections satisfied other surrounding mathematical data. The fact that a spiral length is chosen does not guarantee that the spiral parts will match other existing conditions, as you have just seen.

Test procedures are required in all spiral intersections. The example chosen for this study shows the property line intersecting the spiral within the long tangent line. Property line intersections within the short tangent line are equally possible and are setup in a similar manner by extending the long tangent beyond the spiral PI until it intersects with the property line. A triangle or four-sided figure can be developed where either the law of sines or solving for two unknowns, will produce a method of verifying your results.

Earlier in the chapter, we mentioned that spirals cannot be computed in one direct sequence. Programmable software developers claim they can solve spirals directly but, to this date, this has not been satisfactorily proven.

FIELD LOCATING THE PROPERTY LINE AND SPIRAL INTERSECTION

It was suggested earlier in the chapter that the second method of intersecting a spiral and a tangent could be accomplished in the field. The field method requires a greater amount of time than simply field-tying the centerline to the property line, but it does relieve the office staff from further computations.

FIELD LOCATION METHOD

Assume the mainline tangent, as well as the long and short spiral tangents are field-located, such as in Figure 12.5. Set the instrument on the property line, and extend the line beyond where you estimate the spiral to intersect. Such intersections will be somewhat obvious when the spiral is flat but not quite as obvious when the spiral is sharp.

Set references (nails) along the property line and especially on each side of the estimated intersection with the spiral. Return to the TS, measuring a distance along the imaginary spiral arc to the row of nails, noting one distance that is definitely short of the property line and another distance that is definitely beyond the property line. For this example, assume the rough measurement to the distance short of the property line is 93.00 ft and the rough measurement to the distance on the far side of the property line is 95.00 ft.

SOLVE THE SPIRAL DATA

Complete the spiral computations for each of these distances and accurately locate both points in the field. Use the curve data associated with Figure 12.5, and compute the θ angle, LT, ST, and LC for the 93 ft spiral. The computations are as follows:

$$\theta = \frac{(93.00)^2}{(200)^2} \times 4$$

$$\theta = \frac{8649.00}{40000} \times 4$$

$$\theta = 0.216225 \times 4$$

$$\theta = 0.8649^\circ$$

$$\theta = 0^\circ\ 51'\ 54''$$

$$\text{LT} = 0.66668 \times 93.00 = 62.00 \text{ ft}$$

$$\text{ST} = 0.33334 \times 93.00 = 31.00 \text{ ft}$$

$$\text{LC} = 0.99999 \times 93.00 = 93.00 \text{ ft}$$

Accurately field-locate this partial length of spiral, using the long and short tangent method or using the long chord.

Complete the computation for the spiral that is 95 ft in length and accurately field-locate the end point of this partial spiral.

ESTABLISH THE INTERSECTION

Use two string lines, one to denote the property line and a second string line to denote the arc between the two spiral points, one set at 93 ft and the other set at 95 ft. Where the strings intersect, defines the centerline intersection. Field-measure from the string intersection to the nearest accurately set spiral point, and either add or subtract as appropriate to obtain the exact station.

PREPARING THE SPIRAL FIELD BOOK

This portion of the chapter outlines "field book" requirements and the positioning of data on the page. The following data are required:

1. Station identification for each spiral
2. The deflection angle to each individual station
3. The chord distance between points using a two- or three-person crew

Many consulting engineers contract with outside professional organizations to complete this phase. These outside companies specialize in this type of construction peripherals. Quite often, the engineering company will complete the data in-house, forwarding their results to the highway department for approvals and subsequently submitting copies to the surveyors for their use in the final layout.

Interstate highways and state route construction projects generally move at a rapid pace, and surveyors are constantly under pressure trying to complete multiple tasks in many separate construction areas.

Being under such pressure is a great disadvantage to the surveyor, often leading to computation errors. Errors in construction work quite often place heavy costs on the consulting engineer, contractor, surveyor, or all three, depending on the type of error and who was negligent. Litigation, the end result of construction errors is something all construction and engineering-related professionals must try to minimize or eliminate altogether by searching for survey methods that lead to an error-free product.

One of these methods is to prepare, in advance, all spiral computations under office conditions. Surveyors, however, cover projects where office backup is nonexistent and all field mathematics are computed by the surveyor. Whatever the conditions, make the best of the situation, know you are not alone, and *take your time*. One single error in computation leads to an incorrect layout, the computation must be repeated, and finally, the layout must also be repeated. This, at best, consumes only an amount of time equal to the correction. At worst, construction work could have already commenced before the error was discovered. Then not only is time an added cost, but remedial construction is an even greater added cost.

Privacy, therefore, during computations is absolutely necessary, and the surveyor should always assign other duties to the remaining field personnel while the computations are in process.

Figure 12.6 is an example of the computations used to prepare the field-book entries for spirals. Before beginning the computations, be sure you have an agreement with the governing agency (county, state, or federal) regarding the station layout interval (25 or 50 ft). Review the highway plans for related curve and spiral data, while observing whether the spiral is in its normal position (between a tangent and a curve) or located between two curves. Spirals between two curves are discussed in Chapter 13. The computation process is quite different, as you will see in the next chapter, and an incorrect analysis will lead to computations that are absolutely incorrect.

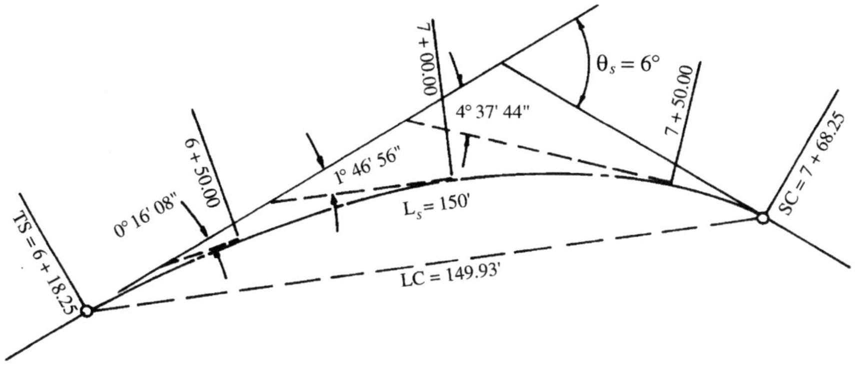

Station required 6 + 50	Station required 7 + 100	Station required 7 + 50
6 + 50.00 –6 + 18.25 L = 31.75 ft	7 + 00.00 –6 + 18.25 L = 81.75 ft	7 + 50.00 –6 + 18.25 L = 131.75 ft
$\theta = \frac{(31.75)^2}{(150)^2} \times 6°$	$\theta = \frac{(81.75)^2}{(150)^2} \times 6°$	$\theta = \frac{(131.75)^2}{(150)^2} \times 6°$
$\theta = 0.2688166°$	$\theta = 1.78215°$	$\theta = 4.6288166°$
$\theta = 0°\ 16'07.7''$	$\theta = 1°\ 46'55.7''$	$\theta = 4°\ 37'43.7''$
$\frac{1}{3}\theta = 0°\ 05'22.6''$	$\frac{1}{3}\theta = 0°\ 35'38.5''$	$\frac{1}{3}\theta = 1°\ 32'34.5''$
LT = 0.66667 × 31.75 = 21.166 ft	LT = 0.66670 × 81.75 = 54.50 ft	LT = 0.666689 × 131.75 = 87.86 ft
ST = 0.33334 × 31.75 = 10.583 ft	ST = 0.33337 × 81.75 = 27.25 ft	ST = 0.333354 × 131.75 = 43.94 ft
LC = 1.0000 × 31.75 = 31.75 ft	LC = 0.99996 × 81.75 = 81.75 ft	LC = 0.99971 × 131.75 = 131.71 ft

FIGURE 12.6 Spiral field-book entries

THOROUGHLY EXAMINE THE CURVE DATA

Make a thorough examination of the curve and spiral data, being acutely aware of "station equations" that might interfere with your computations. When you are confident of the mathematical base information, proceed with the field-book preparation. The number of spiral layout points, as mentioned earlier, is dependent on the severity of the spiral curvature. Usually, sharp curves followed by sharp spirals require minimum centerline locations at 25-ft intervals, and 50-ft intervals are acceptable in flat curves and spirals, as well as mainline tangent alignments.

FIELD BOOK ENTRIES

Figure 12.6 is an example of the computations that must be prepared before the layout data can be entered into the field book. The following centerline curve data will be used to develop these computations:

$$D_c = 8°$$
$$\theta_s = 6°$$
$$L_s = 150 \text{ ft}$$
$$\text{LT} = 100.06 \text{ ft}$$
$$\text{ST} = 50.05 \text{ ft}$$

In Figure 12.6, we will also assume that a layout interval of 50 ft is considered satisfactory, and therefore, with the TS of the spiral located at station 6 + 18.25, the nearest 50-ft interval is station 6 + 50.00, and the difference between stations produces a spiral length of 31.75 ft. The next station is 7 + 00 and the spiral length is 81.75 ft. And the last station is 7 + 50 with a spiral length of 131.75 ft. Remember, all partial spiral lengths are determined by subtracting the TS station from the POS station or by subtracting the POS station from the ST. Spiral distances are never determined by subtracting the POS station from the SC, or the CS.

RECOMMENDED NOTE-KEEPING PROCEDURES

There are many note-keeping procedures; however, the most widely used method is to show the stations vertically in the field book, with the "flat" end of the spiral at the bottom of the page, and entering new stations up-page, regardless of whether the stations are increasing or decreasing.

If you are computing the spiral at the up-station end of the curve, the ST will be the first entry at the bottom of the page, and your stations will be proceeding up-page but in a back-station direction.

FIELD-BOOK OUTLINE

The outline below is the type of notation preferred, whether you are preparing spiral or curve data. The first outline represents information that is appropriate for a two-person survey crew, where the surveyor must double as the rear chain person; while the second outline represents information that is appropriate for a three-person survey crew, where the surveyor remains at one location and the two chain persons are moving forward, setting each new point while measuring from the previous one.

TWO-PERSON SURVEY CREW

Station	Deflection angle	Long chord
7 + 68.25 (SC)	2° 00' 00"	149.92 ft
7 + 50.00	1° 32' 35"	131.71 ft
7 + 00.00	0° 35' 38"	81.75 ft

6 + 50.00	0° 05' 23"	31.75 ft
6 + 18.25 (TS)	0° 00' 00"	0.00
Station	Deflection angle	Long chord

THREE-PERSON SURVEY CREW

Station	Deflection angle	Chord distance between stations on centerline
7 + 68.25 (SC)	2° 00' 00"	
		18.25 ft
7 + 50.00	1° 32' 35"	
		50.00 ft
7 + 00.00	0° 35' 38"	
		50.00 ft
6 + 50.00	0° 05' 23"	
		31.75 ft
6 + 18.25 (TS)	0° 00' 00"	
Station	Deflection angle	Chord distance between stations on centerline

You will note that the deflection angles on both outlines are identical and only the chord distances have changed.

REMEMBER THE BASICS

1. The normal spiral layout commences at the flat end of the spiral and proceeds to the sharp end unless you have specific need to locate the spiral in a backward direction (to be covered later in the chapter).
2. The deflection angle noted in your field-book is one-third theta. This represents the deflection for each of the required stations, while the instrument remains at the TS or ST point.
3. When the survey crew consists of two members moving forward and measuring from the previous point, each distance measured is equal to the difference between stations. In other words, there are no centerline chord corrections for the spiral.

Quite often, however, the surveyor will make a judgment call and, as the spiral approaches the curve, the surveyor will make chord adjustments similar to curves of equal distance. These adjustments are not mandatory, due to the universal understanding that all spiral computation and layout errors are made to be contained within the spiral itself. Remember that the beginning and ending of the total spiral are initially established by using either of two layout methods: the long and short tangent method or by the long chord method.

Although we speak of incremental layout errors as though they are of a considerable magnitude, we find, to the contrary, that the original engineers and mathematicians who first conceptualized the spiral and computed the function tables did so with great forethought. The function tables have been computed so that the majority of position error is eliminated.

SPIRAL STATION POSITION ERRORS

Some simple studies have been made to determine the magnitude of position error that exists at various points along a spiral when centerline arc distances are used to set the point. One study, in particular, used a sample spiral, where the θ angle was 40°, the curve was 9°, and the length of spiral was 500 ft. The results of the study show that, when the spiral is divided into 10 equal parts, the largest position error is located near the sharp part of the spiral, at the SC or CS stations. A tabulation of this position error is shown below:

Station	Distance set by stations (ft)	Distance using function tables (ft)	Difference	Station
6 + 50 (SC)				6 + 50 (SC)
	50.00 ft	49.96 ft	0.04 ft	
6 + 00				6 + 00
	50.00 ft	49.967 ft	0.033 ft	
5 + 50				5 + 50
	50.00 ft	49.983 ft	0.017 ft	
5 + 00				5 + 00
	50.00 ft	49.981 ft	0.019 ft	
4 + 50				5 + 00
	50.00 ft	49.99 ft	0.01 ft	
4 + 00				4 + 00
	50.00 ft	49.991 ft	0.009 ft	
3 + 50				3 + 50
	50.00 ft	49.993 ft	0.007 ft	
3 + 00				3 + 00
	50.00 ft	49.999 ft	0.001 ft	
2 + 50				2 + 50
	50.00 ft	50.00 ft	0.00 ft	
2 + 00				2 + 00
	50.00 ft	50.00 ft	0.00 ft	
1 + 50 (TS)				1 + 50
Station	Distance set by stations (ft)	Distance using function tables (ft)	Difference	Station

A review of the above outline shows that the position error of points along the majority of spirals used in highway design is minimal. Only in unusual situations will the θ angle be 40° and the attached curve be 9°. A second study was made using the illustration in Figure 12.6, and found no station position error greater than 0.007 ft.

Summary

In summary, we have learned the following information about the spiral:

- Eq. (12.4) is used to compute the θ_s angle of the total spiral.
- Eq. (12.5) is used to compute the θ angle of any partial length spiral.
- The function tables (per 1-ft) are used to compute the interior parts of a spiral.

- Field-book entries for centerline layout are generally limited to:
  a. Stations
  b. Deflection angles (one-third θ)
  c. LC or centerline distance between stations
- Spiral layout commences at the flat part of the spiral, where the instrument remains while individual deflection angles are turned. The distances are measured from one instrument set-up point, using the long chord; or the stations can be set using a three-person crew, two people moving forward and locating successive points from the previous point.
- Incremental measuring between spiral stations produce minimal errors in spite of the absence of a chord correction.
- Multiple point spiral layout is cost *ineffective* if the long tangent and short tangent method is used; however, as in most situations, field conditions will vary, making a single instrument setup impossible, and thus the surveyor is forced to use other techniques.
- Cross sections and other radial surveys require the following: Move the instrument forward to the POS, backsight the TS or ST station, and add or subtract two-thirds θ from 90° to locate the radial line.
- Other spiral layout methods are possible, such as using the X and Y distances. Use the function tables to compute these values, which act as a "coordinate" method. The obstacle associated with this technique is that the Y distance is usually very short (less than the focus length of the instrument), and it becomes impractical to use the transit without a double setup.

FIELD-BOOK ENTRIES

Figure 12.7 is an additional example of a spiral layout, showing backup computations supporting the deflection angles and chord distances from which you can prepare the spiral field-book entries. Once the θ angle is determined, using Eq. (12.4) for the full-length spiral, or Eq. (12.5), for the partial-length spiral, compute the remaining parts using the same function tables. Occasionally, the symbols are used interchangeably.

SPIRAL PARTS AS A RATIO

As you review Figure 12.7, notice the length of the LT versus the length of the ST and you probably will have observed that the LT distance (133.37 ft) is roughly twice the length of the ST distance (66.70 ft), and the larger angle in the triangle (2° 40' 00") is twice that of the smaller angle in the triangle (1° 20' 00"). This phenomenon (⅓ : ⅔) is common in all spirals and should alert you when your computations show otherwise.

The following centerline curve data will be used to develop these computations.

Where:

θ_s = 4°
L_s = 200.00 ft
LT = 133.37 ft
ST = 66.70 ft
LC = 199.96 ft

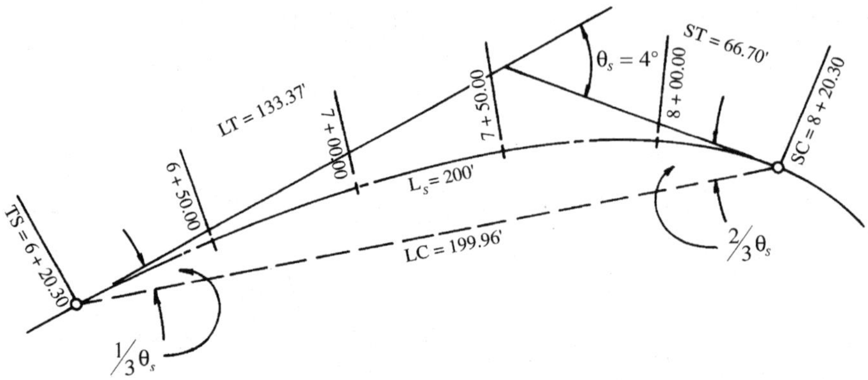

Station 6 + 50.00	Station 7 + 00.00	Station 7 + 50.00	Station 8 + 00.00
$\theta = \frac{(29.70)^2}{(200)^2} \times 4$	$\theta = \frac{(79.70)^2}{(200)^2} \times 4$	$\theta = \frac{(129.70)^2}{(200)^2} \times 4$	$\theta = \frac{(179.70)^2}{(200)^2} \times 4$
$\theta = \frac{882.09}{40000} \times 4$	$\theta = \frac{6352.09}{40000} \times 4$	$\theta = \frac{16822.09}{40000} \times 4$	$\theta = \frac{32292.09}{40000} \times 4$
$\theta = 0.088209°$	$\theta = 0.635209°$	$\theta = 1.682209°$	$\theta = 3.229209°$
$\theta = 0°\ 05'17''$	$\theta = 0°\ 38'06''$	$\theta = 1°\ 40'56''$	$\theta = 3°\ 13'45''$
$\frac{1}{3}\theta = 0°\ 01'45''$	$\frac{1}{3}\theta = 0°\ 12'42''$	$\frac{1}{3}\theta = 0°\ 33'39''$	$\frac{1}{3}\theta = 1°\ 04'35''$

FIGURE 12.7 Locating multiple spiral points

Compute the deflection angles and chord distances for the following stations:

6 + 20.30 (TS)
6 + 50.00
7 + 00.00
7 + 50.00
8 + 00.00
8 + 20.30 (SC)

The computations shown on Figure 12.7 cover the deflection angles for the four interior stations; however, the long and short tangents and the long chord must be computed before the layout can commence. These computations are shown below. The centerline curve data are:

$$\theta_s = 4°$$
$$L_s = 200 \text{ ft}$$

Knowing the above two parts of the spiral (whether full length or partial length) permits you to enter the function tables and compute the remaining parts. The functions, in this example, are computed based on an θ angle of 4°. Knowing that the digits representing this particular function are the equivalent of a 1-ft spiral, and to complete the solution, each function must be multiplied by 200 ft. Therefore,

$$LT = 0.66684 \times 200 = 133.37 \text{ ft}$$
$$ST = 0.33349 \times 200 = 66.70 \text{ ft}$$
$$LC = 0.99978 \times 200 = 199.96 \text{ ft}$$

Remember, the layout of the full-length spiral should be completed before the interior stations are field-located. The above computations are sufficient for this layout, providing you use the long and short tangent and the θ_s angle. If you intend to locate the end of the spiral by using the LC, you must compute the deflection angle that represents this line, which is always one-third θ, or in this example is (4 ÷ 3 = 1° 20' 00").

RECOMMENDED SPIRAL FIELD-BOOK ENTRIES

Station	Deflection angle	Long chord
8 + 20.30 (SC)	1° 20' 00"	199.96 ft
8 + 00.00	1° 04' 35"	179.67 ft
7 + 50.00	0° 33' 39"	129.69 ft
7 + 00.00	0° 12' 42"	79.70 ft
6 + 50.00	0° 01' 45"	29.70 ft
6 + 20.30 (TS)	0° 00' 00"	00.00 ft

SPIRAL LAYOUT, BACKWARD (USING CLOSURES)

Figure 12.8 shows one method of locating spiral stations in a backward direction. Not too many years ago, I was responsible for the layout of approach slabs, abutments, and piers for a bridge crossing the Ohio River. On one side of the river, the bridge was designed with the flat part (TS) of the spiral located on the bridge near the abutment but several feet back onto the deck.

When the preliminary work was being completed and before the embankment for the roadway approach was in place, the surveying of the spiral was routine. I could always occupy the TS and, with the use of the spiral field-book entries, the individual stations could be quickly located. The centerline across the river was established using a Wild, T-2 instrument (1" reading), and all distances were measured with electronic equipment. This meant that all controls were well established and the spiral could be located quite easily until the construction contractor commenced the roadway fill adjacent to the bridge. The bridge, during this particular period in construction, did not yet have falsework or a temporary deck upon which to set the instrument and commence a normal spiral layout from the TS station. You probably can see the problem already. The contractor had compacted approximately 20 ft of fill adjacent to the bridge but not close enough to the bridge to permit the instrument person to occupy the TS where the elevations were common. In other words, the TS was 20 ft, vertically, below the remainder of the spiral.

The contractor required immediate centerline station locations throughout the entire spiral (at 25 ft intervals) to pour a concrete median. This placed me in a position of locating the spiral backward, and to make matters worse, the pouring of a concrete median such as this, was frightening. My knowledge of surveying spirals in a backward direction was several months away (early 1970s), so my company sent a mathematician to the site to find a solution. A quick solution could not be found, except for the method shown in Figure 12.8, and this required computing two unknowns for each 25-ft station within a 400 ft spiral.

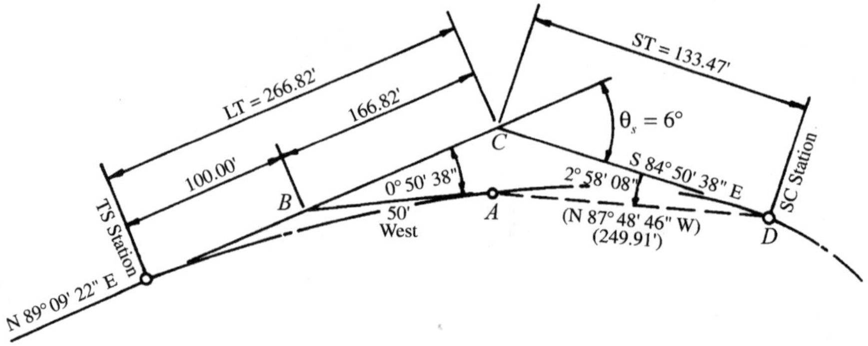

Course	Bearing	Distance	Cosine	Sine	North	South	East	West
A to *B*	West	50.00'	0	2	---------	---------	---------	50.00'
B to *C*	N 89-09-22 E	166.82'	0.014728	0.999891	2.4569	---------	166.8019	---------
C to *D*	S 84-50-38 E	133.47'	0.089869	0.995953	---------	11.9949	132.9299	---------
D to *A*	(N 87-48-46 W)	(249.91')	0.038165	0.999271	(9.538)	---------	---------	(249.7318)

$$\tan \angle = \frac{249.7318}{9.538}$$

$$\tan \angle = 26.182827$$

$$\angle = 87.81276694°$$

$$\angle = 87° \ 48'46''$$

FIGURE 12.8 Locating spirals backward (closures)

Obviously, we did not have the time to field-compute these unknowns, so the mathematician returned to the office and using a programmable computer, solved the various bearings and distances. She then returned to the site where the spiral was eventually completed and the concrete median was poured.

Figure 12.8 shows the backward layout of one point, labeled *A* and located 150 ft from the TS. The original centerline curve data is shown below:

$$L_s = 400 \text{ ft}$$
$$\theta_s = 6°$$
$$\text{LT} = 266.82 \text{ ft}$$
$$\text{ST} = 133.47 \text{ ft}$$

The closure method (Figure 12.8) could have been used to solve our bridge problem had there been sufficient time; however, the problems were solved in the office. When office backup is not available and you must continue with the layout, the following method is one part of the answer. Compute the θ angle for the partial-length spiral:

Where:

$L = 150$ ft

Find the LT, ST, and θ angle:

$$\theta = \frac{L^2}{L_s^2} \times \theta_s \qquad (12.5)$$

$$\theta = \frac{(150)^2}{(400)^2} \times 6°$$

$$\theta = \frac{22500}{160000} \times 6°$$

$$\theta = 0.84375°$$

$$\theta = 0° \; 50' \; 38''$$

Using the function tables:

LT = 0.66668 × 150 = 100.00 ft
ST = 0.33334 × 150 = 50.00 ft

Figure 12.8 includes a sketch of the 400 ft spiral with associated bearings and distances placed in their appropriate locations. This shows the LT (266.82 ft), the ST (133.47 ft), and the θ_s of 6°. The partial-length spiral (150 ft) is also shown on the sketch with its calculated parts, including the LT (100.00 ft), the ST (50.00 ft), and the θ angle of 0° 50' 38". By assembling this information on the sketch (Figure 12.8), a solution for two unknowns can be described as follows:

line *AB*, distance known, bearing assumed to be due west
line *BC*, distance and bearing known
line *CD*, distance and bearing known
line *DA*, distance and bearing *unknown*

Therefore, by solving the bearing and distance of line *DA*, the partial-length spiral can be located from an instrument set up at the SC. However, before the bearing of course *DA* can be located, the surveyor must have prior knowledge of the bearing system or be able to compute the bearing of the radial line through the SC or the bearing of the local tangent at the SC. Be careful to not mix bearings. If you assumed a bearing on one line (such as line *AB*) in the closure (for the ease with which the computation can be made), remember to rotate the closure and bring the two sets of bearings into conformity.

The closure was discussed in Chapter 11, but a review of the process will probably be beneficial.

CLOSURE REVIEW

1. Assign letters, numbers, Roman numerals, or other designations of your choice to each of the corners in the closure. Preferably, leave the unknown bearing and distance as the last course in the closure.
2. Highway plan bearings can be used in many solutions, such as Figure 12.8, because the unknown bearing and distance is on the same line. When plan bearings are used, and the unknown bearing and distance are on two separate lines and the bearings are multi-directional, the problem cannot be solved directly using a "manual" closure. The prob-

lem *can* be solved, however, using programmable computers. Review the subsection "Computational Impasse" located in Chapter 11, which covers these situations more thoroughly.

3. The procedure used in the closure method is simply that of taking a series of traverse lines, as would be used in a random field survey and relating each traverse line to a triangle, where the traverse line becomes the hypotenuse of the triangle. The triangles are assembled on the *outside* of the traverse and form a figure that is rectangular.
4. The abbreviated closure form shown in Figure 12.8 represents the assembling of these triangles into a predetermined form, where ultimately, two unknowns can be computed by the process of elimination.

Note: The sides of a closure (north and south directions) must be equal, and the top and bottom of a closure (east and west directions) must be equal.

PROCEEDING THROUGH THE CLOSURE

As we proceed through the closure in Figure 12.8, line *AB* is assumed to be due west, and the distance of the line is equal to the ST (50.00 ft) of the partial spiral. Line *BC* deflects off line *AB* by the θ angle (0° 50' 38") of the partial spiral, where the length is equal to the difference between the LT of the full-length spiral (266.82 ft) and the LT of the partial-length spiral (100.00 ft), or 166.82 ft. The bearing of line *BC* is calculated, as described in Chapter 3, by setting up two intersecting lines (north and south) to represent the four quadrants at point *B*, and knowing that line *AB* is due west, line *BC*, by observation, must be in quadrant 1.

LINE BC

The bearing of line *BC* is the difference between 90° and the θ angle, or 90° 00' 00" – 0° 50' 38" = 89° 09' 22". All bearings that are less than 90° and located in quadrant 1 are considered northeast, and therefore, the bearing of line *BC* is N 89° 09' 22" E. Add this bearing information into the appropriate cells in the closure form, locate the sine/cosine functions for the bearing, multiply the functions by the course distance, and enter this data onto the form. Remember to line-through the directions not applicable.

LINE CD

Line *CD* deflects off line *BC* by the θ_s angle (6° 00' 00") of the full-length spiral, where the length of line *CD* is equal to the ST (133.47 ft) of the full-length spiral. To determine the bearing of line *CD*, set up the two intersecting quadrant lines at point *C* and approximate the bearing of line *BC*, which is N 89° 09' 22" E in quadrant 1. In this situation, the bearing of line *CD* cannot be determined by either adding or subtracting an angle from 90° in quadrant 1, because the deflection angle of 6° 00' 00" at point *C*, moves line *CD* into quadrant 2. Prepare a sketch at this intersection with line BC in quadrant 1, and line *CD* in quadrant 2. By observation, you can probably determine that both angles must be added together (89° 09' 22" + 6° 00' 00" = 95° 09' 22") and subtracted from 180° to equal the bearing of the line. Therefore, the bearing of line *CD* is S 84° 50' 38" E. Add this information into the appropriate cells in the closure form.

LINE DA

The last course, line *DA*, represents the missing bearing and distance in the traverse but also represents the hypotenuse of the last triangle, whose leg distances were inserted in the closure to complete the balancing of the north-south and the east-west directions. You will note (Figure 12.8) that all cells in the closure are filled except those noted in parentheses and shaded. The distances in the shaded cells are obtained by comparing the lengths of the north-south and the east-west lines and placing those numbers in their appropriate columns. Balance the north with the south and the east with the west.

Example:

The north column has one entry (2.4569 ft) and the south column has one entry (11.9949 ft). Obviously, the north column does not equal the south column. Therefore, you can add this difference (9.538 ft) in the appropriate cell. The east column has two entries and the west column has one entry.

These columns do not equal each other, and by observation, you can determine that the west column is short by a distance of 249.7318 ft.

TRIGONOMETRIC FUNCTIONS CONTRIBUTE TO THE CLOSURE

By adding these last two distances to the closure form, now compute the hypotenuse of a right triangle whose one leg distance is 9.538 ft, and whose second leg distance is 249.7318 ft. Find this hypotenuse based on the following trigonometric functions:

1. The sine of an angle is equal to the opposite distance divided by the hypotenuse, and we do not know the hypotenuse.
2. The cosine of an angle is equal to the adjacent distance divided by the hypotenuse, and we do not know the hypotenuse.
3. The tangent of an angle is equal to the opposite distance divided by the adjacent distance, and we know both distances.

Therefore, the hypotenuse, or bearing of the line, can be determined by dividing the north or south leg into the east or west leg of the closure:

Find:

$$\tan \angle = \frac{\text{opposite}}{\text{adjacent}} \text{ trigonometric function}$$

Substituting:

$$\tan \angle = \frac{249.7318}{9.538}$$

$$\tan \angle = 26.182827$$

$$\angle = 87.8127669°$$
$$\angle = 87° 48' 46''$$

DETERMINING THE DISTANCE

The bearing is N 87° 48' 46" W, but the distance of the line remains undetermined. To complete this solution, one must review the mathematical process by which entries are placed into their respective cells. Remember, these entries were originally determined by multiplying the cosine or sine of the angle by the distance in each course to fill the appropriate cell. By reversing the process, and dividing the cosine or sine of the angle into the triangle leg, you can compute the length of the course.

REVERSING THE PROCESS, MADE SIMPLE

- Think of the distance in each course as a 2.
- Think of the cosine or sine function of each course as a 4.
- Multiplying $2 \times 4 = 8$, and 8 is entered into the cell.
- Reverse the process and divide 4 (the cosine or sine) into 8 (the cell) and the result is 2 (the hypotenuse and missing distance).

Example:

Refer to Figure 12.8 and divide the cosine function into the north leg of the triangle; the result is 249.91 ft. As a check, divide the sine function into the west leg of the triangle; the result is a matching distance of 249.91 ft. Occasionally, the two divisions will not match. If the difference is greater than a few hundreds of a foot, you probably have an error and must review your input data. If the two distances reasonably match, say within ±0.03 ft, use the result obtained by dividing through with the largest number. Division with digits that have a number of zeros immediately following the decimal point will produce greater errors.

SPIRAL LAYOUT, BACKWARD (USING TRIGONOMETRY)

Figure 12.9 is an example of the computations used to layout the spiral in a backward direction from an instrument set up at the SC. This method uses trigonometric solutions that involve relating simple triangles to one another. The triangles are those produced when X and Y distances are computed between adjacent stations. The X is the distance measured along the mainline extended, to a point where a perpendicular will intersect the SC. The Y is the measured distance of this perpendicular.

Example:

In Figure 12.9, the full-length spiral is 150 ft, the X_c of the full-length spiral is 149.835 ft, and the Y_c of the full-length spiral is 5.232 ft. The centerline curve data are:

$$L_s = 150 \text{ ft}$$
$$D_c = 8°$$
$$\theta_s = 6°$$

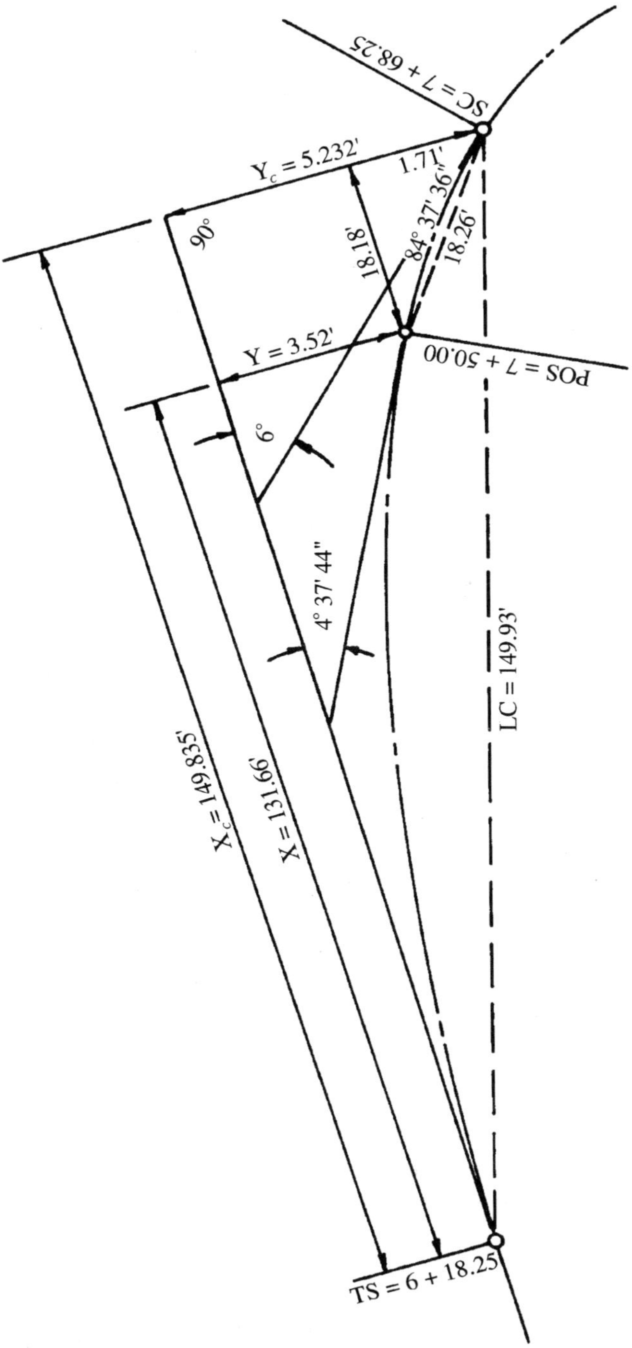

FIGURE 12.9 Locating spirals backward (trigonometry)

Find the θ angle, X distance, and Y distace of a partial-length spiral ending at POS station 7 + 50.00:

$$\begin{array}{l} 7 + 50.00 \text{ (POS)} \\ \underline{-6 + 18.25 \text{ (TS)}} \\ 1 + 31.75 \text{ (length of partial spiral)} \end{array}$$

$$\theta = \frac{(L)^2}{(L_s)^2} \times \theta_s \tag{12.5}$$

$$\theta = \frac{(131.75)^2}{(150)^2} \times 6$$

$$\theta = \frac{17358.0625}{22500} \times 6$$

$$\theta = 0.771469444 \times 6$$

$$\theta = 4.628816667°$$

$$\theta = 4° \ 37' \ 44''$$

To find the X_c and Y_c distance for the full-length spiral, enter the function tables with a θ angle of 6°, and multiply the functions by the length of the spiral. The computations are as follows:

$$X_c = 0.99890 \times 150 = 149.835 \text{ ft}$$
$$Y_c = 0.03488 \times 150 = 5.232 \text{ ft}$$

To find the X and Y distance for the partial-length spiral, enter the function tables with a θ angle (rounded to 4.6°) and multiply the functions by the length of the spiral. The computations are as follows:

$$X = 0.99935 \times 131.75 = 131.664 \text{ ft}$$
$$Y = 0.02675 \times 131.75 = 3.524 \text{ ft}$$

Now that the X and Y distances for both spirals have been computed, subtract the smaller X distance from the larger X distance (149.84 ft – 131.66 ft) and the result is 18.18 ft. This represents one leg of the triangle between the full-spiral at station 7 + 68.25 and the partial-spiral at station 7 + 50 (Figure 12.9). The second leg of the same triangle can be computed by subtracting the smaller Y distance from the larger Y distance (5.23 ft – 3.52 ft) and the result is 1.71 ft. The triangle is a right triangle, and therefore, the included angle is 90°.

THE PYTHAGOREAN THEORY

By employing the pythagorean theory, you can solve for the missing distance in a triangle when two sides and the included angle are known. The pythagorean equation is as follows:

Where:

Z = the hypotenuse of the triangle

18.18 ft = one leg of the triangle

1.71 ft	=	second leg of the triangle
Z	=	$\sqrt{(1\text{st side})^2 + (2\text{d side})^2}$

Substituting:

$$Z = \sqrt{(18.18\ \text{ft})^2 + (1.71\ \text{ft})^2}$$
$$Z = 18.26\ \text{ft}$$

The hypotenuse distance (18.26 ft) represents one part of the layout, but before the layout can continue, the bearing of the hypotenuse must be determined. This can be computed by trigonometry when all three sides of the triangle and the included angle are known as in Figure 12.10.

The angles within this triangle can be solved using either the sine, cosine, or tangent trigonometric equations. Angles α and β are computed as follows:

$$\tan \alpha = \frac{\text{opposite}}{\text{adjacent}}$$

Substituting:

$$\tan \alpha = \frac{18.18\ \text{ft}}{1.71\ \text{ft}}$$

$$\tan \alpha = 10.63157895$$
$$\tan \alpha = 84.62660226^\circ$$
$$\alpha = 84^\circ\ 37'\ 36''$$

$$\sin \beta = \frac{\text{opposite}}{\text{hypotenuse}}$$

Substituting:

$$\sin \beta = \frac{1.71\ \text{ft}}{18.26\ \text{ft}}$$

$$\sin \beta = 0.093647317$$
$$\sin \beta = 5.373469665^\circ$$
$$\beta = 5^\circ\ 22'\ 24''$$

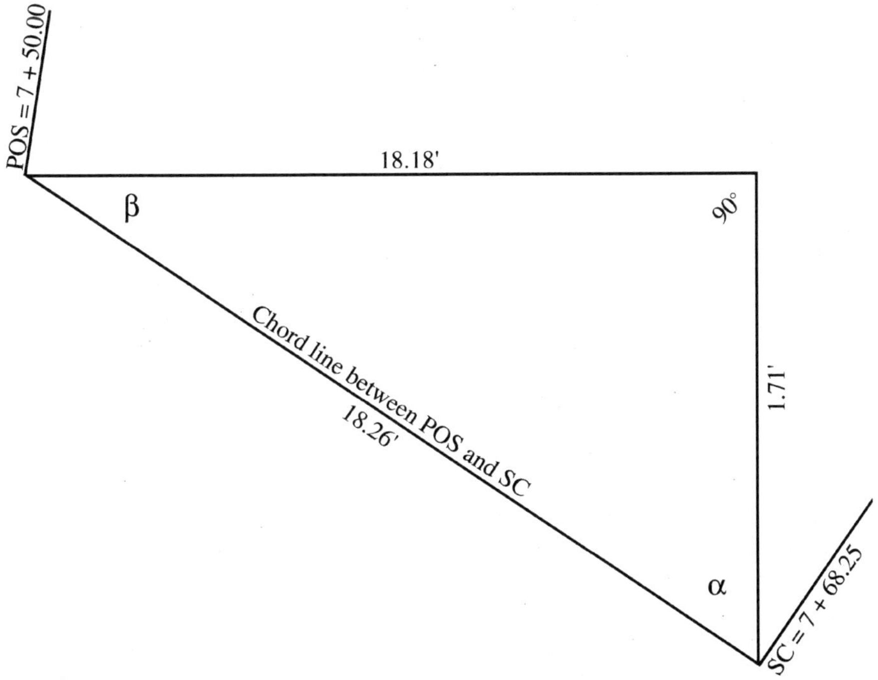

FIGURE 12.10 Backward spiral layout using X and Y distances

Remember to make a summation of the three angles; they must equal 180°.

With the angles and distances completed, we must turn our attention to the layout of the POS station 7 + 50.00 with the instrument set up at SC station 7 + 68.25. There are three field methods that can be used successfully; these are:

1. Occupy the SC point, station 7 + 68.25, backsight the end point of the X_c extended, observe the required angle (to the left) of 84° 37' 36", measure the distance (18.26 ft) and set the point.
2. Occupy the SC point, station 7 + 68.25, backsight along the bearing of the LC to TS station 6 + 18.25. Compute the bearing of the LC, as well as the bearing of the hypotenuse of the small triangle, and turn an angle equal to their difference. Measure the distance (18.26 ft) and set the point.
3. Occupy the SC point, station 7 + 68.25, backsight to any known point on the curve. Compute the bearing along this backsight, and knowing the bearing of the hypotenuse of the small triangle, compute the angular difference. Observe this required angle, measure the distance, and set the point.

Method 1 is probably easier to visualize and easier to complete if the Y_c distance is within the focal length of the instrument. If not, you will probably be required to occupy the end point of the X_c, setting an auxiliary point at 90° to the mainline tangent. Return to the SC point, backsight the auxiliary point, turn the required angle to the left, and set the point.

The second and third methods require you to develop the skills to compute angles from bearings. Under the pressures of construction layout, this often produces field errors. However, you will soon develop sufficient confidence to handle these layouts, and to avoid difficult computations when something easier is available.

CHAPTER 13
TRANSITION SPIRALS BETWEEN TWO CURVES

Introduction

My earliest recollection of a spiral between two curves goes back to the early 1960s. One afternoon during this period, our survey crew returned to the office with a concern. The crew had been trying to field-establish the centerline survey for its initial layout, using the plan curve data and relating this computed data to the original ground control traverse from which the topographic maps were prepared. The centerline could not be made to close with any of the ground controls.

Many hours were subsequently spent in the office searching for some practical reason why this alignment would not close. Eventually, the engineers found the answer: One of the centerline spirals had been designed surrounded on each side by curves, and the centerline alignment had been computed as though the spiral was located in its normal position, between a tangent and a curve.

The error should have become obvious during any one of many plan applications, such as during the plotting of the centerline on the plan-profile sheets, during the preparation of the geometric plan, and during any number of other drawings where the graphics should have shown an alignment discrepancy. The most elementary method of checking for alignment errors, that of computing and comparing deflection angles with bearings, was not done. This comparison would have immediately shown that the deflection angles of the spirals and curves did not equal the deflection angle of the main tangents. The normal checks failed, and until the initial field layout, the error lay dormant.

The plans were nearly complete, the major tangents and certain curves had been field-established; cross sections had been observed, plotted, and used in design; and the plans ready for state submittal. This certainly was not the time to locate such an error, and to make matters worse, the mathematical resources in our office to correct this problem had never been addressed.

Fortunately, the company employed excellent engineers, who analyzed the situation and determined that Barnett's (1940) equations could be used to recompute the spiral.

The method used to correct this problem was to hold the original P_a distance, computing back through the equation and forcing the new spiral to align close to the original location. Thus, the ability to compute the old and new spirals near the same location eliminated a resurvey of cross sections but introduced a station equation due to the difference in spiral lengths. This chapter is devoted to the study of this unusual mathematical and field condition.

During my early professional career, spirals and their computations were generally off limits due to a lack of expertise in this area and were equally difficult to locate in the field for the same reason. Transition curves were beginning to be used more frequently in design as a result of the former Bureau of Public Roads enforcing federal highway standards, which mandated the use of spiral transitions on both sides of a curve, when the curves were greater than 1° 28' 00".

Chapter 12 should have convinced you that the mathematics of spirals are quite easy to compute, providing the function tables are available. The equations and function tables used in this presentation are from Barnett (1940). Figure 13.1 shows the spiral transition between two curves. The following symbols are used when transition spirals are located between two curves. All subscript$_1$ designations refer to the flatter curve, and subscript$_2$ designations refer to the sharper curve.

CS_1 and CS_2	= Common points of curve and spiral
R_1 and R_2	= Radii of the circular curves
L_a	= Length of transition from CS_1 to CS_2
P_a	= Offset distance between the circular curves produced at the common radius (midpoint or half-line)
Δ_1 and Δ_2	= Intersection angles between the tangents at CS_1 and CS_2, and the parallel tangents of the circular curves produced at the common radius
D_1 and D_2	= Degree of curvature of the circular curves
R_a	= Curve radius of length $(D_2 - D_1)$
θ_a	= Equivalent spiral angle
T_1	= Local tangent distance from CS_1 to the intersection, point I
T_2	= Local tangent distance from CS_2 to the intersection, point I
AB	= Distance along the half-line from the intersection of either T_1 or T_2
AI	= Distance from the half-line, along T_1 to point I
BI	= Distance from the half-line, along T_2 to point I

MINIMUM SPIRAL LENGTHS

The transition spiral length between curves is usually chosen during engineering design, where the length is computed to meet the following requirement. The general procedure is as follows (Barnett, 1940):

$$L_s = 1.6\frac{(V)^3}{R_2} - 1.6\frac{(V)^3}{R_1} \tag{13.1}$$

Where:

R_1 and R_2 represent the flatter and sharper radii of the two curves
R_1 = 8° (5729.58 ÷ 8 = 716.1975 ft)
R_2 = 14° (5729.58 ÷ 14 = 409.2557 ft)
V = 40 mi/h
1.6 = constant, based on centrifugal force

Substituting:

$$L_s = 1.6\frac{40^3}{409.2557} - 1.6\frac{40^3}{716.1975}$$

$$L_s = 205.21032 - 142.9773$$

$$L_s = 107.23 \text{ ft}$$

Therefore, the minimum length of spiral between curves of 8° and 14°, with a velocity of 40 mi/h, is ±107 ft; however, this "general procedure" does not automatically set the spiral length.

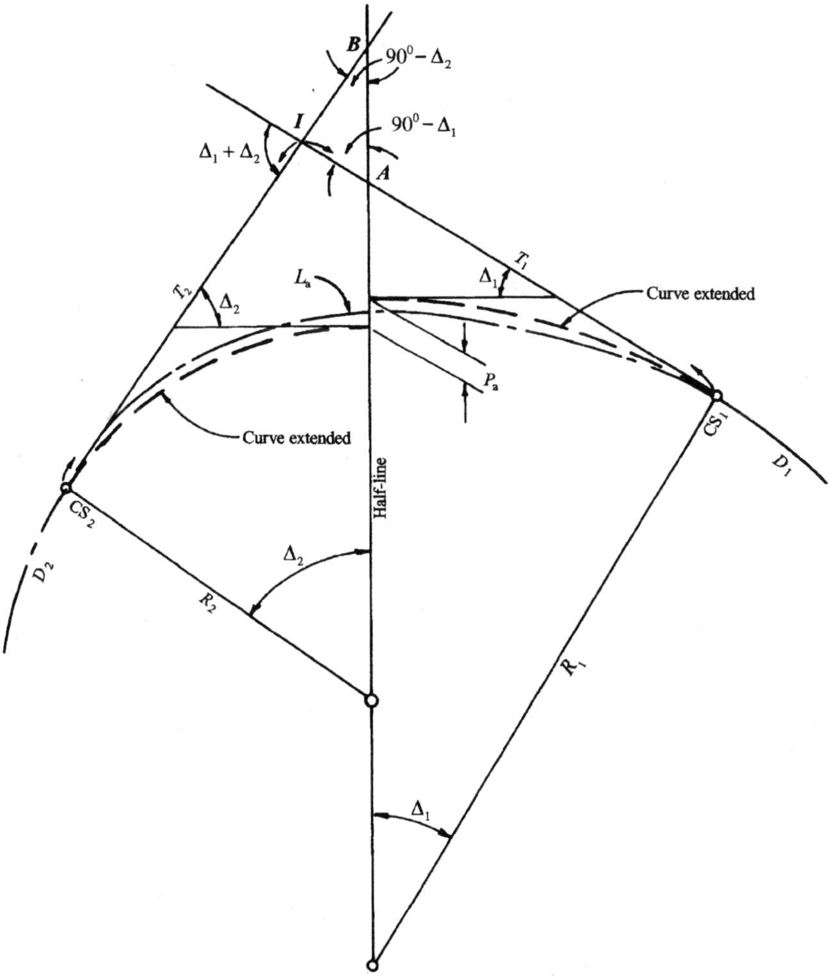

FIGURE 13.1 Spiral transition symbols and location of parts

Superelevation and the rate of change often control whether spirals between compounding curves are required and, if so, what the length might be. Barnett (1940) states:

> Compounding two curves that differ in radius by more than 50 percent of the smaller radius results in a sudden change of curvature which is noticeable if the transition is omitted. A transition of at least 100 feet long should be inserted. Where two curves differ in radius by less than 50 percent and the preceding methods of design result in transitions less than 100 feet long, they may be omitted and such change in superelevation as may be required can be effected on the flatter curve.

However, the final transition length, known as L_a, is usually set by the highway design engineer, or the spiral can be inserted between two existing curves, where the length L_a of the spiral is computed based on the equations that follow.

When the spiral limits are not defined, as in our example (Figures 13.2 and 13.3), you have the option of choosing a length of spiral and adjusting the "beginning" and "ending" of the curves so they match each end of the spiral at common local tangent points. These local tangent points are identified as point 7 on the flatter curve, and point 3 on the sharper curve.

GRAPHICALLY LOCATING THE SPIRAL

After graphically locating each curve on a scaled drawing, choose the approximate "curve to spiral" points (3 and 7), and extend each curve back or forward into the space to be occupied by the spiral, to a line that approximates one-half the distance between CS_1 and CS_2. This length will be initially scaled and later computed to represent the L_a distance. The line is referred to as the half-line or midpoint line and when the computations are completed, the half-line will be shown to mathematically bisect the length of the total spiral.

COMPUTING THE TRANSITION

Where:

L_a = set at 880 ft
D_2 = the sharper curve at 445 ft radius (12° 52' 32")
D_1 = the flatter curve at 555 ft radius (10° 19' 25")

Find the equivalent spiral angle:

$$\theta_a = \text{the equivalent spiral angle} \tag{13.2}$$

$$\theta_a = \frac{L_a}{200} \times (D_2 - D_1)$$

$$\theta_a = \frac{L_a}{200} \times (12°\ 52'\ 32'' - 10°\ 19'\ 25'')$$

$$\theta_a = \frac{880}{200} \times 2.55194444°$$

$$\theta_a = 11.2285555°$$

$$\theta_a = 11°\ 13'\ 43''*$$

* (Note: The equivalent spiral angle is used only to compute the P_a distance from the function tables.)

Find:

R_a = $(D_2 - D_1)$ or radius of a degree of curve equal to the difference between degrees of the two attached curves. (13.3)

R_a = (12° 52' 32" – 10° 19' 25")

R_a = 2° 33' 07"

R_a = 2245.18 ft*

*(Note: The radius (R_a) in a spiral transition has no field locating function.)

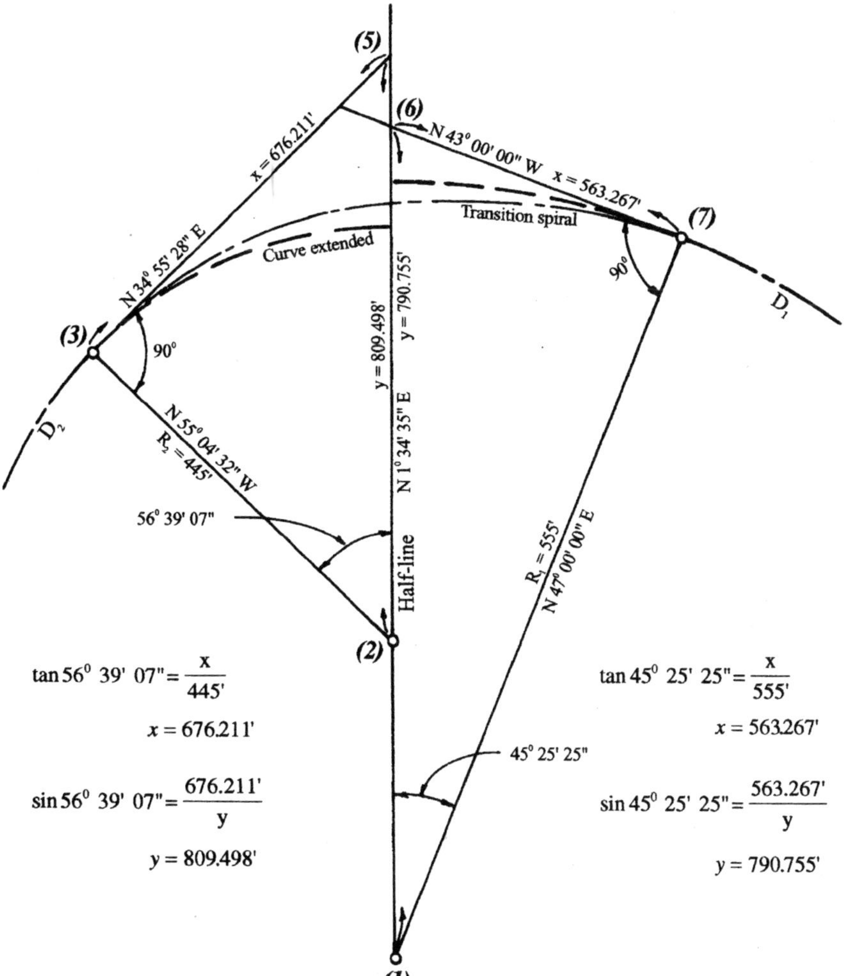

FIGURE 13.2 Spiral transition symbols with bearings and distances

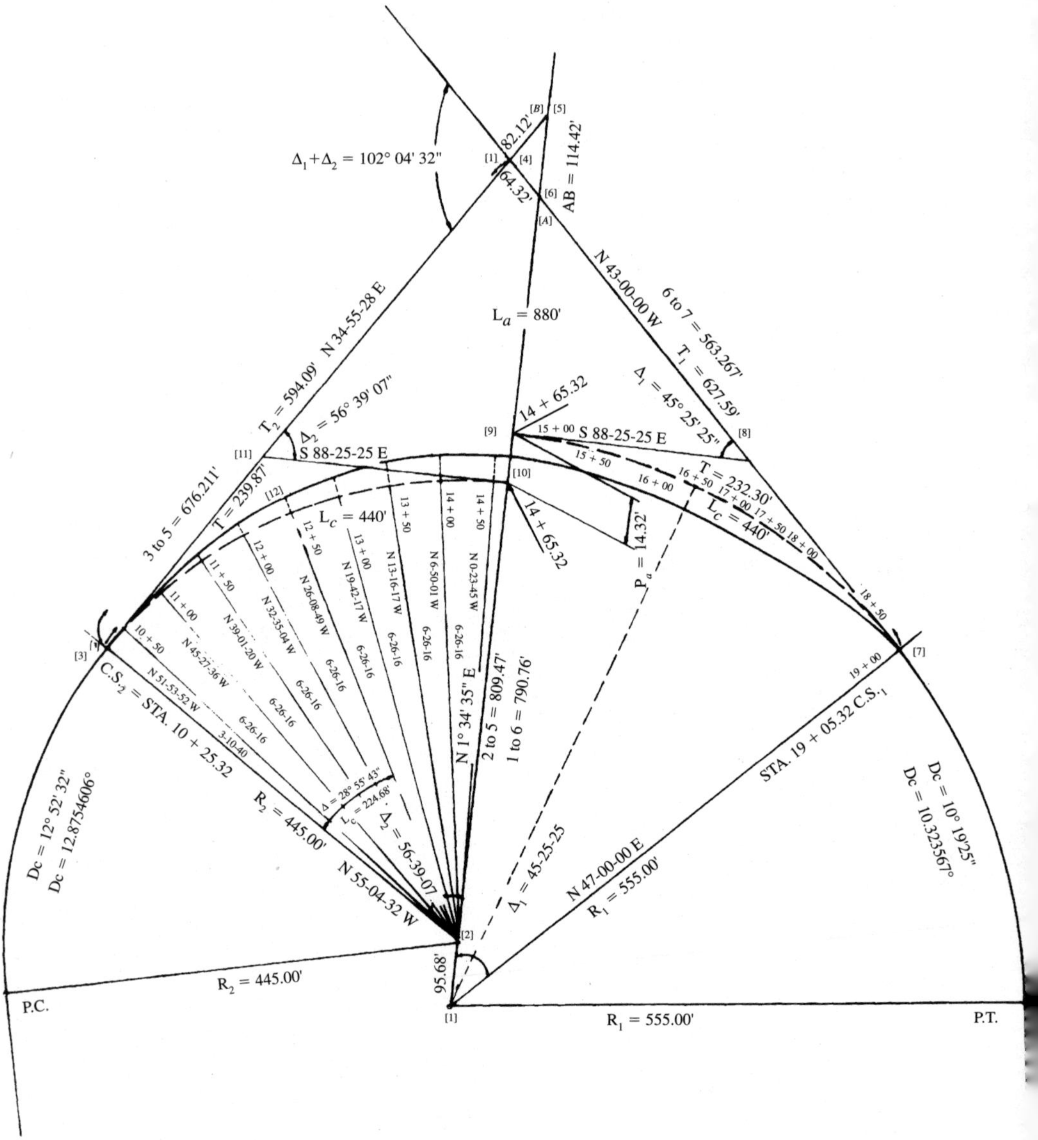

FIGURE 13.3 Drawing of complete spiral calculations

The following two equations are used to compute the Δ angles for each of the extended curves terminating at the half-line. For the flatter curve:

$$\Delta_1 = \frac{L_a}{200} \times D_1 \tag{13.4}$$

Substituting:

$$\Delta_1 = \frac{880}{200} \times 10.32356757$$

$$\Delta_1 = 45.42369731°$$
$$\Delta_1 = 45° \ 25' \ 25''$$

For the sharper curve:

$$\Delta_2 = \frac{L_a}{200} \times D_2 \tag{13.5}$$

Substituting:

$$\Delta_2 = \frac{880}{200} \times 12.87546067$$

$$\Delta_2 = 56.65202695° \qquad (13.5)$$
$$\Delta_2 = 56° \ 39' \ 07''$$

The Δ angles in Eqs. (13.4) and (13.5) should be verified to determine whether each computed angle represents one-half of the L_a distance. The equations found in Chapter 11, can be used to solve the L_c, knowing both the Δ angle and the D_c.

Verifying the length of curve:

$$L_c = \frac{\Delta}{D_c} \times 100 \tag{11.3}$$

Substituting (using the flatter curve data):

$$L_c = \frac{45.42369731}{10.32356757} \times 100$$

$$L_c = 440 \text{ ft}$$

Substituting (using the sharper curve data):

$$L_c = \frac{56.65202695}{12.87546067} \times 100$$

$$L_c = 440 \text{ ft}$$

For field layout purposes, compute the tangent T distances (see Figure 13.3) for each of the curves by using the tangent equation in Chapter 11.

$$T = R \times \frac{\tan \Delta}{2} \tag{11.4}$$

Substituting (using the flatter curve data):

$$T = 555.00 \times \tan 22° \ 42' \ 43''$$
$$T = 555.00 \times 0.41855403$$
$$T = 232.30 \text{ ft}$$

Substituting (using the sharper curve data):

$$T = 445.00 \times \tan 28° \, 19' \, 34''$$
$$T = 445.00 \times 0.53903252$$
$$T = 239.87 \text{ ft}$$

RECAPPING TRANSITION SPIRAL DEFINITIONS

The following transition "parts" were computed on the previous pages, with more to follow:

L_s = Where the minimum length of transition is computed, based on radius, velocity, and a coefficient for centrifugal force; Eq. (13.1)

θ_a = The equivalent spiral angle; Eq. (13.2)

R_a = The radius of a curve, of a degree equal to the difference between degrees of the two attached curves; Eq. (13.3)

Δ_1 = The subtended angle of the flatter curve extended from the CS_1 or CS_2 to the half-line; Eq. (13.4)

Δ_2 = The subtended angle of the sharper curve extended from the CS_1 or CS_2 to the half-line; Eq. (13.5)

CONTINUED SOLUTIONS

Continuing with the solution for other parts of the transition,

Find:

P_a = the distance between the two curves extended toward each other and measured at the half-line. The computations use a combination of functions from Table 12.1 (using the equivalent spiral angle θ_a for the theta) and multiplied by the spiral length to obtain the distance.

$$P_a = \theta_a \text{ (in functions)} \times \text{spiral length} \tag{13.6}$$

Substituting:

$$P_a = 11° \, 13' \, 43'' \text{ (11.2° rounded)} \times 880.00 \text{ ft}$$
$$P_a = 0.01627 \times 880.00$$
$$P_a = 14.32 \text{ ft}$$

You will notice that none of Eqs. (13.1) through (13.6) have established the actual position of the spiral in relation to the local tangents through points CS_1 and CS_2. The final location of these local tangents is subject to their intersection with the half-line, as identified in triangle *ABI* at the top of Figures (13.1) and (13.3). The three legs in the triangle that adjust this positioning are known as lines *AB, AI,* and *BI*. These line designations are computed as shown below:

Find line *AB*:

$$AB = (R_2 \times \text{exsec } \Delta_2) - (R_1 \times \text{exsec } \Delta_1) - P_a \tag{13.7}$$

Find line *AI*:

$$AI = AB \times \frac{\cos \Delta_2}{\sin (\Delta_1 + \Delta_2)} \tag{13.8}$$

Find line *BI*:

$$BI = AB \times \frac{\cos \Delta_1}{\sin (\Delta_1 + \Delta_2)} \tag{13.9}$$

Substituting:

Using the assumed curve data from Figure 13.3, we have the following:

Line *AB*

$AB = (445.00 \times \text{exsec } 56° \; 39' \; 07'') - (555.00 \times \text{exsec } 45° \; 25' \; 25'') - 14.317$ (13.7)
$AB = (445.00 \times 0.8190958) - (555.00 \times 0.4247864) - 14.317$
$AB = 364.497 - 235.756 - 14.317$
$AB = 114.42$ ft

Line *AI*

$$AI = 114.42 \times \frac{\cos 56° \; 39' \; 07''}{\sin (45° \; 25' \; 25'' + 56° \; 39' \; 07'')}$$

$$AI = 114.42 \times \frac{0.54972364}{0.977872579}$$

$AI = 114.42 \times 0.562162854$
$AI = 64.32$ ft

Line *BI*

$$BI = 114.42 \times \frac{\cos 45° \; 25' \; 25''}{\sin \; (45° \; 25' \; 25'' + 56° \; 39' \; 07'')}$$

$$BI = 114.42 \times \frac{0.701859573}{0.977872578}$$

$BI = 114.42 \times 0.717741338$
$BI = 82.12$ ft

Although we have computed the distances in triangle *ABI* (Figure 13.3), the position of the local tangent at points CS_1 and CS_2 cannot yet be determined. There are three separate methods in which these local tangents can be located. They are:

Method 1.

Compute the hypotenuse and opposite leg in right triangle 1,7,6 (Figure 13.3) and add the previously computed distance in line AI to line 7-6, and the remaining distance is equal to T_1.

The computation for this right triangle follows:

Trigonometric function:

$$\tan 45° \; 25' \; 25'' = \frac{\text{opposite}}{\text{adjacent}}$$

$$1.01489723 = \frac{\text{opposite}}{555 \text{ ft}}$$

$$\text{opposite} = 563.267 \text{ ft}$$

Therefore, line 7-6 equals 563.267 ft plus 64.325 ft equals T_1 equals 627.59 ft.

Trigonometric function:

$$\sin 45° \, 25' \, 25'' = \frac{\text{opposite}}{\text{hypotenuse}}$$

$$0.712315337 = \frac{563.267 \text{ ft}}{\text{hypotenuse}}$$

$$\text{Hypotenuse} = 790.755 \text{ ft}$$

Therefore, line 1-6 equals 790.755 ft.

Compute the hypotenuse and opposite leg in right triangle 2, 3, 5 (Figure 13-3) and subtract the previously computed distance in line *BI* from line 3-5. The remaining distance is equal to T_2.

Trigonometric function:

$$\tan 56° \, 39' \, 07'' = \frac{\text{opposite}}{\text{adjacent}}$$

$$1.519575522 = \frac{\text{opposite}}{445 \text{ ft}}$$

$$\text{Opposite} = 676.211 \text{ ft}$$

Therefore, line 3-5 equals 676.211 ft minus 82.127 ft equals T_2 equals 594.09 ft.

Trigonometric function:

$$\sin 56° \, 39' \, 07'' = \frac{\text{opposite}}{\text{hypotenuse}}$$

$$0.835346587 = \frac{676.211 \text{ ft}}{\text{hypotenuse}}$$

$$\text{Hypotenuse} = 809.498 \text{ ft}$$

Therefore, line 2-5 equals 809.498 ft.

Method 2.
Computing the T_1 and T_2 distances direct, using the following equations.

Find:

$$T_1 = R_1 \times \tan \Delta_1 + AI \text{ (same as method 1)} \qquad (13.10)$$

Substituting:

$$T_1 = 555 \text{ ft} \times \tan 45° \, 25' \, 25'' + 64.325 \text{ ft}$$
$$T_1 = 555 \text{ ft} \times 1.01489723 + 64.325 \text{ ft}$$

Therefore, T_1 equals 627.593 ft (same distance as calculated in method 1).

Find:

$$T_2 = R_2 \times \tan \Delta_2 - BI \text{ (same as method 1)} \qquad (13.11)$$

Substituting:

$$T_2 = 445 \text{ ft} \times \tan 56° \, 39' \, 07'' - 82.12 \text{ ft}$$
$$T_2 = 445 \text{ ft} \times 1.519575522 - 82.12 \text{ ft}$$

Therefore, T_2 equals 594.09 ft (same distance as calculated in method 1).

With T_1 and T_2 computed, the distance (627.59 ft) from CS_1 to the intersection, point *I*, can be measured in the field, as can the distance (594.09 ft) from CS_2 to the intersection, point *I*. Note: The distances T_1 and T_2 in this example are considerably long, and the usual construction site terrain and foliage would probably not permit the measuring of these lengths. In lieu of this layout option, a manual or computer-generated closure can be calculated, where the central points of the two radii (points 1 and 2, Figure 13.3) can be located one from the other. Method 3 also acts as a means of verifying other previous computations.

Method 3.

The manual closure is set up using points 1,2,3,11,10,9,8,7 and back to 1 (see Figure 13.3). Note: When all parts of the closure are assembled, as in Fig. 13.3, the only unknown is the distance between points 1 and 2. Remember to avoid solving closures where there is only one unknown. When continuing with such a closure, one of the known bearings or distances is withheld from the closure, and when the final results are computed, one of the two unknowns will match the unit withheld. In the closure below, the bearing of line 1-2 will be withheld.

Course	Bearing	Dist.	Cosine	Sine	North	South	East	West
2-3	N 55-04-32 W	445.00'	0.57249	0.81990	254.7602	---------	---------	364.8586
3-11	N 34-55-28 E	239.87'	0.81990	0.57249	196.6710	---------	137.3243	---------
11-10	S 88-25-25 E	239.87'	0.02750	0.99962	---------	6.59875	239.7790	---------
10-9	N 01-34-35 E	14.32'	0.99962	0.02750	14.3145	---------	0.393938	---------
9-8	S 88-25-25 E	232.30'	0.02750	0.99962	---------	6.39050	232.2119	---------
8-7	S 43-00-00 E	232.30'	0.73135	0.68199	---------	169.8934	158.4281	---------
7-1	S 47-00-00 W	555.00'	0.68199	0.73135	---------	378.5091	---------	405.9013
1-2	(N 1-34-14E)	(95.68')	(0.9996)	(0.0274)	(95.646)	---------	(2.622662)	---------
totals		(95.69')			465.7457	561.3917	768.1372	770.7599

$$\tan \angle = \frac{2.622662}{95.646}$$

$$\tan \angle = 0.02742051$$
$$\angle = 1^\circ\ 34'\ 14''$$

Therefore, the closure bearing of course 1-2 is N 1° 34' 14" E and the distance is 95.68 ft. Notice, the actual bearing of course 1-2, computed using Δ_1 and Δ_2 angles, is N 1° 34' 35" E. The 21" difference is not significant for computing or for layout and is probably caused by the rounding of digits and especially the introduction of the P_a distance into the closure. This distance, remember, is a proportional spiral function and will not produce exact trigonometric results.

SPIRAL TRANSITION LAYOUT

Locating the actual transition arc is accomplished by computing the offset from each curve to the transition. We already know that the separation between the two curves along the half-line is equal to the P_a distance, and the spiral intersects this distance at $\frac{P_a}{2}$. Other offset distances between the two curves and the spiral are computed by multiplying the $\frac{P_a}{2}$ by the cube of the proportion of the distance to the nearest CS.

By viewing Figure 13.3, you can see that offsets from the flatter curve are directed inward toward the center of the circle, while offsets from the sharper curve are directed outward from the center of the circle.

CALCULATING THE PROPORTIONAL CONSTANT

Calculating a proportional constant at the midpoint of the spiral, using the P_a distance in Figure 13.3, and the L_a distance in the same figure, we have the following:

Where:

P_a = 14.32 ft

L_a = 880 ft

$$\text{Proportional constant} = \frac{\frac{P_a}{2}}{\left[\frac{L_a}{2}\right]^3} \qquad (13.12)$$

Substituting:

$$\text{Proportional constant} = \frac{\frac{14.32\text{ ft}}{2}}{\left[\frac{880\text{ ft}}{2}\right]^3}$$

$$\text{Proportional constant} = \frac{7.16}{440^3} = \frac{7.16}{85184000} = 0.000000084$$

Therefore, the proportional constant at the midpoint of the spiral is 0.000000084, and when multiplied by the cube of the distance to the nearest CS ($0.000000084 \times 440^3 = 7.155$ ft, or 7.16 ft rounded), establishes the offset distance from the curve to the spiral at each station. This example is computed at the half-line, and we know that the transition bisects the P_a distance (14.32 ft). Therefore, our computation is correct. Other station offsets are computed in a similar manner, using the cube of the distance to each particular station in the equation. The computations in Table 13.1 show the recommended procedure for a multiple station layout.

The computations in Table 13.1 (on the following page) cover the first one-half of our example (Figure13.3), commencing at station CS_2 10 + 25.32. The offset for the second-half of the spiral is also shown on the following page and is related to the nearest CS, which is station 19 + 05.32.

The computations in Table 13.2 represent multiple stations; however, with the use of the proportional constant the mathematics is made easy, and the offset lengths can be observed for their sequential change.

With the offset distances computed, we can turn our attention to the field layout process, which can be completed using any one of the following three methods:

1. Spiral layout, by deflection
2. Spiral layout, by coordinate computation
3. Spiral layout, by locating points on the extended curve

TABLE 13.1 Offsets Between Spiral and Curve

Station	L from CS_2	L^3	Offsets from curves ($L^3 \times 0.000000084$ in ft)
10 + 25.32	0.0000	0.000	0.0000
10 + 50.00	24.68	15032.647	0.00
11 + 00.00	74.68	416498.007	0.03
11 + 50.00	124.68	1938163.367	0.16
12 + 00.00	174.68	5330028.727	0.45
12 + 50.00	224.68	11342094.09	0.95
13 + 00.00	274.68	20724359.45	1.74
13 + 50.00	324.68	34226824.81	2.88
14 + 00.00	374.68	52599490.17	4.42
14 + 50.00	424.68	76592355.53	6.43
14 + 65.32	440.00	85184000.00	7.16

TABLE 13.2 Offsets Between Spiral and Curve

Station	L from CS^1	L^3	Offsets from curves ($L^3 \times 0.000000084$ in ft)
14 + 65.32	440.00	85184000.00	7.16
15 + 00.00	405.32	66587713.45	5.59
15 + 50.00	355.32	44859968.09	3.77
16 + 00.00	305.32	28462022.73	2.39
16 + 50.00	255.32	16643877.37	1.40
17 + 00.00	205.32	8655532.01	0.73
17 + 50.00	155.32	3746986.65	0.31
18 + 00.00	105.32	1168241.29	0.10
18 + 50.00	55.32	169295.93	0.01
19 + 00.00	5.32	150.57	0.00
19 + 05.32	0.00	0.00	0.00

SPIRAL LAYOUT, BY DEFLECTION

The study thus far has been directed to computing the offsets from the curve to the spiral at known centerline stations. The deflection method permits the surveyor to layout the spiral with an instrument setup at either CS_1 or CS_2 without using programmable computers. Using Figure 13.3 as the example, commence with an instrument setup at CS_2 10 + 25.32 and assume you are to locate station 12 + 50.00. Review the location of the spiral in relation to the curve. Obviously, the sharper curve is inside the proposed spiral, whereas the flatter curve is outside the proposed spiral. This is important because when the spiral deflection angle is computed, corrections (added or subtracted) are made to the curve deflection angle, based on the position of the spiral, that is whether inside or outside of the curve. The deflection method is as follows:

1. Compute the deflection to the point on the curve
2. Compute the deflection to the point on the spiral

Observe Figure 13.3, noticing that the spiral is outside the sharper curve, and therefore, the spiral deflection reduces the curve deflection, assuming your backsight is along the mainline tangent, line 3-I. By contrast, assume the instrument is set up at station 19 + 05.32, locating any known station (within the half-line). In this condition, the spiral deflection increases the curve deflection, assuming your backsight is along the mainline tangent, line 7-6. Observe Figure 13.3 and make sure you understand why the corrections are added or subtracted. The two sets of computations are shown below using the following centerline curve data:

$$D_1 = 10° \ 19' \ 25"$$
$$D_2 = 12° \ 52' \ 32"$$
$$\theta = 11.22855°$$
$$L_a = 880 \text{ ft}$$

The deflection angle to the *curve* is computed by first subtracting stations to obtain the curve length and then computing the subtended angle, known as ϕ (Phi).

$$\begin{array}{r} \text{POC station } 12 + 50.00 \\ \underline{\text{CS}_2 \text{ station } \ 10 + 25.32} \\ L_c = 2 + 24.68 \end{array}$$

Therefore:

$$L_c = 224.68 \text{ ft} \tag{13.13}$$

$$\phi = \frac{L_c \times D_c}{100}$$

$$\phi = 28.928584°$$
$$\phi = 28° \ 55' \ 43"$$

$$\frac{\phi}{2} = 14° \ 27' \ 51" = \text{the curve deflection angle*}$$

* (Note: Remember, the deflection angle to a point on curve is always one-half the subtended angle as discussed in Chapter 11.)

The deflection angle to the *spiral* is computed using the following equation:

$$L_c = 224.\ 68 \text{ ft} \tag{13.14}$$

$$\phi = \frac{L^2}{L_a^2} \times \frac{\theta_a}{3}$$

$$\phi = \frac{(224.68)^2}{(880)^2} \times \frac{11.22855}{3}$$

$$\phi = \frac{50481.1024}{774{,}400} \times 3.74285$$

$$\phi = 0.243986°$$

$$\phi = 0° \ 14' \ 38'' = \text{the spiral deflection angle}$$

By observation, we know the spiral is outside the curve (closer to tangent line 3-I), and the deflection of the spiral must be subtracted from the curve deflection, such as:

$$\begin{array}{r} 14.464166° \\ -0.243986° \\ \hline 14.220180° \end{array}$$

Therefore, 14° 13' 13" equals the adjusted deflection angle to locate spiral station 12 + 50.00 from CS_2 station 10 + 25.32.

FIELD LAYOUT TECHNIQUE FOR THE DEFLECTION METHOD

The field technique for the deflection method is to occupy, as in our example, CS_2 station 10 + 25.32, backsighting forward along the mainline tangent, line 3-I, and with the instrument turn an angle of 14° 13' 13" and measure a chord distance equal to the station difference on the arc. Since station 12 + 50.00 is a significant distance (224.68 ft) from the CS, you will probably want to apply a chord correction. Remember, there are no chord corrections for centerline spiral layout; however, toward the end of this chapter, we will suggest an empirical method for computing chord distances where the centerline points are longer than the usual 50 ft.

Continue to compute the deflection angles for the remaining stations, remembering that each CS instrument setup and each station's respective deflection angle are valid only within the half-line. If you intend to use this method of locating spiral points, short chord distances of 50 ft or less will have little impact on the accuracy of each station. As in all spirals, however, locate the beginning and ending of the spiral, using proper surveying technique, and let whatever error exists be contained within these limits.

SPIRAL LAYOUT BY COORDINATE COMPUTATIONS

This method of spiral layout uses the center of the curve, the length of radius, and the offset distance, combining these three items to establish a coordinate for each station on the spiral. When the coordinates are available at each station, bearings and distances can be determined using programmable computers and inversing between coordinates. When adjacent bearings are computed, which is the result of inversing, the surveyor can turn angles and measure distances from point to point by relating the two adjacent bearings and compute the deflection angle.

Refer to Figure 13.3 from CS_2 station 10 + 25.32 to the half-line, station 14 + 65.32. Each 50 ft interval is shown on the drawing radiating from the central point of the curve, each delta angle is computed based on the length of curve between stations, and each bearing is shown along the radius for the various stations. In Figure 13.3, we have assumed a coordinate of N 5000.00 and E 5000.00 for traverse point 2, and by computation, found the bearing along the half-line to be N 1° 34' 35" E. Compute the bearings for each station using Eq. (11.3) to determine the subtended angle, followed by inputting this data into programmable computers to determine the coordinate at each spiral station point.

Spiral point 12, station 12 + 50 (Figure 13.3) is one example of this computation. Refer back to the offset table where the offset distance for station 12 + 50 is 0.95 ft. This distance is added to the radius (445 ft), making a total distance of 445.95 ft and, with the programmable computer set for traverse mode, or side-shot mode, the coordinates are computed for each spiral station. When two coordinates, such as, points 3 and 12 are computed and inversed, the bearing between these two stations is N 49° 09' 07" E, with a distance of 222.54 ft. If this represented an actual project, you would continue to compute the remaining coordinates for each station, remembering that the length of the radial line changes by an amount equal to the offset distance. The calculation of coordinates for each spiral point, not only provides the option (inversing between any two coordinates) of locating the spiral from one instrument setup, but also provides the opportunity to move forward on the centerline, setting the next adjacent station. Table 13.3 shows the typical computation entries to establish spiral points from one setup.

Table 13.4 shows typical calculation entries when the instrument is moved from one location to another. The field layout procedure for this is:

- Establish the first station (10 + 50.00) as you would using Table 13.3.
- Move forward to this station (10 + 50.00), backsighting the CS_2, relating bearing with bearing, turning the computed angle, measuring the distance, and setting the point.
- Continue this process until you have completed all of the required station locations.

TABLE 13.3 Locate Spiral Points from One Location

Instrument Setup	Deflection Bearing and Distance	Station Destination
CS_2 10 + 25.32	N 36-30-48 E	10 + 50.00
(local tangent bearing)	24.678 ft	
N34-55-28 E	N 39-42-33 E	11 + 00.00
	74.596 ft	
	N 42-52-41 E	11 + 50.00
	124.297 ft	
	N 46-01-28 E	12 + 00.00
	173.651 ft	
	N 49-09-07 E	12 + 50.00
	222.541 ft	
	N 52-15-25 E	13 + 00.00
	270.874 ft	
	N 55-20-34 E	13 + 50.00
	318.566 ft	
	N 58-24-47 E	14 + 00.00
	365.540 ft	
	N61-28-10 E	14 + 50.00
	411.741 ft	
	N 62-24-08 E	14 + 65.32
	425.738 ft	

TABLE 13.4 Instrument Moved from One Location to Another

Setup	Backsight Bearing	Foresight Bearing	Deflection Angle	Distance	Station
10 + 50.00	S 36-30-48 W	N 41-17-12 E	4-46-24	49.98'	11 + 00.00
11 + 00.00	S 41-17-12 W	N 47-36-36 E	6-19-24	49.99'	11 + 50.00
11 + 50.00	S 47-36-36 W	N 53-51-54 E	6-15-18	50.01'	12 + 00.00
12 + 00.00	S 53-51-54 W	N 60-03-46 E	6-11-52	50.05'	12 + 50.00
12 + 50.00	S 60-03-46 W	N 66-10-14 E	6-06-28	50.13'	13 + 00.00
13 + 00.00	S 66-10-14 W	N 72-12-42 E	6-02-28	50.25'	13 + 50.00
13 + 50.00	S 72-12-42 W	N 78-11-58 E	5-59-16	50.41'	14 + 00.00
14 + 00.00	S 78-11-58 W	N 84-06-48 E	5-54-50	50.62'	14 + 50.00
14 + 50.00	S 84-06-48 W	N 87-54-12 E	3-47-24	15.57'	14 + 65.32

Moving the Instrument to a New CS

Each of the above spiral layout computations covered the first one-half of the spiral in Figure 13.3. For a continuance of the table in the second half of the spiral, the *computations* must be computed based on the second radius with a different central radius point (1) and beginning at CS_1 station 19 + 05.32. These computations will not be shown.

Field layout, however, can be from either CS_1 or CS_2 due to the use of *coordinates*. Simply inversing between your setup station and the station to be located will provide the required bearing and distance.

EMPIRICAL CHORD CORRECTION FOR SPIRAL TRANSITIONS

Problem awareness

Practicing surveyors are aware that the location of spiral stations, whether "simple" or "compound," should include a chord correction. However, surveyors are left to their own discretion when choosing a reliable taping method. Barnett (1940) does not mention linear measurement, and therefore, the surveyor often chooses to use curve chord corrections when the spiral stations approach the curve. Chapter 12 mentioned a few studies (using 50 ft intervals) that indicated the spiral station position error, when applying the functions from the tables, have adjusted for this error, and the error is minimal. However, if long chord distances are to be measured, chord corrections must be either computed by coordinates or by the empirical method.

Common practice

The common practice of initially locating each end of the spiral before filling in the interior controls the total error by isolating the problem within the length of the spiral. However, when station locations become very critical, such as mainline bridge piers and abutments, the recommended practice is to compute the location of the pier or abutment by proceeding through the center radii points, following along the length of the radius plus or minus the offset distance, establishing coordinates at each break point, and determining the bearing and distance by inverse, thus locating the point in the most accurate and acceptable manner.

CHORD CORRECTION

Locating single spiral points at distances greater than 50 ft from the instrument setup (assuming coordinates have not been computed) requires a simple ratio calculation (see Figure 13.4) that can be completed in the field. The following "empirical" method is suggested for computing chord lengths. It is most effective for short distances, but is acceptable for long distances. The method has small errors near the spiral half-line, when both

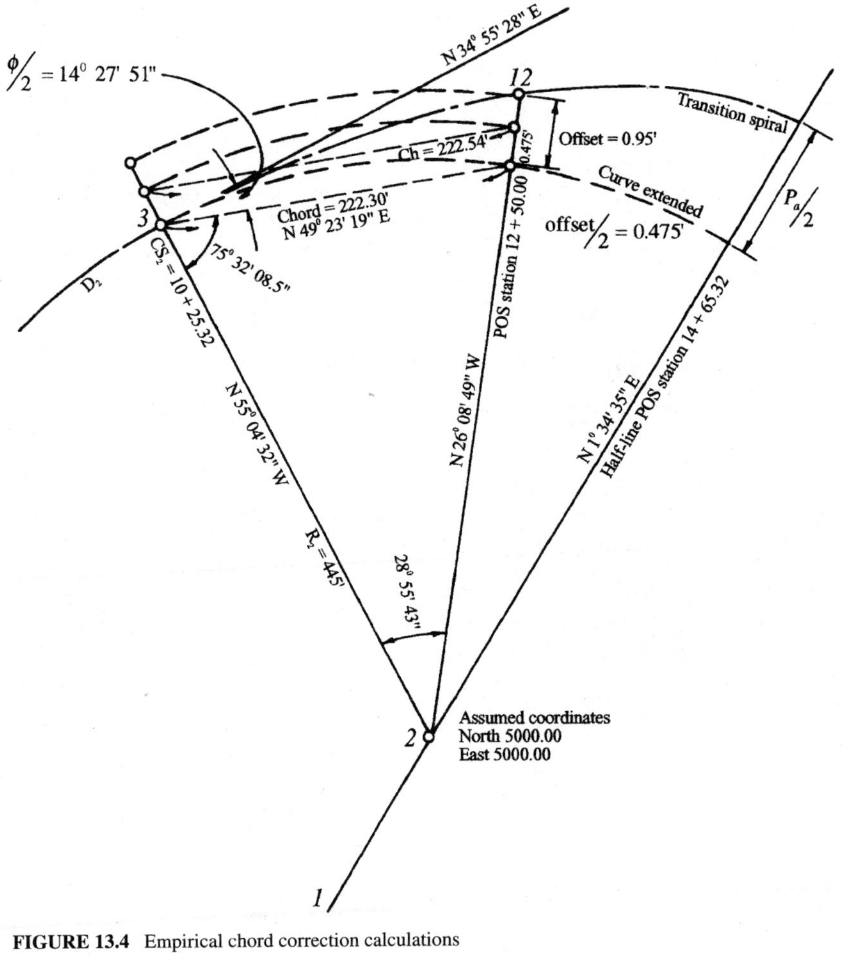

FIGURE 13.4 Empirical chord correction calculations

curves are sharp and when the P_a distance is approximately 10 ft or greater. Commence the angular calculation by using Eqs. (13.13) and (13.14), which are:

- Compute the deflection angle of the simple curve using Eq. (13.13)
- Compute the deflection angle of the transition spiral using Eq. (13.14)
- Relate the deflections to each other and compute the *adjusted deflection angle* as mentioned earlier in the chapter.

Correct the chord distance as follows:

- Compute the normal chord length (see Figure 13.4) between any two stations on the extended curve using Eq. (11.5)
- Compute the spiral chord length by assuming a concentric curve, the length of which is equal to the radius of the curve plus or minus one-half of the offset distance (plus, if the spiral is outside the curve or minus, if the spiral is inside the curve)
- Use this adjusted radius as the length of a concentric curve
- Finally, set up a ratio of the two concentric radii to determine the approximate length of the chord

Example:

Compute the chord distance between CS_2 station 10 + 25.32 and POC station 12 + 50.00.

$$\begin{array}{r} 12 + 50.00 \text{ POS} \\ -10 + 25.32 \text{ CS}_2 \\ \hline L_c = 2 + 24.68 \text{ ft} \end{array}$$

The centerline curve data are:

$$D_c = 12.87546067°$$
$$R = 445 \text{ ft}$$
$$\text{Offset distance} = 0.95 \text{ ft at station } 12 + 50.00$$

Find the curve deflection angle ϕ, Eq. (13.13) and the chord length, Eq. (11.5):

$$\text{Subtended angle of the curve} = \phi = \frac{L_c \times D_c}{100} \tag{13.13}$$

Substituting:

$$\phi = \frac{224.68 \times 12.87546067}{100}$$
$$\phi = 28.928585°$$
$$\phi = 28° \; 55' \; 43''$$
$$\frac{\phi}{2} = 14° \; 27' \; 51'' = \text{curve deflection angle*}$$

*Note: The deflection angle is always one-half the subtended angle whether computing for curves or transition spiral layout.

$$\text{Ch} = 2R \times \frac{\sin \Delta}{2} \tag{11.5}$$

Substituting:

$$\text{Ch} = 2\,(445)\,\frac{\sin 28.928585°}{2}$$

$$\text{Ch} = 890 \times \sin 14.4642925°$$
$$\text{Ch} = 890 \times 0.249776593$$
$$\text{Ch} = 222.30 \text{ ft}$$

Setup the ratio to compute the empirical chord correction:

$$\frac{\text{Chord distance}}{\text{Radius}} = \frac{\text{unknown chord}}{\text{radius} \pm \text{one-half the offset distance}}$$

Substituting:

$$\frac{222.30 \text{ ft}}{445.00 \text{ ft}} = \frac{x}{445.475 \text{ ft}}$$

$$x = 222.54 \text{ ft}$$

Coordinates were used to compare the accuracy of the empirical chord correction with the actual location of the station. The coordinate was previously established on station 12 + 50.00 using the center of the radius point 2, the bearing of the radial line, the length of the radius (445 ft) plus the offset (0.95 ft) to equal 445.95 ft. The deflection angle of the curve and spiral from CS_2 to station 12 + 50.00 was also computed earlier in the chapter, so we applied this information to establish a bearing on the line between CS_2 and station 12 + 50.00.

Station 12 + 50.00	Station 12 + 50.00
Offset method	Chord corrected method
North 5400.3147	North 5400.335
East 4803.4807	East 4803.462

Therefore, this example indicates that, although both adjacent curves are considered sharp, and the difference between station CS_2 and station 12 + 50.00 (224.68 ft) is considered quite a distance to measure without a chord correction, the combination of the adjusted deflection angle and the corrected chord, locates the station in an acceptable position.

Example:

Compute the chord distance between CS_2 station 10 + 25.32 and POS station 14 + 50.00 using the same curve data as in the example above, except the offset distance at station 14 + 50 is 6.43 ft.

$$\begin{array}{r} 14 + 50.00 \\ -10 + 25.32 \\ \hline L_c = 4 + 24.68 \end{array}$$

Find, the curve deflection:

$$\text{Subtended angle of the curve} = \phi = \frac{L_c \times D_c}{100} \tag{13.13}$$

$$\phi = \frac{424.68 \times 12.8754606}{100}$$

$$\phi = 54.67950608°$$

$$\phi = 54° \ 40' \ 46''$$

$$\frac{\phi}{2} = 27.33975304°$$

$$\frac{\phi}{2} = 27° \ 20' \ 23'' = \text{curve deflection angle*}$$

*Note: The deflection angle is always one-half the subtended angle whether computing for curves or transition spiral layout.

Find, the spiral deflection:

$$\text{Spiral deflection} = \phi = \frac{L^2}{L_a^2} \times \frac{\theta_a}{3} \qquad (13.4)$$

$$\phi = \frac{424.68^2}{880^2} \times \frac{11.22855}{3}$$

$$\phi = \frac{180353.1024}{774{,}400.00} \times 3.74285$$

$$\phi = 0.871687253°$$

$$\phi = 0° \ 52' \ 18'' = \text{spiral deflection angle}$$

The spiral is located on the outside of the curve; therefore, the deflection of the spiral must be subtracted from the curve deflection, assuming your control bearing is along the local tangent whose bearing is N 34° 55' 28" E.

Find, the adjusted deflection angle:

$$\begin{array}{l} 27.33975304° \\ \underline{-0.87168725°} \\ 26.46806579° \\ 26° \ 28' \ 05'' = \text{adjusted deflection angle} \end{array}$$

The local tangent bearing at station 10 + 25.32 is N 34° 55' 28" E, and the adjusted deflection angle to station 14 + 50.00 is 26° 28' 05". By relating the local tangent bearing with the deflection angle (in this instance the two angles are added), the chord bearing becomes N 61° 23' 33" E.

Note: The bearing for this same line, computed earlier in the chapter, using coordinates, is N 61° 28' 10" E. The difference in the two bearings is 0° 04' 37", and this is obviously an example where the magnitude of the error along the chord bearing is too substantial to use for layout other than rough grading. Although the chord bearings are quite different, we will continue the comparison by showing the coordinates on the point to determine the displacement. It is also obvious that the closer the surveyor approaches the half-line, the greater the position error of the point.

Find the chord distance along the curve extended between CS_2 station 10 + 25.32 and POC station 14 + 50.00.

$$\text{Ch} = 2R \times \sin \frac{\Delta}{2} \qquad (11.5)$$

$$\text{Ch} = 2(445') \times \sin \frac{54° \ 40' \ 47''}{2}$$

$$\text{Ch} = 890 \times \sin 27° \ 20' \ 23''$$

$$\text{Ch} = 890 \times 0.459265508$$
$$\text{Ch} = 408.75 \text{ ft}$$
$$\text{Spiral offset at station } 14 + 50.00 = 6.43 \text{ ft.}$$

Therefore, the ratio to compute the chord correction is set up as follows:

$$\frac{\text{Chord distance}}{\text{Radius}} = \frac{\text{unknown chord}}{\text{radius}}$$

$$\frac{408.75 \text{ ft}}{445.00 \text{ ft}} = \frac{x}{448.215 \text{ ft}}$$

$$x = 411.70 \text{ ft}$$

Therefore, the bearing of the deflection angle from CS_2 station 10 + 25.32 to POS station 14 + 50.00 (as computed earlier) is N 61° 23' 33" E. This bearing and the corrected chord distance of 411.70 ft provide the coordinate of the point. By comparing this coordinate with the offset coordinate, we have the following:

Station 14 + 50.00	Station 14 + 50.00
Offset method	Chord corrected method
N 5451.41929	N 5451.885
E 4996.88122	E 4996.581

This comparison is a good example of the position error created when two sharp curves are transitioned with a spiral length of 880 ft as in our hypothetical example. This situation will seldom, if ever, be found in highways designed in the late twentieth-century. However, if the situation does arise, the most prudent spiral layout should be based on the offset method, where the coordinates are computed from the center point of each curve, calculating along the radius bearing, after adjusting the radius length to include the offset distance, and finally inversing between coordinates.

As a further example of an acceptable chord correction, the following coordinate comparisons are made for POS station 16 + 50.00, assuming the instrument setup is CS_1 station 19 + 05.32 and foresighting back-station to POS station 16 + 50.00. The length of the spiral in this example is 255.32 ft.

Station 16 + 50.00	Station 16 + 50.00
Offset method	Chord corrected method
North 5422.414257	North 5422.387
East 5192.54167	East 5192.529

The above comparison suggests that the chord corrected ratio produces a layout result that is acceptable when the separation (offset distance) between spiral and curve is less than 10 ft. This method has a greater importance to practicing surveyors for the simple reason it can be computed in the field without programmable computers.

SPIRAL LAYOUT, BY LOCATING THE CURVE INITIALLY

Our field layout and computation study of the transition spiral has covered many interesting mathematical possibilities, the first of which is identified as the spiral layout using the deflection method. This view of a transition spiral being located between adjacent curves, where the curves are extended to the half-line, and a spiral is computed based on the cubed proportion, gives us a sense of the immense study behind Barnett's (1940) calculations. When we also consider that these original calculations were performed before 1940 and presumably without the use of programmable computers, we must pay our respects to this educator whose textbook material has guided engineers and surveyors through the majority of the twentieth-century and soon into the twenty-first century.

Our study has outlined methods of staking the spiral from a single instrument setup point or by moving the instrument from station to station. We have also discussed more than one way of field layout and introduced the empirical chord correction.

With today's advances in computer technology, we have the great advantage of verifying and computing point positioning with a minimum of effort and complete reliability. THE BIG QUESTION IS: *How did our predecessors locate the spiral between two curves during the early decades of this century?* THE ANSWER IS: Early surveyors located transition spirals in those early days just as we do today, using the calculating method available at that period. The simplicity of this early field layout can be completed using the following four steps:

- Locate each station (both directions) on the extended curves, stopping at the half-line
- Turn radial lines at each curve station
- Compute and measure the spiral offsets
- Set the point

APPENDIX
SAMPLE FORMS

INVOICE
(Company name and address)

Project No.:	Invoice No.
Name:	Date:

SERVICE DESCRIPTION	DATE	HOURS	CHARGE
	TOTALS		
	Prev.Balance		
	Tax 6.5 %		
	Bal. Due		

Respectfully submitted

FIGURE A.1 Example of an invoice form

TIME SHEET

(Company name and address)

Name:	
Project:	
Date:	Day of Week
WORK DESCRIPTION	
8:00 a.m.	
8:30	
9:00	
9:30	
10:00	
10:30	
11:00	
11:30	
12:00 noon	
12:30	
1:00	
1:30	
2:00	
2:30	
3:00	
3:30	
4:00	
4:30	
5:00	
5:30	
6:00	
6:30	
7:00	
7:30	
8:00	
Other	

FIGURE A.2 Example of a daily time sheet

CLIENT CONTACT

Name:____________________ Date: ________________ Time: ______________

Address: ________________ Telephone: ___________ Job No: ______________

____________________ Estimate: ______________________________

____________________ Mileage: Out _________ Back ______________

WORK REQUEST: ____

____

____

PROPERTY	TOPOGRAPHIC	CONSTRUCTION
State:	Cross-section	Highway:
County:	Topography:	Bridge:
Township:	Utility:	City Street:
Section:	Elevations:	Water Line:
Square:	Grading Plan:	Sewer Line:
Area:	Scale:	Site Plan:
Town and Range:	Aerial Photo:	Subdivision:
Vol. and Pg:	Contour Interval:	Slope Staking:
Adjacent Owners:	Other requirements:	Ground Control:
Tax Map No:	1)	1)
Auditor's No:	2)	2)
Route No:	3)	3)
Other:	4)	4)
1)	5)	5)

Directions ____

____

____

____

____

FIGURE A.3 Example of a client contact form

FIELD SKETCH

Personnel:	Location:	Sheet Number:

	State:	Date:
	County:	Temp:
	Route No:	Weather:
	County Rd:	Job No:

Sketch:

FIGURE A.4 Field sketch

LEVEL NOTES

Personnel: ______________ Job No. _________ Date: _________ Sht. No. ___________

Bench Mark:

Station	B-S	H.I.	F-S	Exist. El.	Prop. El.	Cut/Fill	Desc.

FIGURE A.5 Example of level note form

SLOPE AND TOPOGRAPHY NOTES

Personnel: ________________ Job No. _________ Date: _______ Sheet No. ___________

Bench Mark Identity: ___________ Bench Elev. ________ B-S ________ H.I. ____________

Sta. or Pt. No.	Vert. Angle	Slope Dist.	Horiz. Dist.	F-S	Ground Elev.	Description

FIGURE A.6 Slope measurements and topography location using two instruments (transit and automatic level).

SLOPE STAKE NOTES

Bench	Mark	Elev. =		Station or B.L.	"Plan" Data	Trial (4) (slope)	Trial (3) (slope)	Trial (2) (slope)	Trial (1) (slope)	Work Hub
B - S	H. I.	F - S	Elev.		Center - line distance to plan break	______	______	______	______	Center - line distance to Work- Hub
1)										
2)										
3)										
4)										
Place	paddle	info.	here		Elevation					Elevation

Bench	Mark	Elev. =		Station or B.L.	"Plan" Data	Trial (4) (slope)	Trial (3) (slope)	Trial (2) (slope)	Trial (1) (slope)	Work Hub
B - S	H. I.	F - S	Elev.		Center - line distance to plan break	______	______	______	______	Center - line distance to Work- Hub
1)										
2)										
3)										
4)										
Place	paddle	info.	here		Elevation					Elevation

Bench	Mark	Elev. =		Station or B.L.	"Plan" Data	Trial (4) (slope)	Trial (3) (slope)	Trial (2) (slope)	Trial (1) (slope)	Work Hub
B - S	H. I.	F - S	Elev.		Center - line distance to plan break	______	______	______	______	Center - line distance to Work- Hub
1)										
2)										
3)										
4)										
Place	paddle	info.	here		Elevation					Elevation

B.L. = Baseline; B - S = backsight; F - S = foresight; H. I. = height of instrument; W/Hub = Work-Hub

FIGURE A.7 Example of slope stake notes as described in Chapter 10

CLOSURE FORM

Course	Bearing	Distance	Cosine	Sine	North	South	East	West

$\tan \alpha =$ ------------------------

$\tan \alpha =$

$\alpha =$

FIGURE A.8 Example of a closure form as described in Chapter 11

BIBLIOGRAPHY

Barnett, Joseph. 1940. *Transition Curves for Highways*. Washington, D.C.: U.S. Government Printing Press.

Breed, Charles B. 1971. *Surveying*. 3d ed. New York: John Wiley & Sons.

Breed, Charles B., George L. Hosmer, and W. Faig. 1958. *Surveying: The Principles and Practice*. 11th ed. New York: John Wiley & Sons.

Brown, Curtis M. 1957. *Boundary Control and Legal Principles*. New York: John Wiley & Sons.

Church, William M. 1956-1970. *Construction Stakes and Their Use*. Columbus, Oh.: Carl M. Geupel Co., Inc.

Clarkeson, John. 1952. *Highway Location and Design*. Scranton, Pa.: International Textbook Co.

Davis, Raymond E., Francis S. Foote, and Joe W. Kelley. 1968. *Surveying: Theory and Practice*. New York: McGraw-Hill.

Ford, Amelia Clewley. 1910. *Ph.D. Bulletin No. 352*. Madison, Wi.: University of Wisconsin Press.

Highway Research Board. 1936. *Safe Side Friction and Superelevation Design*. Washington, D.C.

Ohio State Auditor's Office. Reprint 1995. *The Building of Ohio*. Columbus, Ohio. Mr. Jim Petro, Auditor of State.

Ohio Auditor of State. 1995. *Ohio Lands: A Short History*. Columbus, Oh.

Ohio Department of Transportation. 1985. *Construction and Materials Specifications*. Columbus, Oh.

Sherman, C.E. 1925. *Ohio Land Subdivision*. vol. III. Columbus, Oh.: The Ohio State University Press.

Sperling, Abraham. Ph.D. 1957. *Mathematics*. Greystone Press. New York, New York.

Urquhart, Leonard C. 1959. *Civil Engineering Handbook*. 4th ed. New York: McGraw-Hill.

White, C. Albert. 1983. *A History of the Rectangular Survey System*. Washington, D.C.: U.S. Government Printing Office.

Woods, Kenneth B. 1960. *Highway Engineering Handbook*. New York: McGraw-Hill.

Wright, Paul H., and Radnor J. Paquette. 1979. *Highway Engineering*. 4th ed. New York: John Wiley & Sons.

INDEX

A

acreage
  size of property, 28
  square feet to acres conversion, 28
acreage of various divisions, 19
aerial photography, 8, 96, 132
angle computation, **26**
  angle summation check, 34
  angle-to-bearing calculation, 30
  angular orientation of land, 30
  azimuths, 31
  bearing-to-angle computation, 33-34
  bearings, 31
  interior angle plotting, 32-33
  meridian method of angle plotting, 31-32, **32**
  plotting, 30
  true North orientation, 30
angle-to-bearing calculation, 30
angular measurement: degrees, minutes, seconds, 19
angular orientation of land, deed descriptions, 30
appraising value of land for purchase, 9
architect's scale, 55
arcs, 161, 162, 164-166, **165**
area
  average area calculation, 116-117, **118**
  computerized calculations, 118
Articles of Confederation, ceding of public lands (c. 1780), 12
as-built surveys, 10
automatic level, 111
azimuths, 31

B

bar graphs, scale, 57
Barnett, 81, 209, 211, 261

Illustrations are in **boldface.**

base-point stations, 68-69
baselines, 13, **18**, 19, 20
bearings, 31
  angle-to-bearing calculation, 30
  bearing-to-angle computation, 33-34
  curves, 186, 190
beginning and end point of project, 47
bench marks (BM), topographic maps, 111
benefits of improved highway systems, 5
bidding
  Construction and Material Specification Manual, 74-75
  contractors, 9
bifurcated highways
  curves, 130
  grade, 130
blocks (*see also* quadrangles), 13, **17, 18,** 19
blue flagging, slopes/slope staking, 155
bounds (see metes and bounds)
bridges, 9, 10
Brown, 27
bucking-in, concentric curves, 180-181
Bureau of Public Roads (*see* Federal Highway Administration)

C

calculations sheet, 54
catch points, slopes/slope staking, 154
categories of highways
  Interstate Highway System, 4
  National Defense Highways, 4
  primary-system highways, 4
  secondary-system highways, 4
  Urban System highways, 4
centerline plat, 9, 10, 47, 61-66, **62-63**
  centerline of survey line, 64-65
  curve data, 64, 65

centerline plat *continued*
grade, edge of pavement vs. centerline, 130
horizontal and vertical control points, 65
information shown on plat, 61, 64
locating the centerline of survey, 64-65
points of intersection (PI), 65
profile grade, 75
property plat, 64
reference monuments, 66
referencing, 66
scale, 65-66
slopes/slope staking process, 150
spirals, 65
standard construction detail, 66
state plane coordinate system, 64
typical reference detail, 66
U.S. Geological quadrangle maps for location, 65
vicinity maps, 65-66
chords, 161, 163, 164-166, **165,** 167, 277-282, **278**
Clarkson, 3, 4
Clewley, 12
client-contact form, **288**
climate vs. design, 81
closure form, **293**
closures, curves, 182-185, **183, 184,** 191-192
computational impasse, 193-194
cosines, 195, 203-207
function of zero and one, 195-196
sequence of equations, 202-208
sines, 195, 203-207
closures, spirals, 251-256, **252,** 256-260, **257, 259**
communication and transportation, 3
compaction factors, 122
compass and chain, 25
completion dates, 9
computer-aided design, 9, 30
concentric curves, 177-180, **177, 179**
Congress Lands, Ohio land system, 21
Connecticut Western Reserve, Ohio land system, 12, 20
Construction and Material Specification Manual, 74-75
construction limits, slopes/slope staking, 150, 151
construction title sheet, 48-54, **50-51, 52-53**
contour lines, 96
contour maps (*see also* topographic maps), 96-106
contractors
Construction and Material Specification Manual, 74-75
slopes/slope staking instructions, 157, 159
control elevation computation, curves, 139
conventional signs, 47, 49
coordinates, 30
state plane coordinate system, 64
cosine, curves, 186-189, 195, 203-207
cost comparison studies, 8
county boundaries, 23, 46, 48
course, curves, 185
cross section surveying method, topographic maps, 108, 110
cross sections, 9, 10, 54, 67, 113-122, **114**
average area calculation, 116-117, **118**
commodities, volume calculations, 119-120
compaction factors, 122
Construction and Material Specification Manual, 74-75
cut, **114**
earthwork volume computations, 116-122
fill, **114**
gathering information in the field, 115, **115, 116**
interstate road, typical section, **72-73**
multiple cross sections, 115-116
plotting cross sections, **119**
primary road, typical section, **70-71**
secondary road, typical section, **68**
slopes/slope staking, 147, 154
tapers and volume, 120
cross slopes, 67
crown vs. superelevated section, **79**
crown, crest, or summit condition curves, 132, 138
curves/curve data (*see also*spirals), 64, 65, 125-146, 161-208
aerial photographs, 132
algebraic difference calculation, 134-135, **135**
angular totals within multiple-sided figure, 206
arcs, 161, 162, 164-166, **165**
bearing, 186
bearing, secondary solution, 190-191
bifurcated highways, 130
bucking-in on concentric curves, 180-181
centerline vs. edge of pavement grades, 130
chords, 161, **163,** 164-166, **165,** 167
closure tolerances, 191-192

closures, 182-185, **183, 184**
computational impasse, 193-194
function of zero and one, 195-196
sequence of equations, 202-208
sine and cosine, 195
computerized calculations, 162
concentric curves, 177-180, **177, 179**
control elevation computation, 139
cosine, 186-189, 195, 203-207,
course, 185
crown, crest, or summit condition curves, 132, 138
curve elevation computation, 141-142
deflection angle, **163**
degree of curve, 163, 166, 168-170, **169**
delta angle, **163**
design standards, 130-131
distance, 186
drafting process, 129-130
earthwork balance, 128
elevation at midpoint, 133-134
elevations on tangent lines, 137-138, **137**
equations for curves, 166-167, 181-182
external distance (E), **163**
field layout of curves, 170
field test for degree of curve, 168-170, **169**
grade in and grade out, 132, 139
grade maximum and minimum, 129
grade sign convention, 128
graphics for profile view, 129
highest point calculation, 140-141
highway curves, 162
horizontal curves, 126, 162
inaccessible curves, computing curves from curves, 196-201, **197**
instrument "move-up", line-of-sight breaks, 175-177
intersecting curves, 201-202, **201**
intersection of line and curve, 181, 182, 192-193
length of vertical curve, 131
line-books, 173, 174-175
locating transition spirals, 282-283
lowest point calculation, 140-141, **140,** 144
meander lines, 161, 181
middle ordinate (MO), **163,** 167
midpoint of curve computation, 139-141
minimum allowable vertical curve, 131
minimum safe radius, 80-81
mosaics used in design, 132
multiple-point curve computations, 173-174
multiple-point curve layout, 173
offsets between spiral and curve, 272, **273**
parts of a curve, **163**
plan view, 125, **127**
point of curve (PC), **163,** 171
point of intersection (PI), **163,** 171
point of tangent (PT), **163,** 171
point of vertical curve (PVC), 132, 139
point of vertical intersection (PVI), 132, 139
point on a curve, locating, **172**
preliminary field investigation, 171
preparing sketches for layout, 171-173
profile selection, 127
profile view, 125, **126, 127,** 129
radius, **163,** 164, 166
railroad curves, 162, 164
random (unengineered) curves, 171
sag condition curves, 132, 142-144
setting grade or establishing elevations, 125, **126, 127**
sight distance requirements, 141-142
sine, 186-189, 195, 203-207
spirals, 245
stationing, 167-168
substituting curve for spiral, **213,** 231-234
symbols used in curves, **163**
tangent distances, **163,** 166
tangent offset at points computation, 144-146, **145**
tangent offset computation, 135-139
transition in curve, **213,** 231
transition spirals between curves, 261-283
vertical clearance, 129
vertical curve computation, **128,** 132-135, **133, 134, 135**
cuts (excavations) (*see also* slopes/slope staking), 147, **149**

D

deed descriptions, 26-28
angular orientation of land, 30
description of land, 29
geometric shape of property, 28
graphic presentation of land description, 29-30
intent, 26-27
junior and senior property rights, 27
location of property, 28
metes and bounds, 27
natural objects as reference points, 27, 28
plats, 30

deed descriptions *continued*
plotting deed descriptions, 29-30, 30-37
size of property, 28
title identity, ownership, 27-28
transferring title, 29
degrees, angular measurement, 19
description of land, deed descriptions, 29
design and construction of federal-aid highways, 4-5
distance measurement (*see* linear measurements)
ditch ratios, 147, 149-150
Donation Tract, Ohio land system, 20
drainage, 9, 10, 54
drive details sheet, 54

E

early American surveying, rectangular system, 12
earthen side slope, 77-78
earthwork, 9
earthwork volume computations (*see* volume computations)
edge of shoulder subgrade, 77, 78
Eisenhower, Dwight D., 4
electronic distance measurement (EDM), 25, 37
elevation at midpoint of curve, 133-134
elevation data, topographic maps, 98
elevation level loop, 98-99, **99, 100**
slopes/slope staking, 150
embankments (*see also* slopes/slope staking), 147, **148**
engineer's scale, 54-55
Equator as reference, 19
excavations (*see also* slopes/slope staking), 147, **149**

F

federal funding
applying for funding, 9
Federal Highway Act (1921), 4
Federal Road Act (1916), 3
Federal-Aid Highway System, 4
National Pike construction (c. 1800), 3
primary-system highways, 4
Federal Highway Act (1921), 4
Federal Highway Administration, 8
Federal Road Act (1916), 3
Federal-Aid Highway System, 4
field sketch form, **289**
fills (embankments) (*see also* slopes/slope staking), 147, **148**
Fire Lands, Ohio land system, 12, 20
First and Second Ohio Company Purchase, Ohio land system, 20
foresighting, slopes/slope staking, 155
forms
client-contact form, **288**
closure form, **293**
field sketch form, **289**
invoice form, **286**
level note form, **290**
slope and topography notes form, **291**
slope stake notes form, **292**
slope-staking data form, **152, 154, 155, 156 160**
time sheet, **287**
fraction-to-decimal conversion, 76-77
fractional townships, 19
French Grants, Ohio land system, 21
friction, tire and roadway, 81

G

general notes sheet, 54
general summary sheet, 54
geological conditions, 8
geometric shape of property, deed descriptions, 28
global positioning satellite (GPS), 25
grade in and grade out, 132
grades (*see also* slopes/slope staking), 9, 10
centerline vs. edge of pavement grades, 130
maximum and minimum, in curves, 129
sign convention, 128
graphic presentation of land description, deed descriptions, 29-30
Greenwich Prime Meridian, 19

H

highway curves, 162
highway planning, 3-10
aerial photography, 8
appraising value of land for purchase, 9
as-built surveys, 10
benefits of improved highway systems, 5
bidding process, contractors, 9
bridges, 9, 10
centerline plat, 9, 10
completion dates, 9
computer-aided design, 9
cost comparison studies, 8
cross sections, 9, 10
design and construction of federal-aid highways, 4-5
drainage, 9, 10

earthwork, 9
existing streets evaluation, 8
existing-facilities evaluation, 8
factors affecting route selection, 5, 8
finalizing the route, 8-10
funding the proposed highway, 9
geological conditions, 8
grade evaluation, 9, 10
historical impact of U.S. highways, 3-4
industrial value of land, 5, 9
interchange evaluation, 9
land use evaluation, 5
legislation affecting highways, 4-5
materials cost analysis, 9
Needs Studies, 8
preliminary route study map, **6-7**
purchasing right-of-way land, 9
right-of-way, 9
route selection process, 5-10, **6-7**
site study, geological conditions, 8
title sheets, 41-59, **42-43, 44-45**
topographic maps, 8
traffic analyses, 8
U.S. Geological Survey quadrangle maps, 8
utilities evaluation, 9
width of roadway, 9
zoning factors, 5
history of highways
early American survey systems, 12
historical impact of U.S. highways, 3-4
public lands in U.S., **14-15**
Roman roads, 11-12
horizontal and vertical control points, 65
horizontal curves, 162

I

inch-per-foot to feet-per-foot conversion, 151-153
index of sheets, title sheets, 47, 49, 54
indiscriminate surveys, 25-37
adjusting final angle, 35-37
angle summation check, 34
angular orientation of land, 30
azimuths, 31
bearing-to-angle computation, 33-34
bearings, 31
closing horizontal angles, 35
coordinates, 30
correcting angular errors, 35
deed descriptions, 26-28
description of land, 29
distance measurement, 35-37
doubling horizontal angles, 35
graphic presentation of land description, 29-30
interior angle plotting, 32-33
meridian method of angle plotting, 31-32, **32**
metes and bounds, 25, 27
natural objects as reference points, 27, 28
plats, 30
plotting deed descriptions, 29-30, 30-37
true North orientation, 30
industrial value of land, 9
intent, deed descriptions, 26-27
interchange evaluation, 9
intermediate stations, 58
Interstate Highway System, 4-5, 161
interstate road, typical section, **72-73**
invoice form, **286**

J

Jefferson, Thomas, 13
junior and senior property rights, 27

L

land grants, 20-21, **22**
latitude and longitude, 12, 13, **18**
Equator as reference, 19
Greenwich Prime Meridian, 19
square dimensions of sections, variability, 13, 18
true North orientation, 30
lawsuits (*see* litigation)
legislation affecting highways, 4-5
Articles of Confederation, ceding of public lands, 12
design and construction of federal-aid highways, 4-5
federal funding regulations, 4
Federal Highway Act (1921), 4
Federal Highway Administration, 8
Federal Road Act (1916), 3
Federal-Aid Highway System, 4
Interstate Highway System, 4-5
National Defense Highways, 4
Needs Studies mandate, 8
primary-system highways, 4
public lands in U.S., **14-15**
secondary-system highways, 4
state highway systems, 4
Urban System highways, 4
level loop, elevation, 98-99, **99, 100**
slopes/slope staking, 150

level note form, 290
line-books, curves, 173, 174-175
linear measurement, 19, 35-37, 181
  architect's scale, 55
  chains, 19
  engineer's scale, 54-55
  links, 19
  metric measurement, 64
  mile, 19
  miniature plan, 56
  negative fractional stations, 59
  perch, 19
  pole, 19
  size of property, 28
  station equations, 58-59, 69, 74
  statute mile, 19
  tools used by surveyors, 25
links, 19
litigation, 21-22
  deed descriptions, 26-28
  description of land, 29
  junior and senior property rights, 27
  title identity, ownership, deed descriptions, 27-28
  transferring title, 29
location of property, deed descriptions, 28
lock-levels, 112
longitude (*see* latitude and longitude)
lots, 13, 18

M

maintenance of traffic sheet, 54
materials
  commodities, volume calculations, 119-120
  Construction and Material Specification Manual, 74-75
  cost analysis, 9
mathematics
  adjusting final angle, 35-37
  angle computation, 26
  angle summation check, 34
  angle-to-bearing calculation, 30
  angular orientation of land, 30
  architect's scale, 55
  azimuths, 31
  bearing-to-angle computation, 33-34
  bearings, 31
  closing horizontal angles, 35
  contour elevation computation, 101, **103,** 139
  correcting angular errors, 35
  cosine, 186-189, 195, 203-207
  curves, 125-146, 161-208
  distance measurement, 35-37
  doubling horizontal angles, 35
  elevation at midpoint of curve, 133-134
  elevations on tangent lines, 137-138, **137**
  engineer's scale, 54-55
  fraction-to-decimal conversion, 76-77
  inch-per-foot to feet-per-foot conversion, 151-153
  interior angle plotting, 32-33
  maximum superelevated rate, 77, 81
  meridian method of angle plotting, 31-32, **32**
  midpoint of curve computation, 139-141
  minimum allowable vertical curve, 131
  minimum length of spiral run-out calculation, 85, 88
  minimum safe radius, 80-81
  negative fractional stations, 59
  plotting deed descriptions, 29-30, 30-37
  Pythagorean theory, 258-260
  rate of slope, 149-150, 156-157
  ratio of elevations, topographic maps, 101-105
  sine, 186-189, 195, 203-207, 240, 241
  slope-rate change calculation, 88, 90-92, **90**
  spiral length calculation, 81-83, 85
  square feet to acres conversion, 28
  station equations, 58-59, 69, 74
  tangent offset at points computation, 144-146, **145**
  tangent offset computation, 135-139
  true North orientation, 30
  vertical curve computation, 132-135, **133, 134, 135**
  volume computations, 116-122
maximum superelevated rate, 77, 81
meander lines, 161, 181
median, two-lane road with median, **79**
meridians, 13, **18,** 19, 20, 30
  angle plotting, 31-32, **32**
metes and bounds, 12, 20, 22, 27
metric measurement, 64
Michigan Survey, Ohio land system, 21
mile, 19
miniature plan, 56
minimum safe radius, 80-81
minutes, angular measurement, 19
monuments, reference monuments, 66
monuments, U.S. Geological Survey, 98
mosaics used in designing curves, 132
municipal boundaries, 23

N

National Defense Highways, 4
National Pike (c. 1800), 3
National Rectangular system (*see* rectangular system)
natural objects as reference points, 27, 28
negative fractional stations, 59
North American Vertical Datum (NAVD), 75

O

offset
  station offsets, 57-58
  tangent offset at points computation, 144-146, **145**
  tangent offset computation, 135-139
Ohio land system (*see also* surveying techniques), 11, 20-23, **22**
  baselines, 20
  Congress Lands, 21
  Connecticut Western Reserve, 12, 20
  Donation Tract, 20
  Fire Lands, 12, 20
  First and Second Ohio Company Purchase, 20
  French Grants, 21
  land grants, 20-21, **22**
  litigation resulting from poor surveys, 21-22
  meridians, 20
  metes and bounds, 20, 22
  Michigan Survey, 21
  municipal, county, state boundaries, 23
  Old Seven Ranges, 20
  property line retracement difficulties, 21-22
  ranges, 20
  Refugee Tract, 21
  Symmes Purchase, 20-21
  townships, 20
  Two- and Twelve-Mile Square Reserves, 21
  United States Military District, 21
  Virginia Military Survey District, 12, 20
Old Seven Ranges, Ohio land system, 20
ownership (*see* title identity)

P

paddle notes, slopes/slope staking, 157, **158,** 159
Paquette, 3, 4
perch, 19
plan and profile sheet, 54
planning (*see* highway planning)
plats, 30
plotting deed descriptions, 29-30, 30-37
  adjusting final angle, 35-37
  angle summation check, 34
  angles, 30
  azimuths, 31
  bearing-to-angle computation, 33-34
  bearings, 31
  closing horizontal angles, 35
  coordinates, 30
  correcting angular errors, 35
  distance measurement, 35-37
  doubling horizontal angles, 35
  interior angle plotting, 32-33
  meridian method of angle plotting, 31-32, **32**
  meridians, 30
point of vertical curve (PVC), 132, 139
point of vertical intersection (PVI), 132, 139
point of vertical tangent (PVT), 132, 139
points of intersection (PI), 65
pole, 19
preliminary route study map, **6-7**
primary road, typical section, **70-71**
primary-system highways, 4
profile and cross-slope elevations, 75-77
profile grade, 75
property plat, 64
property rights, 27
public lands in U.S., **14-15**
purchasing right-of-way (*see* right-of-way)
Pythagorean theory, 258-260

Q

quadrangles, 19
quadrangles or blocks of land, 13, **17, 18**

R

railroad curves, 162, 164
railroads, 4
random (unengineered) curves, 171
random-shot surveying method, topographic maps, 106-108
range lines, 13, **17, 18**
ranges, 19, 20
ratio of elevations, topographic maps, 101-105
ravine, **109**
RDS level, 111
rectangular system of surveying, 11-13
  acreage of various divisions, 19
  angular measurement: degrees, minutes, seconds, 19
  baselines, 13, **18,** 19

rectangular system of surveying *continued*
blocks, 19
definitions in rectangular system, 19
early American surveys using rectangular system, 12
Equator as reference, 19
fractional townships, 19
Greenwich Prime Meridian, 19
indiscriminate surveys, 25
latitude and longitude, 12, 13, **18**
linear measurement: chains, links, etc., 19
lots, 13, **18**
meridians, 13, **18,** 19
metes and bounds, 22
municipal, county, state boundaries, 23
origin and history of rectangular system, 11-13
public lands in U.S., **14-15**
quadrangles or blocks of land, 13, **17, 18,** 19
range lines, 13, **17, 18**
ranges, 13, 19
reference lines, 13, **18**
sections, 13, **18,** 19
square dimensions of sections, variability, 13, 18
tiers, 13, 19
towns, 19
townships, 13, **17, 18,** 19
typical description of land using rect.system, 18-19
U.S. (National) Rectangular system, 13, **16**
reference lines, 13, **18**
reference monuments, 66
referencing, 66
reference monuments, 66
standard construction detail, 66
typical reference detail, 66
Refugee Tract, Ohio land system, 21
revision tables, 47
right-of-way, 9, 54
calculations sheet, 54
centerline plat, 61
cross sections sheet, 54
drainage details sheet, 54
drive details sheet, 54
general notes sheet, 54
general summary sheet, 54
maintenance of traffic sheet, 54
plan and profile sheet, 54
right-of-way sheet, 54
schematic plan, 49, 54
station and offset corners method, 61
title identity, ownership, 27-28
title sheet, 48-54, **50-51, 52-53**
traffic control sheet, 54
typical section sheet, 54
Roman roads, 11-12
route numbers, 46, 48
route selection process, 5-10, **6-7**
run-out, minimum length of spiral run-out calculation, 85, 88

S

sag condition curves, 132, 142-144
scale, 47, 54-56
analyzing scale, 56
architect's scale, 55
bar graphs, 57
centerline plat, 65-66
engineer's scale, 54-55
miniature plan, 56
spirals, 238
stations, highway stationing, 56-59
topographic maps, 101
schematic plan, 49, 54
secondary road, typical section, **68**
secondary-system highways, 4
seconds, angular measurement, 19
section numbers, 48
sections (*see also* typical section; superelevated sections), 46
section numbers, 48
sections (land division), 13, **18,** 19
square dimensions of sections, variability, 13, 18
senior property rights, 27
Sherman, C.E., 11,13, 23
shoulder, edge of shoulder subgrade, 77, 78
sight distance requirements, curves, 141-142
signatures, title sheets, 47
sine, curves, 186-189, 195, 203-207, 240, 241
site studies, 8
size of property, 28
slope and topography notes form, **291**
slope stake notes form, **292**
slope-rate change calculation, 88, 90-92, **90**
slopes/slope staking (*see also* grades), 147-160
blue flagging, 155
catch points, 154
centerline location, 150
computer printout of slope elevations, 150
construction limits, 150, 151
contractor instructions, 157, 159

cross sections, 147, 154
cross slopes, 67
cuts, 147, **149**
ditch ratios, 147, 149-150
duplicating stake instructions on form, 159
earthen side slope, 77-78
edge of shoulder subgrade, 77, 78
example: slope staking the mainline, **153**
field data required, 150-151
fills, 147, **148**
foresighting, 155
form for slope-staking data, **152, 154, 155, 156, 160**
fractional rate of slope, 149-150
inch-per-foot to feet-per-foot conversion, 151-153
nonfractional rate of slope, 149-150
North American Vertical Datum (NAVD), 75
note keeping during construction, 154-157
paddle notes, 157, **158,** 159
profile and cross-slope elevations, 75-77
profile grade, 75
rate of slope, 149-150, 156-157
ratio, slope ratio, 149-150, 156-157
side slopes, 147
slope-rate change calculation, 88, 90-92, **90**
spirals, 236-237
stationing points, 150
turning points, 150
"work hub" locations, 150, 151, 154

speed (see vehicle speed)

spirals (*see also* curves), 65, 209-260
beginning of spiral to mainline distance, 212
beginning of spiral to perpendicular distance, 212, 226-227
chord, 212, 226, 243, 249, 277-282, **278**
chord correction, transition spirals, 277-282, **278**
closures, 251-256, **252, 256,** 260, **257, 259**
concentric spirals, 231
coordinate computation layout, transition spirals, 275-277, 282
curve data analysis, 245
curve to the spiral, 213
curve with transition, **213,** 231
definition of a spiral, 210
deflection layout method, transition spirals, 273-275
degree of curve, 210-211, 212, 214, 227-228, 233, 239-240, 246, 265
delta angles, 267
designing spirals, 210
drawing spirals, 238-239
early recognition of spiral phenomenon, 209-210
equations for spirals, 211, 214-215
errors, 244
exsecant values, 227
field book requirements, 244-253
field layout of spirals, 236
functions tables, theta angle, **216-225**
graphical location of spiral, 264
hand-held prisms to establish radial line, 237
history of use, 209-210
inside radial line to spiral, 237
intersecting angle (delta) between tangents, 212, 213
layout, closure method, 251-256, **252**
layout, trigonometric method, 255-260, **257, 259**
length of partial spiral, 212
length of spiral, 81-83, 85, 211, 212, 215, 228, 239-240, 246, 249, 267
long tangent, 212, 226, 241, 242, 243, 246, 249
mainline PI to middle of mainline curve distance, 212
midpoint distance, 212, 226
minimum length of spiral run-out calculation, 85, 88
minimum spiral lengths, 262-264
moving instruments, 277
note-keeping procedures, 246
offsets between spiral and curve, 272, **273**
outline of field-book notation, 246-247
outside radial line to spiral, 237
perpendicular distance from tangent to end, 212, 227
points along spiral, **250**
points-on-spiral, calculations, 234-235, **235,** 236-237
proof or test for spiral calculations, 239-242
property line and spiral intersection, 242-244
proportional constants, 272
Pythagorean theory, 258-260
radial lines, 237
radius, 210, 212, 215, 232, 265
railroad spirals, 209-210
ratio of spiral parts to whole, 249-251
scale for spirals, 238
short tangent, 212, 226, 241, 242, 243, 246, 249

spirals *continued*
sines, 240, 241
spiral to the curve, 213
spiral to the tangent, 213
stations, highway stationing, 244, **245,** 247-249, **248**
substituting curve for spiral, **213,** 231-234
superelevation, 209
symbols for transition spirals, 262, **263**
symbols used in spirals, 211-214
tangent and spiral intersections, 237-238
tangent distances, 267-268
tangent to mainline curve distance, 212, 226
tangent to the spiral, 213
theta angle, 212, 215, 228-231, 239-241, 243, 246, 247, 249, 264
theta angles, spiral functions table, **216-225**
transition curves, 211
transition spirals between curves, 261-283
chord correction, 277-282, **278**
layout, 271-272
layout: coordinate computation, 275-277
layout: deflection method, 273-275
layout: initial curve location, 282-283
offsets between spiral and curve, 272, **273**
proportional constant, 272
computation, 264-271, **266**
trigonometric functions for closure, 255-260, **257, 259**
variables in calculating spirals, 214, 239
vehicle speed, 210
square feet to acres conversion, 28
standard construction detail, 66
standard drawings, 47
state boundaries, 23
state highway systems, 4
state plane coordinate system, 64
station-and-offset corners method, right-of-way, 61
stations, highway stationing, 56-59
bar graphs, 57
base-point stations, 68-69
curves, 167-168
intermediate stations, 58
negative fractional stations, 59
offsets, station offsets, 57-58
right-of-way calculations, 61
slopes/slope staking, 150
spirals, 244, **245,** 247-249, **248**
station equations, 58-59, 69, 74
typical section, 68-69
statute mile, 19
streams, **111**
subgrade
earthen side slope, 77-78
edge of shoulder subgrade, 77, 78
superelevated sections, 83
superelevated sections, 79-93, 209
case study, 83-84, **86-87**
climate vs. design, 81
crown vs. superelevated section, **79**
design features of superelevated sections, 79
diagramming a superelevation, **82**
elevation data table, 84, **84**
errors, table errors, 92-93
friction, tire and roadway, 81
maximum superelevated rate, 77, 81
minimum length of spiral run-out calculation, 85, 88
minimum safe radius, 80-81
slope-rate change calculation, 88, 90-92, **90**
spiral length calculation, 81-83, 85
subgrade, 83
table review, 88, **89**
vehicle speed vs., 80
surveying techniques, 11-23
accuracy of surveying methods, 36-37
adjusting final angle, 35-37
angle computation, **26**
angle summation check, 34
angular orientation of land, 30
azimuths, 31
bearing-to-angle computation, 33-34
bearings, 31
closing horizontal angles, 35
coordinates, 30
correcting angular errors, 35
cross section surveying method, topographic maps, 108, 110
deed descriptions, 26-28
distance measurement, 35-37
doubling horizontal angles, 35
first-order surveys, 36
indiscriminate surveys, 25-37
interior angle plotting, 32-33
latitude and longitude, 12, 13, **18**
litigation resulting from poor surveys, 21-22
meridian method of angle plotting, 31-32, **32**
metes and bounds, 12, 27
metric measurement, 64
municipal, county, state boundaries, 23
natural objects as reference points, 27, 28

Ohio land system, 11, 20-23, **22**
plats, 30
plotting deed descriptions, 30-37
property line retracement difficulties, 21-22
random-shot surveying method, topographic maps, 106-108
rectangular system of surveying, 11-13
reference monuments, 66
second-order surveys, 36
state plane coordinate system, 64
third-order surveys, 36
tools used by surveyors, 25
true North orientation, 30
swale, **109**
Symmes Purchase, Ohio land system, 20-21

T

tangent lines, elevations, 137-138, **139**
tangent offset computation, 135-139
tapers and volume, 120
theta angles (see spirals)
theta angles, spiral functions table, **216-225**
tiers, 19
time sheet, **287**
title identity, ownership, deed descriptions, 27-28
title sheets, 41-59, **42-43, 44-45**
architect's scale, 55
beginning and end point of project, 47
calculations sheet, 54
centerline plat, 47
construction vs. right-of-way title sheet, 48-54, **50-51, 52-53**
conventional signs, 47, 49
county information, 46, 48
cross sections sheet, 54
design designation numbers, 47
drainage details sheet, 54
drive details sheet, 54
engineer's scale, 54-55
general notes sheet, 54
general summary sheet, 54
heading information, 46
index of sheets, 47
index of sheets, 49, 54
intermediate stations, 58
limitations of project, 47
maintenance of traffic sheet, 54
miniature plan, 56
names of preparing personnel, 47
negative fractional stations, 59
offsets, station offsets, 57-58
plan and profile sheet, 54
revision tables, 47
right-of-way sheet, 54
route number information, 46, 48
scale analysis, 56
scale, 47, 54-56
schematic plan, 49, 54
section information, 46
section number information, 48
signatures, 47
significant features information, 46-47
standard drawings, 47
station equations, 58-59
stations, highway stationing, 56-59
traffic control sheet, 54
type of construction information, 47
typical section sheet, 54
utility warnings, 47
vicinity maps, 47
toll roads, historical precedence, 3
tools used by surveyors, 25
topographic maps, 8, 95-112
aerial photography, 96
automatic level, 111
bench marks (BM), 111
computing contour elevations, 101, **103**
concentric vs. non-concentric contours, 105-106, **105**
contour lines, 96
contour maps, 96-106
cross section surveying method, 108, 110
drawing in contour lines, 99, 100-101, **103**
drawing shapes, 105-106, **105, 106, 108, 109, 110, 111**
elevation data, 98
elevation level loop, 98-99, **99, 100,** 150
field survey procedures, 106-112
getting topographic maps, 95
grid preparation, 96-98, **97**
intermediate elevations, 99
interpreting contour map, 100
items appearing on topographic maps, 95
level loop, elevation, 98-99, **99, 100,** 150
lock-levels, 112
preparing a contour map, 96-98
random-shot surveying method, 106-108
ratio of elevations, 101-105
RDS level, 111
scale, 101
symbols used on maps, 95
U.S. Geological Survey and monuments, 98

towns, 19
townships, 13, **17, 18,** 19, 20
  fractional, 19
traffic analyses, 8
traffic control sheet, 54
transferring title, 29
transits, 25
true North orientation, 30
turning points, slopes/slope staking, 150
Two- and Twelve-Mile Square Reserves, Ohio land system, 21
two-lane road with median, **79**
typical sections (*see also* cross sections; superelevated sections), 54, 67-78
  base-point stations, 68-69
  Construction and Material Specification Manual, 74-75
  cross sections, 67
  cross slopes, 67
  crown vs. superelevated section, **79**
  earthen side slope, 77-78
  edge of shoulder subgrade, 77, 78
  interstate road, typical section, **72-73**
  North American Vertical Datum (NAVD), 75
  primary road, typical section, **70-71**
  profile and cross-slope elevations, 75-77
  profile grade, 75
  secondary road, typical section, **68**
  station equations, 69, 75
  stations, highway stationing, 68-69
  two-lane road with median, **79**

U

U.S. (National) Rectangular system (see rectangular system), 13
U.S. Geological Quadrangle maps, 8, 65
U.S. Geological Survey and monuments, 98
United States Military District, Ohio land system, 21
Urban System highways, 4
utilities, 9, 47

V

valley, **110**
vehicle speed vs. design, 80, 210
vertical clearances, 129
vertical curve computation, 132-135, **133, 134, 135**
vicinity maps, 47
  centerline plat, 65-66
Virginia Military Survey District, Ohio land system, 12, 20
volume calculations, 116-122
  average area calculation, 116-117, **118**
  commodities, volume calculations, 119-120
  compaction factors, 122
  computerized calculations, 118, 120-122, **121**
  plotting to calculate volume, 119, **119**
  tapers and volume, 120

W

White, 13, 23
width of roadway, 9
Woods, 4
work hubs, slopes/slope staking, 150, 151, 154
Wright, 3, 4

Z

zoning, highway planning, 5, 8

ABOUT THE AUTHOR

Robert C. Vernon's surveying, engineering and design experience spans thirty-nine years, including twenty-nine years in teaching. He began at the consulting engineering firm Beiswenger, Hoch, Arnold & Associates (now U.R.S. Consultants) Akron, Ohio. He was then employed as right-of-way engineering plan reviewer and plan reading instructor for the Ohio highway department's central office College Training program in Columbus, Ohio, and as assistant right-of-way coordinator for Division 10, Marietta, Ohio. Mr. Vernon commenced private surveying practice in 1969 and presently continues this practice as President of Right-of-Way Consultants & Associates/Robert C. Vernon, Inc., and its successor corporation, Robert C. Vernon Enterprises, Incorporated.

Teaching has always been a part of Mr. Vernon's career. Following receipt of a master's degree, Mr. Vernon was instructor at the Kansas City Conservatory (now part of the University of Missouri); at Park College, Parkville, Missouri; and taught in the Willard City Schools, Willard, Ohio. He received his Registered Professional Surveying license in 1963.

In addition to teaching the plan reading and surveying courses while employed at the Ohio Department of Transportation, he also taught a route surveying course (curves and spirals) annually for twenty consecutive years. This course was part of the yearly Surveyor's Refresher Program, originally sponsored by the Ohio Society of Professional Engineers, then later sponsored by the Professional Land Surveyors of Ohio.

Mr. Vernon has published several articles in the American Right-of-Way Association magazine and penned numerous course books for the Ohio Department of Transportation. He has also been affiliated with the National and Ohio Society of Professional Engineers, Professional Land Surveyors of Ohio, the West Virginia Association of Land Surveyors, the American Congress on Surveying & Mapping, and is licensed to practice in Ohio, Pennsylvania, and West Virginia.